D1116798

Radiation Curing of Coatings

Joseph V. Koleske

ASTM Stock Number: MNL45

ASTM International
100 Barr Harbor Drive
PO Box C700
West Conshohocken, PA 19428-2959

Printed in the U.S.A.

Library of Congress Cataloging-in-Publication Data

Koleske, J. V., 1930–
Radiation curing of coatings/Joseph V. Koleske,
p. cm.—(Manual 45 in ASTM International's manual series)
"ASTM Stock Number: MNL45."
Includes bibliographical references.
ISBN 0–8031–2095–8
1. Radiation curing. 2. Plastic coating. 3. Radiation—Industrial applications. 4. Ultraviolet radiation—Industrial applications. I. ASTM Committee D-1 on Paint and Related Coatings, Materials, and Applications. II. Title. III. ASTM manual series; MNL 45.

TP156.C8 K66 2002
668.4—dc21 2002016401

NOTE: This manual does not purport to address all of the safety concerns, if any, associated with its use. It is the responsibility of the user of this manual to establish appropriate safety and health practices and determine the applicability of regulatory limitations prior to use.

Printed in Bridgeport, NJ
April 2002

Foreword

THIS PUBLICATION, *Radiation Curing of Coatings*, was sponsored by Committee D01 on Paint and Related Coatings, Materials, and Applications. This is Manual 45 in ASTM International's manual series.

Contents

Preface

IMAGINE PLACING a layer of a low viscosity liquid mixture made up of monomeric and oligomeric compounds on a substrate, shining a beam of ultraviolet radiation on the liquid, and then in less time than it takes to snap your fingers—effectively, instantaneously—having the entire liquid mass turn into a solid, cross-linked, hard, tough coating with both functional and decorative properties. Sound impossible? Sound magical—like Mary Poppins (to the older generation) or the X-Files (to the younger or younger thinking generation)? Certainly it does! It sounds too good to be true, or as if it were magical in nature. However, it is merely a brief, popularized description of the coating technology known as "radiation curing" or the more limiting term, "photocuring." This technology that deals with using the substrate as your polymerization vessel will be described in detail, but in an understandable manner, in this book. Reading it will give one a good understanding of this topic and enough knowledge to begin formulating radiation-curable inks, coatings, and sealants.

Radiation Curing

Introduction

BEFORE THE EARLY 1970S, there was little apparent concern or even awareness of the amount of solvent that entered the atmosphere from coatings and allied products. As a result, in the early 1970s over 90% of applied industrial coatings were low solids in nature. That is, they contained from 10% to 20% by weight solids, excluding pigments or other suspended materials, and the remainder was solvent. Industrial coatings at that time were typified as being "low solids and solvent borne." Architectural coatings, commonly known as house paints and varnish, were also solvent borne, and latexes were really just coming into commercial reality. In those days, many people said that latexes did not have the quality of oil-based paints—many latex paints did not. Today, latex house paints are prevalent, but because of their excellent application characteristics and properties, oil-based paints still have a solid place in the market. However, their slow drying nature and odor remain as deficiencies. Even latex paints are relatively slow drying.

One would like to think industry could readily see the many advantages of changing from a single, high-polluting technology to a multitechnology arrangement that would provide forward thinking. Thinking that would lead to cost savings as well as resource conservation, while being ecologically friendly and simultaneously providing a healthier work and community environment. Actually, at that time the slow moving coating industry could not or would not see the future—sales and use of solvent dominated the picture.

How much solvent was used in a coating formulation in the early 1970s? When the total solids content was 10%, nine pounds of solvent were used with one pound of final coating material. When the total solids content was 20%, four pounds of solvent were used with one pound of final coating material. These large quantities of relatively inexpensive solvent resulted in excellent application characteristics, flow, and leveling of the liquid coating, and in high gloss, thin, continuous films of the relatively expensive, final solid polymeric coating. Pleasing decorative and functional coatings resulted. But huge quantities of solvent were spewing into the atmosphere, and both industry and the public seemed oblivious to this fact.

At this time there were several reasons for the state of affairs in the coating industry being one of solvent-borne, low solids. Petroleum-derived solvents and the energy, which is also fossil-fuel based, needed to evaporate them were inexpensive. This is easy to realize if one recalls that, prior to the early 1970s, oil and petroleum products had not yet seen the effect of cartel pricing. At the time, if someone had proposed a scenario in which the price of oil would increase by a factor of 10 to 25 times, they would have been told to quit wasting time and pose something realistic. No one wanted to imagine the effect such pricing would have on the coating industry or on individuals. Unless one lived through it, it is difficult to visualize the pressures that were brought to bear on the general public and industry through dictated petroleum shortages that ended up as long lines at gasoline stations and markedly increased prices. Not only did consumers feel the stress of waiting in gasoline lines, they felt the even sharper stress of wondering if they would be able to purchase fossil fuel-based products at any price to heat their homes and drive to their place of employment or on vacation. In the coating industry, companies that had large drying ovens became concerned with the problem of what to do if they couldn't operate their ovens—they knew heating of homes would come first. There was enough oil and natural gas, but it was only available at a high, cartel-set price. And the price of oil went up and up in what seemed a never-ending spiral.

A complicating effect superimposed on the price/availability picture was the fact that solvents were an excellent medium for reducing the viscosity of high molecular weight polymers to an application viscosity. At that time, most polymers used for coatings had high molecular weight. This was a necessary requirement to achieve the high-quality performance characteristics expected from the liquid coatings that were most often used as lacquers. This factor was the major reason large quantities of solvent were used in formulations. Then, almost overnight, oil cost and availability became a serious problem. To satisfy ecological concerns of certain groups, the matter was exacerbated when the government set requirements on the coating industry. The industry was to quickly formulate 80% solids coatings. That is, they were to change from using about four or more pounds of solvent for each pound of final coating to one-fourth of a pound of solvent per each pound of final coating—a factor of 16 or more decrease!

It was in this crucible of market pressures, the Clean Air Act of 1970[1], and existing old know-how that the new technologies of water-borne, high-solids, powder, and radiation-curable coatings, along with the various sub-

[1] The USA Environmental Protection Agency's Clean Air Act of 1970 was amended in 1977 and 1990, but at the time under discussion, it was the original act that was the driving force.

categories such as two-package high solids, water dilutable coatings, higher-solids coatings, and the like, were born.

As might be expected, there were markedly different opinions regarding each of these new technologies. Many thought solvent-borne systems would never be replaced, and in the long run, they were losers. Others thought water-borne coatings were the answer. These were usually people who were in the architectural latex-coating business, and they felt it would be easy to develop industrial latex coatings. High-solids coatings seemed an easy way to meet the demands of the Clean Air Act. All one had to do was decrease molecular weight and high solids could be achieved. They temporarily forgot or just did not know molecular weight-property relationships, and they found that systems to be satisfactorily cross-linked needed functionality suitably located on each molecule to ensure they would be incorporated into the final coating film network. Others thought powder and radiation would be important, but these were more difficult to achieve and often did not receive as much attention from the industry at that time.

In the end all the new technologies found a place in industry. The only parts of industry that did not gain from the changes brought about by the new technologies were those who refused to be forward-looking and clung to the old ways. New companies came into being, names changed, and consolidations took place as the new technologies found their niche in the marketplace. Today, radiation curing is well established and appears to be poised to make great strides into this market [1].

What is Radiation Curing and Why is it Used?

Radiation curing of coatings, inks, adhesives, and sealants[2] is the effectively instantaneous conversion of a usually liquid, low viscosity formulation of ingredients that has been applied to a substrate and exposed to a radiation source into a polymerized and cross-linked mass with decorative and/or functional properties. This process usually results in an effectively complete conversion of the liquid formulation into a solid, useful product, i.e., radiation-cured coatings are essentially 100% solids in nature. Radiation as used in the radiation curing industry is usually ultraviolet radiation (photons or photocuring) or electron beam energy (electrons or E-beam curing), however, coherent or laser radiation and visible radiation[3] or light are also used.

[2] Coatings, inks, adhesives, and sealants can be and are cured with radiation. In this book, the word "coatings" will be taken to be generic to all of these products, unless one is specifically dealing with one of the others. Currently, radiation curing of sealants has limited use. This is primarily due to the need for curing through a considerable thickness of sealant material. Readers interested in radiation curing of sealants might consider the information in the topic "Thick Section Curing" found in Chapter 11.

[3] Webster's New Collegiate Dictionary defines light as that part of the electromagnetic spectrum of radiation that is visible to the human eye.

"Radcure" is an adjective derived to describe anything that has to do with radiation-curing technology—applications, equipment, materials, and so on. When a material is exposed to radiation, it is "irradiated." Although radiation and its nature will be discussed in a following section, a few words about it are appropriate at this time.

The electromagnetic spectrum is made up of electromagnetic waves of radiation that are identified by fluctuations of electric and magnetic fields. It comprises the total range of electromagnetic wavelengths, λ, and extends from essentially zero to infinity. If numbers are to be placed on these wavelengths, it could be said that they range from about 10^{-13} to 10^3 meters. Cosmic rays, gamma rays, and hard X-rays are on the low end of the range, and microwave, radio, radar, television, and electric waves are on the high end of the range [2,3]. Contained within the electromagnetic spectrum is the visible portion of the spectrum, wavelengths from 400 to 760 nanometers or from violet to red, that is known as "light," and the invisible portion of the spectrum just beyond the violet end of the visible region that is known as "ultraviolet" and often referred to as "ultraviolet light." The ultraviolet region of wavelengths extends from about 1 to 400 nanometers, and it contains a region from about 15 to 280 nanometers known as "Ultraviolet C," another from 280 to 320 nm known as "Ultraviolet B," and another region from 320 to 400 nm known as "Ultraviolet A"[4]. Ultraviolet B is radiation that causes the phenomenon known as "sunburn" and prolonged exposure to it over many years can cause skin cancer. Ultraviolet C radiation is very powerful and is used to sterilize surfaces because it has the ability to kill bacteria and viruses. The range from 800 to about 30 000 nanometers is known as the infrared region of the spectra. Above this begins the microwave region and so on. RadTech International has somewhat different, but similar, designations of the ultraviolet radiation ranges. These can be found in the RadTech Glossary of Terms at the end of this volume.

Why does industry use the radiation curing process? Many of the attributes of this process can be summarized as:

- Low capital investment
- Low energy consumption and cost
- High solids and elimination of solvents
- Plastics and other heat sensitive substrates can be coated and cured
- Rapid cure and high productivity
- Multiple operations are possible
- Space requirements are low
- Use of existing equipment when retrofitting radiation equipment
- High degrees of cross-linking and, therefore, potential for improved abrasion, chemical, and stain resistance.

Capital investment is low for ultraviolet curing or photocuring equipment. The equipment usually can be designed to meet specific needs, and no "dead" or unused curing areas need be involved. If electron beam technology is involved, capital investment can be high, but it is used when high, large-volume production coupled with very high line speeds is involved. New small-scale electron beam systems have expanded the electron beam technology into the realm of relatively modest cost.

Radiation-curing systems have low power requirements, and little heating of the substrate takes place. No large ovens are required for evaporation of solvent or water. These factors result in low operating costs, a factor that takes on ever increasing importance as the price of fuel undergoes marked increases, as it did in 2000. An additional benefit is that the systems run cool, and this improves the working environment for employees by making ambient temperatures quite reasonable and easier to control. It is also easier and less time-consuming to change over a finishing line from one formulation or design pattern to another, which results in increased cost savings and productivity.

As mentioned earlier, radiation-curing systems are either 100% solids or very high solids in nature. VOC[4] is near zero, since all of the formulation reacts or remains effectively with the final film. For particular end uses, there are specialized formulations to which solvent is added for viscosity reduction, but it is usually easy to stay within regulatory limits or to fit the equipment with a solvent recovery system. Water-borne radiation curable systems are also used.

Heat-sensitive substrates such as plastics, printed circuit boards and assemblies, paper, wood, and the like can be coated and cured with radiation technology. The reason for this is that there is little buildup of heat in the substrate when thin coating films are involved. If thick films are prepared, judicious selection of formulation ingredients—particularly photoinitiator level—and line speed can be used as controls to remove concerns about heat buildup. This factor can be of particular utility in the electronics industry.

Improved economics and efficiency result from the "instantaneous" curing characteristics of radiation-cured systems. Depending on the nature of the formulation being cured, line speeds of hundreds of feet per min can be easily achieved [7]. This results in increased production. When this is coupled with the fact that multiple operations can be accomplished in a single pass through a finishing line, plant output is often doubled, tripled, or more on a line that is no longer and often shorter by 100 to 200%, or more than a conventional finishing line. That is, a substrate can be printed in one or more colors, the ink cured, over coated with a clear finish, cured, turned over in a

[4] VOC is an abbreviation for Volatile Organic Compounds that is used by the USA Environmental Protection Agency (EPA). VOC is defined as "Any organic compound that participates in atmospheric photochemical reactions; that is, any organic compound other than those the Administrator designates as having negligible photochemical reactivity." Details about VOC and how it is measured can be found in the literature [5, 6].

"Ferris wheel" arrangement, and then coated and cured on the reverse side. Of course, elimination of solvent also represents a calculable cost savings. Savings result because there is no need for solvent, handling solvent, evaporating solvent, recovering solvent, and for safely and ecologically sound disposing of recovered solvent in some manner.

In general, space requirements are much less, often 25–50% less, for a radiation-curing line than for a conventional coating line. The main reason for this is that no large ovens are required for evaporation of solvent and/or water. In addition, there is no need for a solvent recovery system. An additional space-saving feature of radiation-curable coatings is their 100% solids nature. Less drums of formulated coating are required when solvent is eliminated, which also results in decreased handling.

At times, existing coating lines can be retrofitted with radiation-curing equipment. This can be accomplished with minor adaptations and costs, particularly with photocuring equipment, which can be very versatile in nature.

It can be readily understood that radiation-curing lines can run much faster than conventional coating lines. When this is coupled with the multiple operations that can be accomplished on a single pass through the line, marked increases in productivity can be attained. For these and the other reason discussed, radiation curing is a technology that is currently in vogue and that will be experiencing growth of about 10% compounded in the future [8].

Radiation

In the above, there was a brief introduction to radiation and its relationship to the electromagnetic spectrum. Radiation is the propagation and/or emission of rays through space, which is effectively a void, or through a medium, such as air, in the form of either waves or particulate emission. Just what does this mean?

Most people think of activating molecules by using thermal energy as the normal way for chemical phenomena to occur. That is, temperature elevation is used to push molecules up a hill or over a barrier and allow them to attain a different, higher energy state. For example, the conversion of something from the liquid state to the gaseous state is usually accomplished with the use of thermal energy. In the coating industry, many thermal processes or thermally activated reactions are used. Such phenomena usually use thermal energy, whose source is an increase of temperature, to increase one or more of the rotational, translational, and vibrational energy states from a ground state to the upper range of their energy distribution [9]. One of the simplistic processes used in the coating industry is evaporative drying of a lacquer that usually is accomplished with supplied heat or thermal energy. The liquefaction and flow of powder coatings is also accomplished with thermal energy. Another specific example is the crosslinking reaction of hexamethoxymelamine with hydroxyl functionality that is accomplished by supplying thermal energy to the system. Various other thermal processes used in the coating industry should be readily apparent to readers.

Thermal processes such as these are inefficient in terms of energy usage and of the amount of final coating applied to a substrate for a given unit of energy. A more efficient way to supply and utilize the energy necessary to effect a change to an activated or reactive state is to bring the molecules directly into contact with "packets" of energy supplied by sources such as neutrons, β-rays or high velocity electrons, α-particles, or the quanta of electromagnetic energy termed "photons" and supplied by that portion of the electromagnetic spectrum known as ultraviolet radiation or by light. A distinct advantage of photochemically induced reactions over thermally induced reactions is selectivity. Different and particular reactions of the same substance can be promoted by irradiating it with different frequencies of ultraviolet or visible radiation. Investigation of the reactions caused by such energy-supplying processes is termed "radiation chemistry." The process used to effect cure of coatings, inks, adhesives, and sealants with radiation chemistry is known as "radiation curing." If only ultraviolet or visible radiation is used, the particular process is known as "photochemistry." In industry, the two main methods of effecting radiation curing are electron beams and ultraviolet radiation. Equipment for curing with these two methods will be discussed in Chapter 2.

An electron beam is a stream of electrons that are emitted by a source. The electrons all move at the same speed and in the same direction. An electron beam, which is similar to a cathode-ray tube, produces the stream of electrons. The cathode is heated to generate electrons, a high-voltage system then directs the electron stream, and electromagnets guide the stream to the substrate. This technique was mainly used for high volume production because of equipment cost and size. In a general sense, selection between ultraviolet and electron beam curing is dependent on factors such as film thickness, pigmentation or opacity, production scale or volume, and similar items.

Photochemistry is a subdivision of radiation chemistry, and it is the science of the chemical changes that are initiated by ultraviolet and visible radiation on molecules. The particular wavelength region of ultraviolet that is of interest in the coating area is from about 100 to 400 nm, but more practically from about 200 to 400 nm. It should be understood that incident ultraviolet radiation can be transmitted, absorbed, or refracted, *but it is only absorbed radiation that can produce a chemical change* in molecules. This is the first law of photochemistry, which is known as the Grotthuss-Draper Law. Certain compounds can absorb particular ultraviolet wavelengths or energy packets and are completely transparent to other ultraviolet rays. Another way of stating this is that if a molecule does not absorb radiation of a particular frequency, it will not undergo a photochemical change when contacted with radiation of that particular frequency. The amount of radiation absorbed by a compound is proportional to the thickness of the compound's surroundings. The second law of photochemistry, or Stark-Einstein Law, tells us that only one molecule of a radiation absorbing system will be activated for each photon of radiation absorbed by a chemical system. The quantity of photochemical reaction that takes place is determined by the product of time of exposure and the radiation intensity.

The fact that radiation is energy can be simply and readily exemplified. The following examples quickly and vividly demonstrate that ultraviolet and visible radiation are energy. It is well known that sunlight contains wavelengths from both the ultraviolet and the visible spectra. Almost everyone has taken a magnifying lens and focused the sun's rays onto paper or a similar material. When this is done, it is observed that the focused beam of broad-spectrum radiation quickly raises the temperature of the paper to the kindling point, and smoke begins to rise. Another example is the fact that when the ultraviolet rays in sunlight bombard our skin, they cause reddening or tanning as well as other less apparent, long-term changes. The fact that ultraviolet rays are not absorbed or reflected by the same media as other forms of energy, such as visible or light energy, is apparent from the "sunburn" one can obtain on a cloudy day. The ultraviolet rays are transmitted through the cloud moisture, whereas much of the other sun's rays are absorbed or reflected by the cloud, and a darkening or dimming effect and a cooling effect are observed.

Other commonly known, but perhaps not realized as such, photochemical reactions include photography wherein light converts silver bromide to metallic silver, vision in which a protein is isomerized by radiation in the retina of the eye, and photosynthesis wherein photons from sunlight are absorbed by chlorophyll and make energy available for the formation of various chemicals such as carbohydrates, vanillins, phenolics, and so on. Thus, it is readily apparent that ultraviolet and visible radiation, as well as other portions of the electromagnetic spectrum, are energy. However, at this point the magnitude of the energy, its nature, and its mode of transmission are not apparent.

Simplified Classic to Modern Light/Radiation/Energy Theory

In the Seventeenth Century, there were two concepts used to describe the nature of light and ultraviolet radiation. In one of these concepts, Newton posed a corpuscular or particle theory for such radiation. This theory was generally accepted until almost the end of the Eighteenth Century. At this time, Huygens posed the second concept, and it considered such radiation to be wave-like in nature. Studies concerned with the diffraction, interference, reflection, and refraction phenomena that occur with light took place in the early Nineteenth Century, and the results of these studies could be best explained by invoking the Huygens wave theory. Further credence was given to the wave-like character theory when it was found that, in a vacuum, light travels at the same velocity as that of other electromagnetic radiation, namely 3×10^{10} cm · s^{-1}.

In the 1860s, Maxwell presented his electromagnetic theory, in which radiation energy is described as a combination of oscillating electrical and magnetic fields positioned at right angles or normal to each other. Then the energy propagates through otherwise empty space or a medium with the velocity of light in the direction of propagation. Maxwell's theory gave further acceptance to the wave theory of Huygens. However, even though such phenomena could be explained by only invoking the wave theory, the photoelectric effect[5] and other light characteristics could not be explained with only the wave theory. To understand these phenomena, it was necessary to bring into play the corpuscular or packet theory.

Studies by Planck [10] at the beginning of the Twentieth Century indicated that a body radiating light with a frequency ν would radiate energy only in whole number multiples of hν, where h, Planck's constant = 6.624 $\times 10^{-27}$ erg · s, is the proportionality between energy, *E*, and frequency with:

$$E = h\nu = hc/\lambda$$

where *c* is the velocity[6] of light in a vacuum and λ is the wavelength. Planck's hypothesis also stated that a body could only absorb energy in even number multiples of hν. A few years later Einstein [12] developed his relativity expression,

$$E = mc^2,$$

wherein *m* is mass and *c* is as previously defined. His efforts extended Planck's theory, and they indicated that electromagnetic radiation is quantized and made up of discrete quanta or photons each of which, as a unit of energy, has a value $h\nu$. Thus, a photon is a quantum of electromagnetic energy of a single wavelength, i.e., of a single mode of direction and polariza-

[5] When a photon impinges onto a surface and is absorbed, an electron may be emitted from the surface. This emission of an electron is the means by which photons can be detected and is known as the photoelectric effect.

[6] Usually the velocity of light is thought of as a constant; however, in January 2001, it was reported that physicists were able to stop light and then send it on its way at 3×10^{10} cm·s^{-1}. Atoms of gas were magnetically chilled to within a few millionths of a degree from absolute zero. Two laser beams that operated at different frequencies were directed into the gas, wherein it imprinted a pattern into the orientation of the spinning atoms of sodium and rubidium. As the intensity decreased, the probe light dimmed and then vanished, but still was imprinted on the metals where it was frozen or stored. When the control beam was restored, the light stored in the spinning atoms was reconstituted and continued on its way [11].

tion.[7] From these investigations, the modern concept of radiation emerged as energy with a dual nature. Radiation moves through space in wave packets (corpuscles) that have a spectrum of energies (i.e., wavelengths) with wave-like properties.

Earlier it was pointed out that an advantage of curing with ultraviolet or visible radiation is selectivity. For a change to take place, the product of the frequency of the radiation that is absorbed by a molecule and Planck's constant must exactly match the energy that separates the ground state of the molecule, E_g, and the excited state, E_e.

$$E = h\nu = (E_e - E_g)$$

If the frequency is changed, the magnitude of E changes; and if a state $E_e - E_g$ equal to the new magnitude of E exists for the molecule, the molecule will be transformed to this new and different state. Thus, by varying the frequency of radiation impinging on a molecule, it is possible to get photochemical reactions that are completely different from each other, assuming a number of excited states exist for the molecule. As was mentioned earlier, the electromagnetic spectrum encompasses a very broad range of wavelengths that are between about 10^{-13} to 10^{3} meters [18].

Magnitude of Ultraviolet and Visible Radiation Energy

With this picture of radiation and how it functions, we can now turn to calculation of the magnitude of the energy associated with ultraviolet and visible radiation. As described above, Planck's constant is the proportionality constant between the energy of a photon and wavelength, and the energy at

[7] This has been a simplistic discussion of the relationship between radiation and energy. It is beyond the scope of this book to deal with the transition from this information into details of the various theories involved. However, a short discussion may help direct interested readers in understanding radiation and provide literature references that are pertinent to the topic.

Rutherford [13] theorized that the atom was a dense, charged nucleus with surrounding orbiting electrons. This picture required the orbiting electrons to lose energy through radiation, and thus, spiral into the nucleus. Bohr [14] altered Rutherford's picture by placing the electrons into fixed orbits that had different energy levels. His theory of the hydrogen atom had the sole electron in this atom occupying particular, discrete energy levels with the energy state determined by a principle quantum number, n, with n having a whole integer value. The theory worked well for single electron materials, but it was not capable of dealing with atoms that contained more than one electron. When the electron moved from a higher energy level, E_2, to a lower level, E_1, a quantum of radiation was emitted. The frequency of the emitted radiation was given by $E = E_2 - E_1 = h\nu$.

About two decades later, in 1924, de Broglie [15] suggested that the dual nature of radiation might also be true for matter. That is, any moving corpuscle, radiation or matter, has a wavelength associated with it and that a similar relationship exists between momentum and the wavelength of bodies other than light quanta with $\lambda = h/m\nu$. This was later confirmed for electrons by Thomson and others [16,17].

any particular wavelength, $E_{(\lambda)}$, can be calculated from

$$E_{(\lambda)} = h\nu = \left[\frac{hc}{\lambda}\right] = \left[\frac{6.6256 \times 10^{-27}\ \text{erg} \cdot \text{sec}}{\text{quantum}}\right]\left[\frac{c}{\lambda}\right]$$

This energy is the specific energy associated with the wave packet that has a wavelength λ. Quite often, chemists are interested in the energy that is associated with a mole of material or with the energy associated with a mole of quanta. This quantity of energy has been termed an "Einstein" and is equal to Avogadro's number, 6.023×10^{23}, times the energy associated with a particular wavelength. So, if we make this change and substitute for the velocity of light, the expression becomes

$$E_{(\lambda)} = \left[\frac{6.6256 \times 10^{-27}\ \text{erg} \cdot \text{sec}}{\text{quantum}}\right] \times \left[\frac{6.023 \times 10^{23}\ \text{quantum}}{\text{mole}}\right]\left[\frac{2.9979 \times 10^{10}\ \text{cm}}{\text{sec}}\right]\left[\frac{1}{\lambda}\right]$$

or

$$E_{(\lambda)} = \left[\frac{1.1963 \times 10^{8}\ \text{erg} \cdot \text{cm}}{\text{mole}}\right]\left[\frac{1}{\lambda}\right]$$

with the wavelength expressed in centimeters. Since the wavelength is usually expressed in nanometers, and since one joule, J, is equal to 1×10^{7} ergs, the expression for energy can be expressed as

$$E_{(\lambda)} = \left[\frac{1.1963 \times 10^{8}\ \text{erg} \cdot \text{cm}}{\text{mole}}\right]\left[\frac{10^{7}\ \text{nm}}{\text{cm}}\right] \times \left[\frac{1\ \text{joule}}{1 \times 10^{7}\ \text{ergs}}\right]\left[\frac{1}{\lambda\ \text{nm}}\right] = \frac{1.1963 \times 10^{8}}{\lambda}\ \frac{\text{joules}}{\text{mole}}$$

in terms of joules per mole and as

$$E_{(\lambda)} = \left[\frac{1.1963 \times 10^{8}\ \text{joules}}{\lambda\ \text{mole}}\right]\left[\frac{1\ \text{kcal}}{4184\ \text{joules}}\right] = \frac{2.8592 \times 10^{4}}{\lambda}\ \frac{\text{kcal}}{\text{mole}}$$

in terms of kilocalories per mole. Values of $E_{(\lambda)}$ at various wavelengths are given in Table 1.1. The energy required to rupture a double bond and form a

TABLE 1.1. Magnitude of Energy at Various Wavelengths.

WAVELENGTH, λ	ENERGY PER MOLE	
	JOULES × 10^{-5}	KILOCALORIES
100	11.96	286.00
200	5.98	143.00
300	3.99	95.3
400	2.99	71.5
500	2.39	57.2
700	1.71	40.9

free radical depends on the particular chemical molecule receiving the energy. It is usually on the order of 10–80 Kcal per mole ($0.42\text{–}3.35 \times 10^5$ J/mole)) [19], which may be compared with the values in Table 1.1.

Photon Production

A photon is the quantum of electromagnetic energy of a single wavelength. Photons are produced in the ultraviolet and visible radiation regions by

- spontaneous or stimulated radiation transitions between electronic energy levels in atoms or molecules,
- recombination of two radiative bodies as in electron-ion recombination, and
- electron retardation wherein radiation is emitted when an electron is forced to deviate from its normal path as occurs when an electron passes near a positive ion.

If it is possible for an atom to exist in two energy states, E_1 and E_2 with $E_2 > E_1$, there are three possible radiation processes that can connect the two energy levels. There can be a simple, uncomplicated absorption in which the atom at energy level E_1 is placed in a radiation field that contains wavelength λ, where $\lambda = hc/(E_2 - E_1)$. Here, it may absorb a photon and be excited to a higher energy level E_2. Induced or stimulated emission occurs when an atom that exists in energy state E_2 is placed in a radiation field that contains wavelength λ, where again λ is given by $hc/(E_2 - E_1)$. A photon may impinge on the atom and cause it to emit a photon that is coherent with the impinging photon. When this happens, the energy level drops to E_1. Spontaneous emission is a third method of photon generation. In this process, an atom in a higher energy state E_2 can spontaneously emit a photon and drop to energy level E_1. These mechanisms lead to the photons or quanta that activate photoinitiators that, as we shall see in ensuing chapters, cause polymerization of reactive species in the formation of radiation-cured coatings.

Electrons

It is important to point out that electrons are not included in the electromagnetic spectrum. Electrons differ from photons in that they are particulate and not wave-like in nature, and they have mass. Because of their mass, accelerated electrons as formed by an electron beam apparatus have the ability to penetrate matter. These phenomena, as well as the depth of penetration into materials, are further discussed in Chapter 2. Photons, which are included in the electromagnetic spectrum, can also penetrate matter. The depth of penetration depends on their energy, wave, and particulate nature.

References

[1] Koleske, J. V., "RadTech 2000," *Paint and Coating Industry Magazine*, Vol. 16, No. 6, June 2000, p. 34.

[2] Kinstle, J. F., "Radiation Polymerization," *Paint and Varnish Production*, Vol. 63, No. 6, June 1973, p. 17.

[3] Lake, R. T., "Ultraviolet Curing of Organic Coatings," *Radiation Curing*, Vol. 10, No. 2, 1983, p. 18.

[4] Davis, T. W., "Ultraviolet Radiation," *Microsoft® Encarta® Online Encyclopedia 2000.*

[5] *Glossary for Air Pollution Control of Industrial Coating Operations*, EPA-450/3-83-013R, Environmental Protection Agency, Washington, DC, December 1983.

[6] Brezinski, J. J., *Manual on Determination of Volatile Organic Compounds in Paints, Inks, and Related Coating Products*, MNL4, ASTM International, West Conshohocken, PA, 1993.

[7] Schaeffer, W. B., "UV Curable Materials Response and its Relationship to Power Level and Lamp Spectra," *Proceedings, Volume 1, RadTech '90–North America*, March 25–29, 1990, p. 29.

[8] Lawson, K., "Status of UV/EB Curing in North America–2000," *RadTech 2000*, Baltimore, MD, April 12, 2000.

[9] Moore, W. J., *Physical Chemistry*, 2nd ed., Prentice Hall, Inc., NJ, 1955.

[10] Planck, M., *Annalen der Physik*, Vol. 1, 1900, p. 69; *Annalen der Physik*, Vol. 4, 1901, p. 553.

[11] "Physicists Bring Light to a Stop Then Send It on Its Way," Sci-Tech, *www.foxnews.com*, January 19, 2001.

[12] Einstein, A., *Annalen der Physik*, Vol. 17, 1905, p. 132.

[13] Rutherford, E., *Philosophical Magazine*, Vol. 21, 1911, p. 669.

[14] Bohr, N., *Philosophical Magazine*, Vol. 26, 1913, p. 476.

[15] de Broglie, L., *Philosophical Magazine*, Vol. 47, 1924, p. 446.

[16] Thompson, G. P., *Proceedings Royal Society*, A117, 600, 1928.

[17] Davisson, C. and Germer, L. H., *Physics Review*, Vol. 31, 1927, p. 705.

[18] Commission Internationale de l'Eclairage, "International Lighting Vocabulary," CIE, Paris, 1970.

[19] Ridyard, Andrew, "Accurate, High Power UV Spectral Measurement," *Proceedings of RadTech 2000*, Baltimore, MD, 9–12 April, 2000, p. 136.

2 Curing Equipment

IN THE PREVIOUS CHAPTER, the nature of radiation in certain portions of the electromagnetic spectrum was discussed. As was seen, radiation is composed of wave packets called photons with characteristic and discrete energy levels. The radiation associated with ultraviolet curing or photocuring is known as "nonionizing" or "actinic" radiation. It is radiation that usually does not produce electrically charged particles, free electrons, or ions, and it is electromagnetic energy that is capable of producing photochemical activity. Photons with sufficient energy can produce electrons; however, in photocuring such photons are not usually available. Radiation that is associated with electron beam curing is "ionizing" radiation, and it is energy that can directly or indirectly form ions and generate electrons while traveling through a substance. Other energy sources used today include lasers and light or visible radiation. Although these latter sources are not broadly used in commerce today, they may well be of future importance and will be briefly discussed at the end of this chapter.

Selection of electron beam or ultraviolet radiation as the means of curing is a balance between the versatility of the energy generated, the particular curing chemistry to be used, and the equipment complexity and cost. A generalized comparison of the two curing methods is given in Table 2.1. When reviewing the information in this table, one should be certain to remember that equipment cost usually is the overriding factor. Both technologies involve line-of-sight curing, since neither electron beams nor ultraviolet radiation can "see" around corners. Electron beam curing depends on penetration of the material involved with energized electrons and on the density of the material or materials through which they are passing. The electrons generate free radicals that are capable of initiating the polymerization of ethylenically unsaturated compounds such as acrylates. Usually, one does not associate electron beam curing with the cure of epoxides, however, recent literature indicates that epoxides [1–4], carbazoles [5], and vinyl ether/cellulose blends [6] can be cured with this technology. In addition to the coatings area, electron beams are finding important utility in the curing of fiber-reinforced composites [7–9]. Electron beam cured composites have been found to be

TABLE 2.1. Comparison of Electron Beam and Ultraviolet Radiation Systems.

CHARACTERISTIC	ELECTRON BEAM	ULTRAVIOLET RADIATION
Energy Consumption	Lower	Higher
Versatility	Less	Greater
Capital Expenditure	Much Greater*	Less
Maintenance Cost	Less	Greater
Efficiency	Greater	Less
Productivity	Greater	Less
Substrate Selection	Greater	Less
Ease of Operation	Less	Greater

* Small size, narrow width units have markedly decreased cost, but cost is still noticeably greater than for ultraviolet radiation units.

generally comparable to thermally cured products, and the important advantages of instantaneous cure coupled with environmentally friendly chemistry can mean cost savings to manufacturers. This technology has been of most interest to the aircraft and aerospace industries, but the timesaving and ecological aspects of the technology have spurred the interest of the ground transportation industry.

Ultraviolet radiation curing depends on the radiation transmitting properties of the coating formulation, the dominant wavelengths generated by the ultraviolet radiation sources, the particular type of photoinitiator used (free radical or cationic), and its radiation absorption characteristics. Photolysis or degradation of the photoinitiator results in the generation of free radicals that can initiate cure of ethylenically unsaturated compounds such as acrylates or of cations that can initiate cure of compounds such as cycloaliphatic epoxides. Currently used cationic photoinitiators generate both cations and free radicals when exposed to ultraviolet radiation.

In general, curing with ultraviolet (UV) radiation is more widely used than electron beam (EB) curing. This is mainly due to equipment cost and complexity for large-scale applications. Compact electron beam curing systems suitable for research and for small-lot applications have been known for some time [10,11]. An example of such usage is semiconductor metallization applications. Recently, a new, small-size, versatile EB unit that will handle a continuous moving web has been described [12], and it should be useful for moderate sized lots. This compact device operates in the low voltage range of 80 to 125 keV and is finding utility in the coatings and graphics areas. More specifically, it has market acceptance in curing opaque pigmented inks, coatings for flexible substrates, metals, and fiber composites, and adhesives for foil-based laminates. This small-size electron beam is moderately priced at less than $150,000 [13] and includes an electron beam, power supply, and controller.

Electron beam technology has been known for many end uses and today is regarded as an accepted method of curing for a number of them [14]. These include graphic arts [15] on offset web presses [16], laminating and pressure sensitive adhesives [17], and inks [18,19]. As with other radiation-cure technologies, electron beam curing is environmentally friendly, offers energy savings and efficiency, and minimizes waste and need for recycling chemicals [20]. Cost, performance, and needs have been compared for solvent, water, ultraviolet radiation, and electron beam radiation in light of current markets [21].

UV lamps can be easily installed at low cost to the end of an existing manufacturing line, or individual lamps can be configured in a number of ways to accommodate articles of various sizes and shapes. To cure three-dimensional objects, the coating line may be configured in such a manner that the object can be rotated or spun while in the radiation zone. Such objects have asymmetries, impressions, undercuts, and other characteristics that make uniform curing difficult. Optimized positioning of ultraviolet radiation lamps for cure of 3D items has been described and a large-scale UV test facility coupled with computer simulation of the cure process has been developed [22].

The coating industry usually uses ultraviolet radiation in the 250 nm to 400 nm range for curing. At shorter wavelengths than these, X-rays exist at about 0.1 nm and below that gamma rays exist at about 0.0001 nm. These shorter wavelengths can create ionizing radiation within a material and effect polymerization without the use of initiators. However, it is more common to generate ionizing radiation using accelerated electrons as are formed from electron beams.

Electron Beams

An electron beam is a stream of electrons[1] emitted from a single source, an electron beam gun, moving in the same direction and at the same speed. Almost all substances absorb atomic particles such as these high energy, negatively charged particles. Because of their small size and high energy, electrons can penetrate into and through a coating, through a substrate to reach a curable adhesive, and through pigment particles to cure opaque liquids.

If one considers the dualism of wave and particulate behavior, accelerated electrons, which are not a part of the electromagnetic spectrum, are treated as having mass ($\sim 9 \times 10^{-31}$ kg) and behaving as particulates. As mentioned earlier, accelerated electrons, which travel near the speed of light or with photon energy, have the capability of penetrating matter, with the depth of penetration proportional to the electron accelerating potential or beam voltage, as described in Figure 2.1. From the data in this figure, it is ap-

[1] An electron is an elementary, charged particle in an atom. It has an at-rest mass of 9.10956×10^{-31} kg and a charge of 1.602192×10^{-19} coulomb. An electron's spin quantum number is 1/2. Positrons are the positive counterparts of electrons, and, except for the reversal of charge, these possess the same characteristics.

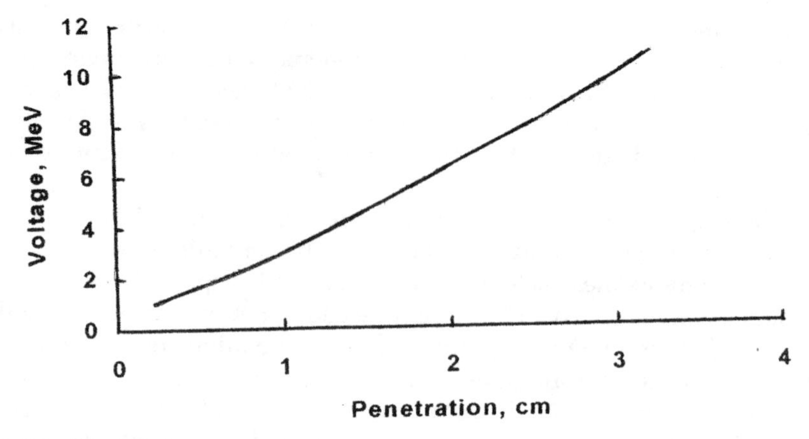

FIG. 2.1—Electron beam penetration as a function of accelerator voltage (Figure courtesy of A. Berejka)

parent that accelerators that operate at up to 10 MeV can penetrate up to 3.2 cm of unit-density material. Since electrons are stopped only by mass, their penetration decreases as the density of the material impacted increases. For example, a 10 MeV electron beam will penetrate 2.4 cm of a 70% carbon fiber composite material that has a density of 1.35 g/cm^3. In the coating area, electron beam accelerators are self-shielded units that operate at 70–300 kV. They are capable of penetrating into unit density material of up to 430 μm (17 mils or 0.017 in.). Thus, electron beams are able to penetrate through pigmented coatings, foils used in laminates, and the like, and they do not rely on photoinitiators or other initiators to cure coatings or adhesives.

In a simplified sense, the stream of electrons, which are key to this technology, is formed with an electron gun [23,24]. This device consists of an emitting cathode and an anode with an aperture for passage of a portion of the formed electrons that move in a straight line at a uniform velocity. The cathodic element, which is in the form of a tungsten wire or filament, is heated to a high temperature of about 2400° K in an acceleration chamber. At this high temperature, electrons are generated around the cathode. These are exposed to a high voltage electrical field and accelerated to a high speed that is dependent on the voltage used. The acceleration area must be under a vacuum or the electrons will lose directionality because of rapid scatter of nitrogen, oxygen, and other gas molecules. As a result of such scatter, there would be no velocity increase. The accelerated electrons that were stripped from the cathodic element are directed through a metallic foil aperture or "static window" that is in proximity to the uncured coating. The window must be durable and be able to withstand substantial heat and mechanical stress. A far more detailed description of this technology can be found in the literature [25].

When the stream of electrons, which is ionizing radiation, is passing through the material to be cured, individual electrons have an opportunity to collide with an electron that belongs to an atom in the uncured material. This collision can remove an electron from the molecule and thus generate a free radical. In the presence of ethylenic unsaturation, these free radicals can initiate polymerization and ultimately cause cross-linking to occur in the usual sequence of initiation, propagation, and termination, as will be discussed later. A photoinitiator, the source of free radical or cationic initiators in an ultraviolet radiation cured system, is not needed in formulations cured by electron beams.

The depth to which the electrons can penetrate into a material is dependent on the density of the material being irradiated and electron energy [26–29]. By proper selection of these parameters, radiation energy usage can be optimized, and any damage to the substrate by electrons can be minimized. Prior to 1970, electron beam equipment used very high voltages of about 500 kV to accelerate the electrons [30]. At this high voltage, substantial shielding was required to contain the X-rays that were formed in the process region. In addition, the electrons had much more penetrating power than was needed, and this resulted in energy inefficiency. Self-shielded electron beam equipment that operated in the 150 kV to 300 kV range was developed in the early 1970s. These devices were less expensive and more energy efficient than the earlier 500 kV units. These electron beam units were designed for industrial use and were self-shielded in nature [31,32]. The optimum operating voltage for these units is 165 kV for curing thin (~ 0.01 mm) coatings and inks. Recently, a new generation of electron beam units that operate at 80 kV to 110 kV has been described [30]. The high-energy electrons developed in these units are capable of breaking bonds and creating ions when they impact on molecules. Such ions convert into free radicals, and these free radicals initiate polymerization of ethylenically unsaturated molecules.

An electron beam's intensity is determined by the beam current. There are well-established protocols that exist for calibrating and determining electron beam dose as, for example, described in ASTM E1818, "Standard Practice for Dosimetry in an Electron Beam Facility for Radiation Processing at Energies Between 80 and 300 KeV." Dose is expressed in a Standard International unit, the Gray, with 1 kiloGray being equal to one Joule per gram mass of absorbed energy. For electron beam processing, dosimetry is traceable to national reference standards that are maintained in the United States by the National Institute of Standards and Technology [13].

There are two generic types of electron beam systems—scanned or swept beam, and curtain or filament beams. In the scanned beam type processor, a relatively narrow beam of electrons is generated and accelerated in a vacuum that is maintained in the high voltage section of the device. An electric or magnetic field is then used to move or scan the beam of electrons back and forth across a metal foil window. The beam of electrons passes through the window into the process area, where it contacts the liquid coating and induces polymerization. The filament beams are single or multiple linear-filament emitters placed normal to product movement direction or multiple emitters with the filaments placed parallel to product movement direction. In this type of electron beam processor, a broadband or curtain of electrons is generated and sent directly to the process area where polymerization is initiated, propagated, and terminated [33,34]. Inerting of the process area is important for eliminating oxygen inhibition of polymerization and to obtain good surface and through cure. Methods for accomplishing this have been described [35–37]. Also, equipment for high line speeds is available [38].

Gamma-irradiation is also used to generate electrons. Typical γ-irradiation systems comprise a radiation source that is usually ^{60}Co, a radiation shield, a system for containing the radiation source, and a material conveying system. Such systems are not used in the coating industry, but they are used in medical device sterilization and food irradiation.

Ultraviolet Radiation Systems

As indicated above, one of the distinctions between electron beam curing and ultraviolet radiation curing is the use of a photoinitiator, though a photoinitiator is useful when epoxides are cured with ionizing radiation. Photoinitiators are compounds that absorb the radiation and then degrade in some manner to form a species that is capable of initiating polymerization—usually free radicals or cations. These compounds will be discussed later. In general, electron beam systems do not require a photoinitiator in the formulation, since the generated electrons cause free radical formation. However, in the electron beam cure of epoxides, iodonium and sulfonium salts are used to improve cure and cure rate.

Ultraviolet radiation curing systems for flat stock can be relatively simplistic in nature. They require an ultraviolet radiation source, a means of cooling the lamps that comprise a power supply, radiation source, reflectors, and a conveyor [39]. The power supply is the controlling device within the curing system. A detailed comparison of energy and intensity for stepped and variable ballast systems has been made [40]. This real-time study concluded that the new solid-state ballast designs will become the dominant factor in ultraviolet radiation curing power supplies.

The conveyor or conveyor system is the means of continuously delivering coated items to the radiation source and of reproducibly exposing the items to the radiation. When flat stock is involved, the surface to be cured is most easily exposed to the radiation source at an angle that is normal to the usually flat conveyor system. The system can be readily designed in such a manner that scattering of the radiation is kept to a minimum and all portions of the surface are exposed to the same, or essentially the same, radiation intensity. When three-dimensional objects are involved, it is important that the coated objects be rotated during irradiation, so that all coated surfaces will be exposed to the radiation source. Proper design of the conveyor line will ensure that all portions of the object receive essentially the same amount of radiation. One can envision alternate schemes in which the source is rotated or moved, and while these may be useful, they are more complex in nature.

The reflectors have a variety of shapes that include elliptical, parabolic, diverging, converging, angled, doubled, and other specially designed geometries. The reflectors may focus the radiation and concentrate it to a small area of high intensity, or they may be interfocused wherein the radiation is spread over a large area. Focused beams are low in cost, but can cause overheating of substrates, require accurate focus, and cause a need for multiple lamps if other than flat surfaces are to be cured. Interfocused lamps do not require critical focus, minimize any heating of the substrate, and allow large areas to be cured effectively. Reflector smoothness is very important to minimize radiation scatter, and the surface is usually polished to a roughness of 100 nm [41]. Recently, a very smooth glass reflector system that is comprised of a borosilicate glass dichroic reflector has been described [42]. The dichroic coating on the reflector absorbs infrared and reflects ultraviolet radiation. This feature makes the unit very useful for curing on plastic surfaces. If complex shapes, such as three-dimensional objects, are to be cured, special configuration of the lamps, reflectors, and/or a means of rotating the object or the bulbs in the radiation area are employed [43–48]. Bottles, yogurt cups, tubes, automobile headlamp bezels, and similar objects exemplify such shapes. Fibers, wires, and cables have been coated in an elliptical tube with a reflecting surface wherein the lamp is positioned near one end of the ellipse and the object to be coated is near the other. In this manner, all portions of the coated fiber can be cured.

The energy output of an ultraviolet radiation lamp can be expressed as effective radiance or integrated energy [49–51], each of which are expressed as watts/cm^2. In addition to ultraviolet radiation, which comprises about 28% of the lamp output, visible radiation (~21%), infrared radiation (~33%), and heat losses (~18%) are other lamp outputs [52]. The effective irradiance or peak intensity is the energy to the reactive chemicals on a substrate, and it is a function of the power input to the lamp and lamp geometry. Integrated energy, which is more commonly known as dose, is the radiant energy output of the lamp that arrives at the reaction surface. It is a function of exposure time and is the time integral of irradiance. These quantities are important for reliable quality control and evaluation of the condition of lamps and the curing unit [53]. Commercial radiometers that measure the photon power available, which is also known as irradiance, are available [54], and ways of positioning sensors and recording output have been described [55]. Radiometry can be used for process control, design, and monitoring, and these have been discussed along with the types of radiometers, main optical factors to measure, sources of error in making the measurements, and ways to report data [56–59]. Devices to measure both intensity and temperature in a curing device have been described [60]. Data from the instrumentation is sent to a computer for storage and analysis. Basically, radiometers consist of an ultraviolet sensitive diode positioned behind a band-pass filter and a radiation collecting optical system that will provide a narrow or wide acceptance angle. This system generates a signal that is sensed by an electronic integration and amplification circuit. The modified signal is then interpreted and displayed as a digital readout of irradiance in watts/cm^2 [54]. Although these radiometers give information about energy output, they do not give information about the spectral distribution of wavelengths the lamp is emitting. Another difficulty is the fact that the devices are calibrated with a particular lamp. If they are used to measure output energy from a lamp that emits different wavelengths, errors will arise.

Spectroradiometers can do away with the difficulties that may be encountered with radiometers, but they are markedly more expensive [54]. These devices function by taking the received radiation through a collecting optical system. It is then split into its spectrum of wavelengths by use of interposing optics and a diffraction grating. After passing through the grating, the radiation is sensed by photodiodes placed near the exit and the collected data is displayed as energy as a function of wavelength spectrum of the lamp.

Visual, pressure sensitive adhesive backed, dosimetry indicators are available [61]. When exposed to actinic or ionizing radiation, these label indicators change color from an unexposed green to a deep magenta. The color changes can be "read" with the naked eye or with a colorimeter or other suitable instrument, if precise numbers are needed.

Ultraviolet radiation sources [62,63] are of two general types—mercury vapor lamps and xenon lamps—with the former type being by far the most common radiation source. Mercury has been used in such sources for a long time because, when vaporized and ionized, it is one of the richest sources of ultraviolet radiation. There are three kinds[2] of mercury vapor lamps available [64]:

- Low pressure
- Medium pressure
- High pressure

Operating characteristics of these lamps are given in Table 2.2. Medium pressure lamps, electrode and electrodeless, are the most widely used industrial

[2] It should be pointed out that in the field of actinic radiation or photochemistry, lamps used are referred to as being "low pressure," "medium pressure," and "high pressure" in nature. However, in the field of illumination, these same three types are referred to as "low pressure," "high pressure," and "very high pressure."

TABLE 2.2. Typical Operating Characteristics of Mercury Vapor Bulbs.

CHARACTERISTIC	LOW-PRESSURE BULBS	MEDIUM-PRESSURE BULBS	HIGH-PRESSURE BULBS
Pressure	~10^{-2} to 10^{-3} Torr	~20 to10^2 Torr	~10 to 100 atm
Operating Temperature, °C	~40	~700	~700
Output, watts/cm	~0.2 to 0.5	~40 to 120	Hi Intensity

ultraviolet radiation sources. High-pressure lamps are mainly used in photolithography, but not in the curing of coatings and inks. However, high-pressure lamps are used in spot curing devices that deliver radiation to small areas as may be encountered in the electronics, medical, and dental areas.

Low and medium pressure lamps are sealed, transparent, quartz tubes that contain mercury and, in certain instances, traces of compounds that can be used to alter the transmission spectra emanating from the bulb. The lamps are available in various lengths and energy output. When electrode lamps are energized with an electrical current, an arc is struck between electrodes located at the ends of the tube, and the contained mercury is vaporized. The vaporized mercury is ionized through collisions with electrons, e.

$$Hg + e \rightarrow Hg^* + e$$

$$Hg^* + e \rightarrow Hg^+ + 2e$$

$$Hg^* + Hg^* \rightarrow Hg^+ + Hg + e$$

$$Hg^* + Hg^* \rightarrow Hg_2{}^+ + e$$

In the case of low-pressure lamps, a small amount of argon is introduced into the bulb. The inert gas does not alter the output spectra, but it does enter into the start-up of photon generation.

$$Ar + e \rightarrow Ar^* + e$$

$$Ar^* + Hg \rightarrow Hg^* + Ar + e$$

The excited mercury atoms, Hg*, move from a variety of transition states to different, but lower, transition states. In this process, energy is released in the form of photons. The emitted photons have a wavelength (energy) that is dependent on the particular transition that takes place. Although many transitions are possible with mercury, particular transitions and, therefore, wavelength or energy outputs can be favored by the conditions (compounds or gases) that exist within the tube.

The electrodes are usually made of tungsten, and they are the electrical-current coupling devices that send energy to the electrodes and, thereby, cause the discharge in the tube. The discharge envelope or lamp body, which is often termed the "discharge envelope," is formed from a rock crystal form of quartz, which is a stable, high-purity, crystalline form of silicon dioxide, that is melted (~1700° C) and cooled to form transparent vitreous silica. This vitreous silica, often termed merely "quartz" in the industry, will transmit ultraviolet radiation, whereas common glass, as is used in windowpanes, light bulbs, etc., is opaque to it [65]. Also, vitreous silica is capable of withstanding the high operating temperatures (700° C) of the system and has a lower coefficient of expansion, as well as lower electrical conductivity than glass. Quartz purity and the presence of trace amounts of other compounds in the quartz will affect transmission properties and can be used to control these properties. If silica sand is used, the vitreous silica is translucent due to the presence of numerous, very tiny air bubbles in the product, which results in a poorly transmitting glass. Both the type of rock crystal used and the method of melting have an effect on the properties of transparent vitreous silica [66].

When trace amounts of compounds are added to the bulb, the process is referred to as doping. Halides of antimony, cobalt, gallium, indium, iron, lead, manganese, or silver alone or in combination are used to alter lamp spectral output [64, 67–70]. Particular lines can be added to the output spectrum, lower wavelengths can be decreased in intensity, or certain lines can be enhanced. For example, if a mixture of cobalt, iron, and nickel iodide is added, a portion of the input power can be converted into radiation in the range of 310–350 nm. Silver halides markedly decrease ozone formation and health hazards that arise from short-wave ultraviolet radiation. Such specially designed lamps can be useful in particular areas, such as the pressure sensitive adhesive area [71,72]. Since only molecules that absorb energy of the particular frequency emitted by the irradiator will react, there are very practical aspects to wavelength and its selection [73]. The three predominant, available bulbs are the standard mercury bulb with strong emission lines or bands at 254, 300–320, and 365 nm, the iron-filled bulb with enhanced output in the 320–390 nm region, and the gallium-filled bulb with enhanced output in the 390–410 nm region [74].

A novel bulb that has enhanced emission in the "titanium dioxide window" has been developed [75]. This bulb with enhanced emission in the 390–420 nm region is a good match with the absorption spectra of phosphine oxide photoinitiators, and coatings cured with it have improved cure speed, pendulum hardness, and solvent resistance in comparison to other bulbs designed for use with pigmented coatings. The bulb offers manufacturers the potential for benefits in white-pigmented coatings.

Electrodeless lamps are vitreous silica discharge tubes that contain mercury and/or other metals or materials. Rather than having electrodes that are supplied electrical current to vaporize the contained metals, electrodeless lamps use microwave energy that is directed by means of wave guides in a microwave chamber that houses the lamp. Exposure to the controlled microwaves results in the emission of ultraviolet radiation. Fusion Systems Corporation made successful application of this technology to high-powered, medium pressure mercury lamps in the mid 1970s [76–80]. These lamps are very popular today, and The Nordson Corporation has become a second supplier of air-cooled electrodeless lamps that are available in widths of 4, 6, 8, 10, and 13 in., with power outputs of up to 2540 watts per cm (1000 watts per in). [81]. Although electrodeless lamps are popular, conventional bulb technology is also widely used in industry.

Electrodeless lamps are more complicated than conventional lamps operated by electrodes, but they have several advantages. These include:

- Essentially instant on and off capabilities.
- Lamp lifetime is longer since there are no electrodes to deteriorate.
- A wide range of additives including metal halides can be used since there are no electrodes with which they might interact.
- The lamps deliver less infrared radiation than conventional bulbs.
- The absence of electrodes makes essentially the entire length of the lamp functional for delivering radiation.

The instant on/off ability is due to the very short warm-up times needed. The warm-up time is measured in seconds rather than the minutes required for lamps with electrodes. This feature also decreases or eliminates the need for shutter or shielding systems to protect sensitive substrates during production line stoppages or other holdups. Since less infrared radiation is generated, the electrodeless lamps run cooler than conventional lamps. This is an advantage when heat sensitive substrates such as plastics and electronic components are involved. With the entire length of the lamp functional, the lamp systems may be manufactured as modules that can be connected in a variety of designs and give an essentially uniform radiation pattern.

The main use for low-pressure mercury vapor lamps is in the common fluorescent lamp used in work places and homes. The tubes in these lamps are coated with a fluorescent compound that absorbs the 185 and 254 nm radiation emitted by mercury under these conditions and re-emits it in the visible region. The tubes are also used without the fluorescent coating to destroy bacteria and mold in sterilization applications and to pre-gel coatings in the wood industry. These lamps have a number of advantages, including low operating temperature, low cost, and long lifetimes. However, they have not been extensively used in ultraviolet radiation curing applications. The short wavelengths involved generate ozone, and this is a difficulty that must be considered when such lamps are used. The low lamp intensity results in only shallow penetration of a liquid coating. Thus, photoinitiator photolysis takes place at a slow rate and mainly in the surface regions of the coating. As a result, ineffective cures take place unless long radiation time periods are involved. In addition, with free-radical photoinitiator systems that are deleteriously affected by oxygen, and if an air atmosphere is present, the oxygen will quickly negate the generated free radicals.

Medium-pressure mercury-vapor lamps are widely used in the photocure of coatings and inks. Because these lamps operate at high temperatures, it is necessary to cool systems that involve their use. Fans or circulating water systems cool the lamps to maintain lamp-operating temperature over a reasonably narrow range. This results in a spectral output that is uniform in nature. Special heat-reducing dichroic filters that block much of the heat producing infrared can also be used to control operating temperature. Conventional medium-pressure lamps require a warm-up time of 10–30 min. to vaporize the mercury completely and to bring the lamp up to peak operating efficiency. Operating lifetime for conventional lamps is 1000 h, but they often are used for 2000 h or longer. Electrodeless lamps have a lifetime warranty of 5000 h for metal halide or "H" or mercury bulbs and 3000 h for special gas-fill bulbs. The decrease in radiation as a function of operating time is gradual, and good maintenance is important to consistent output and curing. In multibulb units, bulbs are often replaced on a rotation basis. Lamps can become fouled during use, and this will decrease output and can slow cure or require higher energy use than should be necessary to effect cure. A test for examining fouling by silicone-based additives was developed, and it pointed out that such additives can be a serious cause of fouling [82]. Reducing or eliminating use of such products in the formulation mitigated or eliminated the fouling.

When focused reflector systems are used, the lamps can generate a very high radiation intensity made up of ultraviolet, visible, and infrared radiation. While a production conveyor is moving, the thermal energy generated by the infrared portion of the radiation doesn't cause any difficulties. However, if the production conveyor or line is stopped for any reason, this thermal energy can be sufficient to ignite paper, to damage a coating, or to damage a part being coated. If a lamp and reflector system can cause this difficulty with the equipment, coating, and substrate being used, the substrate must be protected from high radiation intensity when the line stops. The most usual methods for providing such protection are [63]:

- Placing a shield or shutter that is interlocked with the conveyor between the lamp and the conveyor.
- Devising a system through which the reflector and lamp system are rotated in a manner that the thermal energy is directed away from the substrate.
- Using an interlocked power decreasing system that will automatically decrease the power level to a safe value and then have the capability to allow rapid return to the full power level.

Xenon Lamps

Inert gases are of particular interest as filling for discharge tubes. Their very nature provides an advantage because inert gases will not attack electrodes and/or the fused silica containing-tubes. They can be used at a variety of pressures and require no warm-up period, as can be required by mercury lamps. In addition, because of their atomic nature, there are no molecular processes, such as rotation and vibration, that will use and dissipate energy that might have been radiated. On the disadvantage side, the complete outer shells of the inert gases result in high ionization potentials, compared with metals [83]. As is apparent from the data in Table 2.3, the energy of the lowest excited state of the inert gases is much higher than that of mercury. Of the noble gases, xenon has the lowest energy value excited state, 8.3 eV, in comparison to that of mercury, 4.7 eV. It is the most efficient inert gas for converting electrical energy into radiation [84]. Because of the higher energy levels in xenon, operating lamp temperatures are greater than those obtained with mercury lamps.

Xenon lamps will produce continuous radiation or pulsed radiation. The pulsed lamps have extremely high peak radiation intensity that is produced instantaneously, as would be produced by an electronic flash lamp used for photography. The on/off character possible with xenon lamps makes them very useful for providing high intensity, pulsed radiation that is useful for spot curing of adhesives, sealants, and coatings. The continuous radiating lamps are mainly used for lighting purposes. Xenon arcs produce a continuous spectrum that has a color close to that of noonday sunlight, and thus, are used in searchlights, large-scale illumination, photographic lighting sets, lighthouses, and similar end uses.

Pulsed or flashed xenon lamps have been used in the flash photographic area for more than 70 years [85], and this topic has been reviewed by Carlson and Pritchard [86], as well as others at about the same time. The lamps have been used in specialty lighting situations [87], such as motion and still ultra-high-speed photography, stroboscopes, for nighttime aerial reconnaissance in the Second World War and as a means for generating short-lived species by flash photolysis in laboratories. The high-speed photography gave us pictures of bullets in flight, light bulbs shattering, etc. In about the 1970s, xenon flash tubes began to find a place in ultraviolet radiation curing [88–91].

TABLE 2.3. Ionization Potential of Inert Gases and Selected Metals [18].

ELEMENT	IONIZATION POTENTIAL OF NEUTRAL ATOM, eV*	IONIZATION POTENTIAL OF SINGLY IONIZED ATOM, eV*
Helium	24.59	54.42
Neon	21.56	40.96
Argon	15.76	27.63
Krypton	14.00	24.36
Xenon	12.13	21.21
Mercury	10.44	18.76
Iron	7.90	16.19
Sodium	5.14	47.29

* 8065.541 cm^{-1} = 1 eV (conversion of wave number to electron volts).

Xenon flash tubes are constructed similarly to mercury vapor lamps. The electrodes are polarized, so there is a difference between the anode and the cathode. An emissive material may be coated onto the electrodes to alter the radiation spectra produced. The tube body is made of vitreous silica or borosilicate glass for the same reasons described earlier for mercury lamps, namely high transmission characteristics, low coefficient of expansion, and low electrical conductivity in comparison to glass. In operation, a pulsing circuit is used. A capacitor in the lamp system is charged to a particular voltage and is the source of stored energy. An ignition voltage is applied to ionize atoms of gas in the tube. Then the stored energy is discharged to ground to create a plasma in the flash tube. During discharge, the energy is converted into circuit and tube losses (30–40%) and into radiation (60–70%) that is capable of curing photocure formulations. The radiation is about 25% ultraviolet, 50% visible, and 25% infrared, and the emission spectrum generated is dependent on factors such as [92]:

- Charging voltage
- Capacitor storage size
- Gas type and pressure
- Glass type and additives
- Tube length and cross-sectional area

Xenon lamps were used to investigate the cure of gray-pigmented coatings suitable for marine coatings that would function as touch-up coatings when a ship is underway [93]. Factors, such as the ability to adhere to cold, poorly prepared steel surfaces, were studied. Most problems were solved, except for yellowing that took place during a salt fog test.

Hybrid Xenon/Mercury Lamps

Hybrid lamps of this type have mercury added to the xenon gas in the quartz envelope. This enhances the xenon lamp output in the area encompassing the mercury spectral lines. Systems designed for curing coatings on three-dimensional objects have been described [63]. These incorporate a medium-pressure mercury vapor lamp and a pulsed xenon lamp in a single quartz envelope. Xenon is used to fill the tube, but enough mercury is added to allow a continuous plasma arc in the tube. In this system, small amounts of iron and beryllium iodide are included as dopants. These lamps result in a mercury spectral emission with peaks at 254 and 365 nm that has the broad-band, high intensity xenon, which is rich in short wavelength radiation, superimposed.

The hybrid systems are also used as spot curing devices. Such devices are available with the usual lamps that emit from low wavelengths to 450 nm and are available with ozone-free lamps [94]. Most of these units provide radiation through the ultraviolet spectrum and the visible spectrum [95]. The use of filters allows selection of particular wavelengths when needed to activate particular photoinitiators.

Excimer Radiation

"Excimer" is a term that stands for excited dimer. The dimer is a diatomic molecule that is usually composed of an inert gas atom and a halide atom. These atoms are only bound into the diatomic molecule when they are in an excited state. The excited state and, therefore, the excimer has only an extremely short lifetime, that is on the order of nanoseconds, after which the atoms dissociate and release the excitation energy as photons of ultraviolet radiation that have a high selectivity. That is, the radiation emitted is quasi-monochromatic in nature. Thus, the photons are formed in a manner similar to that formed by excimer lasers. The main difference is that excimer lamps produce incoherent radiation, and thus, can be used for large area application processes [96].

Different wavelengths can be obtained by selecting different gases for the quartz tubes. The inert gas-halide systems that have been found to be most efficient along with their radiation emission frequency are [64]: ArF* (192 nm), KrCl* (222 nm) KrF* (248 nm), XeBr* (282 nm), XeCl* (308 nm), and XeF* (351 nm). If only an inert or rare gas is present in the discharge tube, it has been found that excimers can form when excited rare gas atoms interact with other gas atoms. The emission frequency of such excimers is: Ar_2* (126 nm), Kr_2* (146 nm), and Xe_2* (172 nm). Certain commercial units have lamps that are optimized at wavelengths of 172 nm, 222 nm, and 308 nm [96].

Excimer lamps are designed differently than other ultraviolet-radiation-producing lamps. The electrodes are not located in the gas medium, but are separated from it by a dielectric barrier that is usually quartz. The inner electrode is solid in nature, but the outer electrode is a metal grid that transmits the radiation generated by the excimers' dissociation. A decided advantage of these lamps is that they do not generate infrared radiation in the discharge process, and therefore, operate without raising the substrate temperature any appreciable amount. The lamp surface reaches temperatures of about 35 – 40° C [96]. This is a distinct advantage when heat-sensitive substrates are involved. In addition to curing coatings and adhesives, these devices are used in the medical field where they are capable of removing extremely thin layers of tissue without causing damage to adjacent layers.

Visible Radiation, Light

Light, which is the visible radiation that occurs between 400 and 750 nm [97], has advantages in certain applications over ultraviolet radiation. These applications usually involve those in which a patient and operator could suffer physiological harm by repeated exposure to an unshielded or improperly shielded ultraviolet-radiation source as in dentistry and orthopedic cast or device areas. The energy from light is less powerful than that from ultraviolet radiation sources (See Table 1 of Chapter 1), and thus, in general, cure times are longer. This can be useful in encapsulation or potting applications, as are found in opto-electronics where there is a need to fix alignments of active components quickly and accurately by one of these techniques [98]. Another advantage of light curing is that there is less scatter of the longer wavelengths than those involved in the ultraviolet range, and thus, greater depth of cure can be achieved.

Light sources are inexpensive to acquire and to operate and are readily available in many power intensities. Suitable light sources include incandescent spotlights (400–600 nm), tungsten-halogen lamps, daylight fluorescent tubes (350–450 nm and higher), blue light (400–500 nm), sunlight, and other sources that provide radiation in the 400–760 nm range. If the source emits both light and ultraviolet radiation, as is the case with sunlight, the ultraviolet radiation can be removed, if it is undesirable, by including a suitable filter as part of the radiation delivery system. An additional source that might be considered is sodium metal: 580–600 nm [99].

As with other radiation sources, the rate of curing with light is dependent on lamp intensity, distance of the lamp from the material to be cured, and the thickness to be cured. Incandescent spotlights and tungsten-halogen lamps are usually of 150-watt capacity [100,101]. The latter lamps are incandescent sources that have been optimized for long lamp lifetimes, a factor that decreases lamp maintenance. All incandescent lamps are based on a tungsten filament that acts as a radiator that produces light and other radiation when heated to a high temperature that is on the order of 3000° K. The quantity of light generated depends on the surface area of the filament, and the spectral properties of the light depend on filament design and operating temperature.

Spot curing devices that utilize visible radiation are based on high-pressure mercury vapor lamps [102]. The lamps emit both visible and ultraviolet radiation, but the ultraviolet radiation can be removed with a variety of filters that permit output refinements for specific cure conditions. For example, filters for 250–450 nm, 320–390 nm, 320–500 nm, 365 nm, and 400–500 nm are commercially available.

References

[1] Patent: Irving, E., Banks, C. P., Ciba-Geigy Corporation, "Method of Forming Images," U. S. 4,849,320 (1989).

[2] Patent: Crivello, J. V., Polyset Corporation, "Electron Beam Curable Epoxy Compositions," U. S. 5,260,349 (1993).

[3] Janke, C., Oak Ridge Center for Composites Manufacturing Technology, *www.ornl.gov/gov/publications/labnotes/oct96.html*.

[4] Palmese, G. R., Ghosh, N. N., and McKnight, S. H., International SAMPE Symposium, 45, 1874, 2000.

[5] Shu, J. S., Covington, J. B., Lee, W., Venable, L. G., and Varnell, G. L., Texas Instruments Company, 1985.

[6] Patent: Lapin, S. C., Allied Corporation, "Semi-interpenetrating Polymer Networks," U. S. 4,654,379 (1987).

[7] Berejka, A. J. and Eberle, C., "Electron Beam Curing of Composites in North America," Oak Ridge National Laboratory International Meeting on RADIATION PROCESSING, XII Paper, 2001.

[8] Patent: Janke, C. J., Dorsey, G. F., Haven, S. J., Lopata, V. J., Lockheed Martin Energy Systems, Inc., "Toughened Epoxy Resin System and a Method Thereof," U. S. 5,726,216 (1998).

[9] Goodman, D. L. and Byrne, C. A., "Composite Curing with High Energy Electron Beams: Novel Materials and Processes," Technical Paper, *28th International SAMPE Technical Conference*, Seattle, WA, November 4–7, 1996.

[10] Livesay, W. R., "A New Compact Electron Beam Curing System," *Proceedings of RadTech '90*–North America, Chicago, IL, Vol. 2, 25–29 March, 1990, p. 195.

[11] Comello, V., "E-Beam Systems Combine Durability and Precision," *R&D Magazine*, 40, 11, 12, October 1998.

[12] Felis, K., Avnery, T., and Berejka, A., "Innovative Energy Efficient Low Voltage Electron Beam Emitters," Technical Paper, 12th International Meeting on Radiation Processing, Spring 2001.

[13] Berejka, A., private communications, August and December 2001.

[14] Maguire, E., "EB Equipment," *Proceedings of End User Conference, RadTech 2000*, Baltimore, MD, 9–12 April, 2000, p. 287.

[15] Bean, A. J., "Electron Beam Curing in Graphic Arts," *Proceedings of End User Conference, RadTech 2000*, Baltimore, MD, 9–12 April, 2000, p. 296.

[16] Nelson, P., "EB Compatibility with Presses," *Proceedings of End User Conference, RadTech 2000*, Baltimore, MD, 9–12 April, 2000, p. 334.

[17] Full, J., "EB Coatings and Adhesives," *Proceedings of End User Conference, RadTech 2000*, Baltimore, MD, 9–12 April, 2000, p. 301.

[18] Duncan, D. P., "EB Inks—Progress, Performance, and Potential," *Proceedings of End User Conference, RadTech 2000*, Baltimore, MD, 9–12 April, 2000, p. 327.

[19] Rodriguez, J., "EB Inks and Coatings," *Proceedings of End User Conference, RadTech 2000*, Baltimore, MD, 9–12 April, 2000, p. 339.

[20] Greise, Elmer, "Green Benefits of EB Technology," *Proceedings of End User Conference, RadTech 2000*, Baltimore, MD, 9–12 April, 2000, p. 318.

[21] Creighton, M., "How Stable are EB Coatings?" *Proceedings of End User Conference, RadTech 2000*, Baltimore, MD, 9–12 April, 2000, p. 313.

[22] Schneider, M., Klein, W., and Schröder, C., "Optimized Positioning of UV Lamps for the Treatment of 3D Workpieces," *Proceedings of RadTech Europe '99*, Berlin, Germany, 8–10 November, 1999, p. 711.

[23] Kardashian, R. and Nablo, S. V., "Electron Beam Curing Equipment," *Adhesives Age*, Vol. 25, No. 12, December 1982, p. 48.

[24] Karmann, W., "Radiation Curing Equipment," Technical Paper FC83-269, Society of Manufacturing Engineers, Lausanne, Switzerland, May 9–11, 1983.

[25] Mehnert, R., Pincus, A., Janorsky, I, Stowe, R., and Berejka, A., *UV & EB Curing Technology & Equipment*, Vol. 1, John Wiley & Sons, 1998.

[26] *Radiation Dosimetry: Electron Beams with Energies between 1 and 50 MeV*, International Commission on Radiological Units and Measurements, ICRU Report 35, 1984.

[27] McLaughlin, W. L., "Dosimetry for Low-Energy Electron Machine Performance and Process Control," *Proceedings of RadTech '90-North America*, Vol. 2, Chicago, IL, March 25–29, 1990, p. 91.

[28] Desrosiers, M. F., "A New Real-Time Telecalibration Service Traceable to NIST," *Proceedings of RadTech 2000*, Baltimore, MD, April 9–12, 2000, p. 129.

[29] Aaronson, J. N. and Nablo, S. V., "Standard Diagnostic Techniques for Determining Electron Processor Performance," *Proceedings of RADCURE '86*, Baltimore, MD, 8–11 September, 1986, pp. 17–47.
[30] Läuppi, U. V. and Rangwalla, I., "Advantages of Low Voltage Electron Beam Processing (80 kV–110 kV)," *Conference Proceedings in RadTech Europe '99*, Berlin, Germany, 8–10 November, 1999, p. 357.
[31] Quintal, B. S., "Electron Beam Equipment–The Next Generation," *Proceedings of RadTech '90–North America,* Vol. 1, Chicago, IL, 25–29 March, 1990, p. 138.
[32] Grishchenko, A. I. and Salimov, R. A., "Selfshielded Low Energy Electron Accelerator," *Proceedings of RadTech '90-North America,* Vol. 1, Chicago, IL, 25–29 March, 1990, p. 163.
[33] Weisman, J. and Tripp, E. P., III, "Electron Beam Curing of Coatings and Adhesives," *Radio Frequency/Radiation Plasma Process*, P. N. Chermisinoff, O. G. Farah, and R. P. Ouellette, Eds., Technomic Publishing Co., Inc., Lancaster, PA, 1985.
[34] Hiley, J., Watt, H., and Avnery, T., "High Power Electrocurtain Processors and Their Applications," *Proceedings of RadTech '90-North America,* Vol. 1, Chicago, IL, 25–29 March, 1990, p. 459.
[35] Patent: Coleman, G. E., PPG Industries, Inc., "Apparatus for Irradiation in a Controlled Atmosphere," U.S. 3,676,673 (1972).
[36] Patent: Nablo, S. V., Energy Sciences Inc., "Method of and Apparatus for Shielding Inert-Zone Electron Irradiation of Moving Web Materials," U. S. 4,252,413 (1981).
[37] Nablo, S. V. and Rangwalla, I., "Inerting Techniques in Electron Processing," *Proceedings of RadTech '90–North America,* Vol. 1, Chicago, IL, 25–29 March, 1990, p. 171.
[38] Zimmermann, T., "High Speed Line With UV and EB Curing," *Conference Proceedings of RadTech Europe '99,* Berlin, Germany, 8–10 November, 1999, p. 367.
[39] Gaven, T., "Choosing a UV System," *Proceedings of RadTech 2000*, Baltimore, MD, 9–12 April, 2000, p. 944.
[40] Davis, G. B., "A Comparison Study of UV Energy and Intensity Between Stepped and Continuously Variable Wattage Ballast Systems," *Proceedings of RadTech 2000*, Baltimore, MD, 9–12 April, 2000, p. 960.
[41] Schwarz, B., "UV-Systems Technology in Comparison," *Conference Proceedings, 347, RadTech Europe '99*, Berlin, Germany, 8–10 November, 1999.
[42] Nordson New Product Bulletin, "UV MAC10 Microwave-Powered Ultraviolet Curing System," Nordson Corporation, Amherst, OH, 2000.
[43] Stowe, R. W., "UV Curing of Solid Objects and Non-Flat (3D) Surfaces," *Conference Papers of RadTech '88–North America,* New Orleans, LA, 24–28 April, 1988, p. 532.
[44] Blake, D. E., "UV Curing of 3D Objects Using Near-Point Source Optics," *Conference Papers of RadTech '88–North America*, New Orleans, LA, 24–28 April, 1988, p. 539.
[45] Lake, R. T., "Finishing With 3-Dimensional UV Coating," *Conference Papers of RadTech '88–North America*, New Orleans, LA, 24–28 April, 1988, p. 550.
[46] Riedell, A., "3-Dimensional Finishing of Wood Furniture," *Conference Papers of RadTech '88–North America*, New Orleans, LA, 24–28 April, 1988, p. 556.
[47] Arnold, H. S., Sr., "3D UV Curing in Furniture and Wood Products Now and in the Future," *Conference Papers Addendum of RadTech '88–North America*, New Orleans, LA, 24–28 April, 1988, p. 80.
[48] Stowe, R. W., "UV Curing of Solid Objects and Non-Flat (3D) Surfaces," *Proceedings of the ACS Division of Polymeric Materials: Science and Engineering*, Vol. 60, 108, Dallas, TX, Spring Meeting 1989.
[49] Whittle, S., "Process Control–The UV Equipment Manufacturers Contribution," *Proceedings of RadTech Europe '99*, Berlin, Germany, 8–10 November, 1999, p. 691.
[50] Stowe, R. W., "Practical Aspects of Irradiance and Energy Density in UV Curing," *Proceedings of RadTech Europe '99*, Berlin, Germany, 8–10 November, 1999, p. 339.
[51] Stowe, R. W., "Power Play, Analyzing the use of UV Light as the Energy Source for Drying," *Modern Paint & Coatings*, Vol. 91, No. 3, March 2001, p. 33.

[52] Starzmann, O., "UV—Technology—Status Quo," *Proceedings of End User Conference, RadTech 2000*, Baltimore, MD, 9–12 April, 2000, p. 250.
[53] Beying, A. and Baudoui, Y., "UV Lamps—Performance and Control," *Proceedings of End User Conference, RadTech 2000*, Baltimore, MD, 9–12 April, 2000, p. 266.
[54] Ridyard, A., "Why Measure UV," *Proceedings of RadTech Europe '99*, Berlin, Germany, 8–10 November, 1999, p. 697.
[55] Beying, A., "UV Monitoring–Why and How," *Proceedings of RadTech Europe '99*, Berlin, Germany, 8–10 November, 1999, p. 685.
[56] Stowe, "Radiometry and Methods in UV Processing," *Proceedings of RadTech 2000*, Baltimore, MD, 9–12 April, 2000, p. 113.
[57] Ridyard, A., "Accurate, High Power UV Spectral Measurement," *Proceedings of RadTech 2000*, Baltimore, MD, 9–12 April, 2000, p. 136.
[58] Persson, E., "Making and Interpreting Power Measurements on Arc Lamps and Ballasts," *Proceedings of RadTech 2000*, Baltimore, MD, 9–12 April, 2000, p. 156.
[59] May, J. T., "UV Measurements in Radiation Curing," *Proceedings of RADCURE '86*, Baltimore, MD, 8–11 September, 1986, pp. 17–23.
[60] May, J. T., "Instrument to Map UV Intensity and Temperature in UV Machine," *Proceedings of RadTech '88- North America*, New Orleans, LA, 24–28 April, 1988, p. 133.
[61] Siegel, S. B. and Berejka, A. J., "Visual Dosimetry Indicators," *Addendum Papers for RadTech '88-North America*, New Orleans, LA, 24–28 April, 1988, p. 54.
[62] Phillips, R., *Sources and Applications of Ultraviolet Radiation*, Academic Press, ISBN 0 1255388 04, 1983.
[63] Keough, A. H., "UV Sources and Applications," Technical Paper FC85-427, Society of Manufacturing Engineers, Basel, Switzerland, 6–8 May, 1985.
[64] Heering, W., "Efficient Generation of UV Radiation," *Proceedings of RadTech Europe '99*, Berlin, Germany, 8–10 November, 1999, p. 333.
[65] *Handbook of Chemistry and Physics*, 76th ed., David R. Linde, Editor-in-Chief, 10–305, 1995–96.
[66] Koller, L. R., *Ultraviolet Radiation*, 2nd ed., John Wiley & Sons, New York, 1965.
[67] Hagood, D., "Understanding UV-Curing Techniques and Processes," *Industrial Paint & Powder*, Vol. 77, No. 7, July 2001, p. 22.
[68] Patent: Langer, A., Genz, A., Deisenhofer, M., Keile, W., Lewandowski, B., Reichardt, J., Treuhand-Gesellschaft fur Elektrische, "Quartz Glass with Reduced Ultraviolet Radiation Transmissivity, and Electrical Discharge Lamp Using Such Glass," U. S. 5,572,091 (1996).
[69] Patent: Yakub, Y., Azran, Y., Lamptech Ltd., "Metal Halide Lamp Including Iron and Molybdenum," U. S. 5,594,302 (1997).
[70] Patent: Watzke, E., Kloss, T., Brix, P., Ott, F., Schott Glaswerke, "Reducing Melt Borosilicate Glass Having Improved UV Transmission Properties and Water Resistance and Methods of Use," U. S. 5,610,108 (1997).
[71] Fisher, R., "Window of Opportunity," *Adhesives Age*, Vol. 43, No. 10, October 2000, p. 40.
[72] Foreman, P., Eaton, P., and Shah, S., "The Best of Both Worlds," *Adhesives Age*, Vol. 44, No. 9, September 2001, p. 50.
[73] Schaeffer, W. R., "The Practical Effects of Wavelength Selection," *Proceedings of RadTech '88-North America*, New Orleans, LA, 24–28 April, 1988, p. 127.
[74] Whittle, S., "Developments in UV Curing Equipment," RADTECH Report, Vol. 13, No. 1, January/February 2000, p. 15.
[75] Bao, R. and McCartney, R., "The Applications of VIP 397/418 Bulbs in Free Radical White Pigmented Coatings," *Proceedings of RadTech 2000*, Baltimore, MD, 9–12 April, 2000, p. 983.
[76] Patent: Spero, D. M., Eastlund, B. J., Ury, M. G., Fusion System Corporation, U. S. 3,872,349 (1975).
[77] Patent: Spero, D. M., Eastlund, B. J., Ury, M. G., Fusion System Corporation, U. S. 3,911,318 (1974).

[78] Patent: Ury, M. G., Eastlund, B. J., Braden, R. S., Wood, C. H., Fusion Systems Corporation, U. S. 4,042,850 (1977-08-16).
[79] Finnegan, E., "Equipment for Curing Ultraviolet Curable Adhesives and Release Coatings," *Journal of Radiation Curing*, Vol. 9, No. 6, July 1982, p. 4.
[80] Schaeffer, W. R., Technical Paper FC75-768, Society of Manufacturing Engineers, Detroit, MI, 1975.
[81] Nordson Corporation, Spectral® UV Systems, "DCF UV Cure Units," Product Specification Sheet, 2000.
[82] Ross, J. S., Malkowski, E. A., Farby, J. C., and Leininger, L. W., "A Simple Technique for Prediction of Lamp Fouling in UV Cure Systems," *Proceedings of End User Conference, RadTech 2000*, Baltimore, MD, 9–12 April, 2000, p. 279.
[83] *Handbook of Chemistry and Physics*, 76th ed., David R. Linde, Editor-in-Chief, **10**-207–209, (1995–1996).
[84] Hoyt, G. D. and McCormick, W. W., *Journal of Optical Society of America*, Vol. 40, 1950, p. 658.
[85] Laporte, M., *Comptus Rendus*, Vol. 204, 1937, p. 1240.
[86] Carlson, F. E. and Pritchard, D. A., *Illumination Engineering*, Vol. 42, 1947, p. 235.
[87] Edgerton, H. E., *Electronic Flash, Strobe*, McGraw-Hill, New York, 1970.
[88] Panico, L. R., Technical Paper FC75-326 (1975), Technical Paper FC76-499 (1976), Technical Paper FC83-601 (1983), Society of Manufacturing Engineers, Detroit, MI.
[89] Michalski, M., Technical Paper FC75-337 (1975), Society of Manufacturing Engineers.
[90] Bordzol, L., Technical Paper FC75-338 (1975), Society of Manufacturing Engineers.
[91] Panico, L. R., *Paperboard Packaging*, Vol. 61, No. 6, 1976, p. 50; "Pulsed UV Curing Provides User-Friendly Solutions to Tough Problems," Vol. 40, No. 11, January 1997, p. 34.
[92] Stöhr, A. and Renschke, J., "UV-Flash Curing and its Applications," *Proceedings of RadTech Europe '99*, Berlin, Germany, 8–10 November, 1999, p. 353.
[93] Howell, B. and O'Donnell, M., "Radiation Curing of Coatings With Xenon Lamps," *Modern Paint and Coatings*, Vol. 85, No. 4, April 1995, p. 38.
[94] Lightningcure LC5 and 200 Series Brochure TLSX1033E01, Hamamatsu Photonics K.K., Shizoka-ken, Japan, November 2000.
[95] Hood, R., "High Intensity Ultraviolet Spot Curing," *Proceedings of RadTech '90–North America Radiation*, Chicago, IL, Vol. 1, 25–29 March, 1990, p. 159.
[96] Roth, A., "Excimer Lamps—A Cold Curing Alternative in the UV Range," *RadTech Report*, Vol. 10, No. 5, September/October 1996, p. 21.
[97] Kosar, J., *Light Sensitive System*, J. Wiley, New York, 1966.
[98] Rogers, S. C., "Visible Light Curing Provides Low Shrinkage and Good Depth," *Adhesives Age*, Vol. 31, No. 4, April 1988, p. 20.
[99] Day, H. R. and Leppeimeier, E. T., "Photopolymers—Principles, Processes, and Materials," SPE Technical Paper, Mid-Hudson Section, November 1, 1967.
[100] Patent: Buck, C. J., Johnson & Johnson Products, Inc., U. S. 4,512,340 (1985).
[101] Patent: Dart, E. C., Perry, A. R., Nemcek, J., Imperial Chemical Industries, Ltd., U. S. 3,874,376 (1975).
[102] "Novacure® Ultraviolet/Visible Spot Cure System," (brochure) EFOS Inc., Mississauga, Ontario, Canada.

3

Free-Radical Photoinitiators and Initiation Mechanism

Introduction

RADIATION POLYMERIZATION follows the general scheme for any polymerization. Suitable reactive monomers must be first initiated to form a species that is capable of being polymerized, then be propagated so as to form a polymeric chain, and finally, the growing polymer chains must be terminated in some manner. These radiation polymerizations are effected or initiated through one or the other of two general mechanisms: free radical initiation or cationic initiation. Free-radical initiation is achieved either by use of an electron beam or other suitable means that generates ionizing radiation capable of generating free radicals from suitable ethylenically unsaturated monomers, or by use of ultraviolet radiation and a photoinitiator that will photolyze (degrade) in the presence of such radiation and produce free radicals. Cationic initiation is achieved by photochemical means and requires the use of a photoinitiator that will photolyze to form Lewis or Bronsted acids. Species that generate Bronsted or protonic acids are currently in commercial use. In addition to generating an acid, the photolysis of cationic photoinitiators also results in the generation of free radicals. This allows dual or hybrid cures of cycloaliphatic epoxides and acrylate esters, as well as other combinations.

Free radicals will initiate polymerization of a variety of compounds that contain ethylenic unsaturation, and these will propagate by an addition process. Cations will initiate polymerization of cyclic compounds, such as cycloaliphatic epoxides, by a ring opening, rearrangement process and certain unsaturated compounds, such as vinyl ethers, by an addition process. The mechanisms through which a variety of monomers will polymerize are given in Table 3.1. Although the monomers given in Table 3.1 will polymerize by the indicated mechanisms, marked differences exist among the different types of monomers in their rate of polymerization. Thus, it should be understood that because a particular monomer is listed in the table, it does not necessarily follow that the monomer can be practically cured with a radiation process. For example, ethylene is a gas and would not be manageable in such a coating process.

TABLE 3.1. Mechanism for Polymerization of Various Monomers (Adapted from [1]).

	POLYMERIZATION MECHANISM			
MONOMER TYPE	FREE RADICAL	CATIONIC	ANIONIC	COORDINATION
α-Olefins	No	No	No	Yes
Acrylates	Yes	No	Yes	Yes
Dienes	Yes	No	Yes	Yes
Epoxides, cycloaliphatic	No	Yes	Yes	No
Epoxides, glycidyl	No	Yes	Yes	No
Ethylene	Yes	Yes	No	Yes
Isobutylene	No	Yes	No	No
Methacrylates	Yes	No	Yes	Yes
N-vinylcarbazole	Yes	Yes	No	No
N-vinylpyrrolidone	Yes	Yes	No	No
Nitroethylenes	No	No	Yes	No
Styrene	Yes	Yes	Yes	Yes
Vinyl halides	Yes	No	No	Yes
Vinyl esters	Yes	Yes	No	Yes
Vinyl ethers	No	Yes	No	No
Vinylidine halides	Yes	No	No	Yes

As would be expected, certain types of monomers polymerize more rapidly than others, and these are the monomers that are of importance to curing with ionizing or nonionizing radiation. For example, it is well known in the industry that vinyl groups are far less responsive to ultraviolet radiation curing than acrylates, and that the following cure rate hierarchy exists:

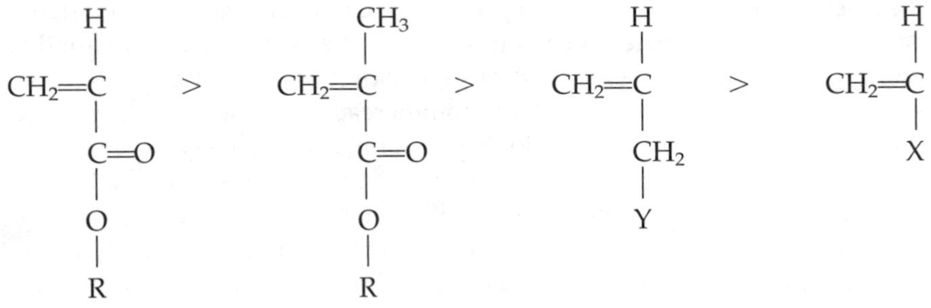

$$\underset{\text{Acrylate}}{CH_2{=}C(H)-C({=}O)-O-R} > \underset{\text{Methacrylate}}{CH_2{=}C(CH_3)-C({=}O)-O-R} > \underset{\text{Allyl}}{CH_2{=}C(H)-CH_2-Y} > \underset{\text{Vinyl}}{CH_2{=}C(H)-X}$$

wherein R might be an alkyl group, Y an hydroxyl group, and X a halide, phenyl, or alkyl ester. Although both acrylate and methacrylate monomers will polymerize by a free radical process, the former polymerize markedly faster than the latter, and they are the preferred monomers in free-radical initiated, radiation-cure processes. Although vinyls are indicated to be slowest of those monomers shown above, styrene is used in combination with unsaturated polyesters to form systems with a desirable cost/property balance that are cured with high-energy, ionizing radiation. Cycloaliphatic epoxides, which are invisible to ultraviolet radiation, cure much more rapidly in cationic systems than glycidyl epoxides that contain aromatic structures, such as the diglycidyl ethers of bisphenol A. The aromatic structures in the latter epoxides block radiation, and this affects depth of cure and cure response.

Photoinitiators

Photoinitiators are compounds that absorb radiation and are thereby raised to an excited state. From their radiation-excited state, photoinitiators photolyze or degrade directly or indirectly into free radicals or cations. These become the initiating species and cause very rapid polymerization of photocurable formulations, based on a variety of chemistries. The main chemistries that involve the use of photoinitiators are:

- Unsaturated polyesters/styrene—free radical initiation
- Polyenes/thiols—free radical initiation
- Acrylates/methacrylates—free radical initiation
- Cycloaliphatic epoxides/polyols/vinyl ethers—cationic initiation

This chapter is concerned with free radical initiation and the polymerization of acrylates. The next chapter will deal with cationic initiation and the polymerization of cycloaliphatic epoxides and related formulating ingredients.

Free Radical Photoinitiators

There are two main types of free-radical generating photoinitiators. Actually, as will be seen later, there are three types, since certain cation-generating photoinitiators also generate free radicals during their photodecomposition. The photodecompositions leading to free radicals [2] are often classified as Norrish Type I [3, 4], Norrish Type II [5], and Norrish Type III reactions [2].

The two main types of photoinitiators are those that form an active species by a fragmentation process or by a hydrogen abstraction process. The photoinitiators that undergo Norrish Type I reactions photolyze through a homolytic fragmentation mechanism or α-cleavage and directly form free radicals capable of initiating polymerization. The absorbed radiation causes bond breakage to take place between a carbonyl group and an adjacent carbon. The photoinitiators that undergo Norrish Type II reactions are activated with radiation and form free radicals by hydrogen abstraction or electron extraction from a second compound that becomes the actual initiating free radical. An obvious advantage of homolytic fragmentation over hydrogen abstraction is the necessity of a bimolecular reaction taking place in the latter case. Norrish Type III reactions are intramolecular, nonradical processes that involve a β-hydrogen atom. The process leads to formation of an olefin and an aldehyde through a carbon-carbon bond scission next to a carbonyl group [2].

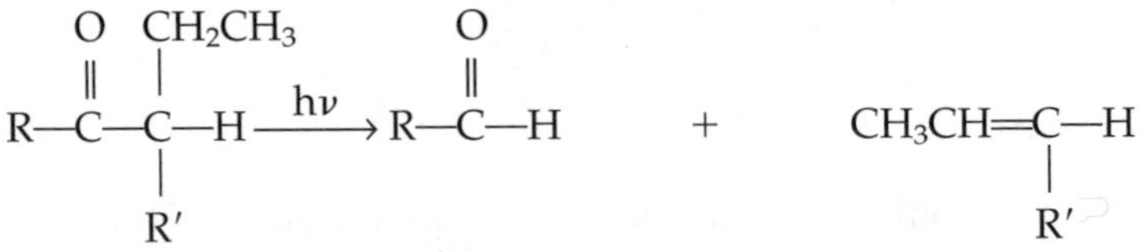

$$R{-}\overset{\overset{\displaystyle O}{\|}}{C}{-}\underset{\underset{\displaystyle R'}{|}}{\overset{\overset{\displaystyle CH_2CH_3}{|}}{C}}{-}H \xrightarrow{h\nu} R{-}\overset{\overset{\displaystyle O}{\|}}{C}{-}H \quad + \quad CH_3CH{=}\underset{\underset{\displaystyle R'}{|}}{C}{-}H$$

Norrish Type III Reaction

Homolytic Fragmentation Type

Photoinitiators that function by means of a homolytic fragmentation or α-cleavage mechanism produce two radicals when they photolyze. Ultraviolet radiation is absorbed by the photoinitiator, and it is raised to an excited state. The excited molecule then spontaneously fragments into two free radicals, as exemplified by the photolysis of a benzoin alkyl ether.

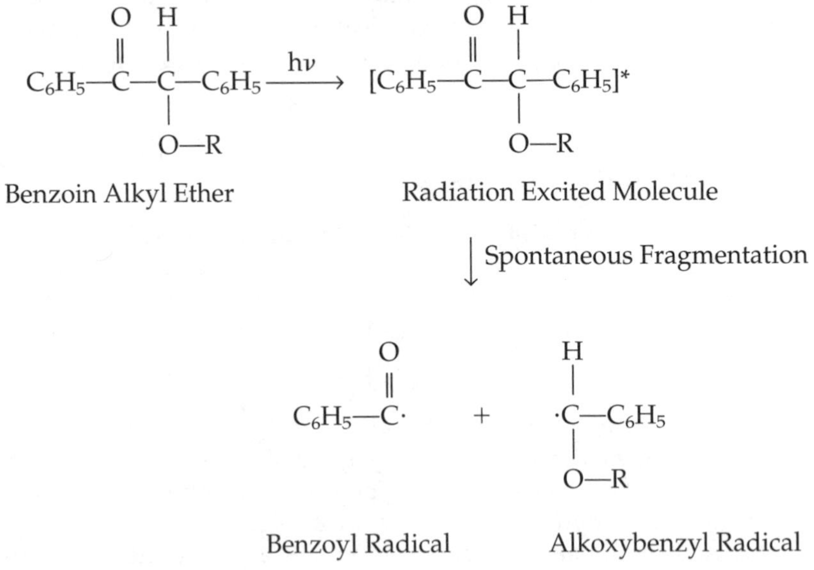

$$C_6H_5{-}\overset{\overset{\displaystyle O}{\|}}{C}{-}\underset{\underset{\displaystyle O{-}R}{|}}{\overset{\overset{\displaystyle H}{|}}{C}}{-}C_6H_5 \xrightarrow{h\nu} [C_6H_5{-}\overset{\overset{\displaystyle O}{\|}}{C}{-}\underset{\underset{\displaystyle O{-}R}{|}}{\overset{\overset{\displaystyle H}{|}}{C}}{-}C_6H_5]^*$$

Benzoin Alkyl Ether Radiation Excited Molecule

↓ Spontaneous Fragmentation

$$C_6H_5{-}\overset{\overset{\displaystyle O}{\|}}{C}\cdot \quad + \quad \cdot\underset{\underset{\displaystyle O{-}R}{|}}{\overset{\overset{\displaystyle H}{|}}{C}}{-}C_6H_5$$

Benzoyl Radical Alkoxybenzyl Radical

The radicals formed may or may not have the same capability of initiating polymerization.

Although the photolysis reaction of benzoin had been investigated as early as 1901 [6], it was much later, in 1941, that Christ [7] applied for a patent that dealt with ways of cementing poly(methyl methacrylate) bomber-turret halves with a solution of methyl methacrylate, benzoin, and lauroyl peroxide. The result was a clear, seamless juncture that was accomplished with ultraviolet radiation in 10 min. This was a marked improvement over the 8-h cure period required for the conventional cure schedule that was in use at the time. Other patents granted in the early 1940s dealt with similar systems and the same photoinitiator [8,9]. No doubt at least two factors were responsible for the slow cure rate (10 min.) mentioned above. First, benzoin is a poor photoinitiator, and second, methacrylates do not polymerize as rapidly as acrylates.

In 1946, a patent by Renfrew [10] pointed out that prior art in the radiation curing area was not commercially successful because of the slow rate of polymerization. In general, the processes were too slow and offered no marked advantage over other known art, unless special needs were present as in the cementing of bomber turrets. Renfrew found that the benzoin ethers, such as benzoin ethyl ether, gave cure rates almost 2.5 times faster than those of benzoin. Here again the studies involved methacrylates or styrene and very broad-spectrum radiation of 180 to 700 nm. The results were carried out in bulk and confounded by the presence of lauroyl peroxide, which markedly enhanced the bulk polymerization as the temperature increased from the heat of reaction to the peroxide's decomposition point.

Later studies confirmed the early results and found that benzoin alkyl ethers were the most effective benzoin derivatives for photopolymerization. The rate constant for photocleavage of the alkyl ethers was found to be greater than 10^{10} s^{-1}, whereas it was only about 10^9 s^{-1} for benzoin [11]. The difference in rate is thought to be related to the ability of the benzylic substituent to weaken the bond by stabilizing an adjacent positive charge. Thus, electron-donating substituents such as alkyl ethers are more effective than electron-accepting substituents such as the alkyl esters. Studies by Osborn and Sanborn [12] demonstrated that there are reactivity differences between the various butyl ether derivatives of benzoin with n-butyl ether and i-butyl ether being more reactive than s-butyl ether, which is more reactive than the t-butyl ether derivative.

A number of homolytic fragmentation photoinitiators are listed in Table 3.2. Such photoinitiators are usually used at a $<1\%$ to 10% by weight concentration with about 2% to 5% most common. A broad list of photoinitiators, as well as their structure, absorption spectra, and physical properties, is available from a commercial source [13].

A particular class of homolytic fragmentation photoinitiators, the acylphosphine oxides, have been said to have two particular positive features. These initiators have the ability to cure thick, pigmented coatings, and they have a low yellowing tendency allowing good white coatings to be obtained [14,15]. They are also useful for clear coatings [16] and have been used in dentistry [17]. In the past, the only acylphosphine oxide compound available was 2,4,6-trimethylbenzoyldiphenylphosphine oxide (TMPO), which

TABLE 3.2. Examples of Homolytic Fragmentation-Type Free-Radical Generating Photoinitiators.

COMPOUND	APPEARANCE	MAIN ABSORPTION, nm
2,2-Diethoxyacetophenone	White solid	242–325
Benzil dimethylketal	White solid	220, 254–335
1-Hydroxycyclohexylphenyl-ketone	White solid	208, 245
2-Ethoxy-2-isobutoxyacetophenone	—	—
2,2-Dimethyl-2-hydroxyacetophenone	Clear liquid	265–280 320–335
2,2-Dimethoxy-2-phenylacetophenone	Off-white solid	330–340
n-Butylbenzoin ethers, mixture	Light yellow liquid	246–325
i-Butyl benzoin ether	Yellow liquid	~250
iso-Propyl benzoin ether	Off-white solid	~250
2,2,2-Trichloro-4-t-butylacetophenone	Solid	—
2,2-Dimethyl-2-hydroxy-4-t-butylacetophenone	—	—
Sulphonyl chloride derivatives of benzophenone	Solid	—
1-Phenyl-1,2-propanedione-2-O-ethoxycarbonyl ester	—	—
1-Phenyl-1,2-propanedione-2-O-benzoyl oxime	—	—

photolyzes by first passing through an excited state and then very rapidly [18] fragments as described below.

CH3 O O C–P CH3 CH3 —hν→ CH3 O C• CH3 CH3 + O •P

TMPO Trimethylbenzoyl radical Phosphinoyl radical

The phosphinoyl radical is very reactive for initiating the polymerization of ethylenically unsaturated molecules such as acrylates, methacrylates, vinyl acetate, and styrene [19]. Although the benzoyl radical will initiate polymerization, the phosphinoyl radical is two to three times more effective in initiating polymerization of these compounds [20].

TMPO has a maximum absorption at 379 nm and is colored in nature. However, the photochemical reaction has a bleaching effect that destroys the light-absorbing chromophore in the compound, rendering it colorless. Surface curing characteristics in an air atmosphere are enhanced by addition of a compound such as hydroxydimethylacetophenone. Because of its liquid character and miscibility with acrylates, it provides an easier to use photoinitiator system.

It has been possible to sensitize the phosphine oxides with optical brighteners and obtain an eight-fold increase in functional photo speed of imaging compositions in graphic arts applications [21]. The optical brighteners used are compounds such as 2,2′-(2,5-thiophenediyl)bis(t-butyl benzoxazole). The phosphine oxides have also been used in radiation-cured, exterior, waterborne coatings [22].

Hydrogen Abstraction Type

When exposed to ultraviolet radiation, hydrogen abstraction photoinitiators are taken from a ground state $[PI\text{-}F]^o$ to an excited state, $[PI\text{-}F]^*$.

$$[PI\text{-}F]^o \xrightarrow{h\nu} [PI\text{-}F]^*$$

Although photoinitiators of this type, Table 3.3, are excited with ultraviolet radiation, they will not spontaneously fragment or photolyze and generate free radicals capable of initiating polymerization as the homolytic fragmentation type did. These photoinitiators require the presence of a synergist that will interact with the excited molecules and form free radicals by means of electron transfer and hydrogen abstraction.

Synergists are compounds that contain a carbon atom with at least one hydrogen atom in the alpha position to a nitrogen atom. Conventional synergists are tertiary amines as described below. These and others are given in Table 3.4.

Alpha Carbon Atom	Alpha Carbon Atoms
↓	↓
$(CH_3)_2\text{—N—}CH_2CH_2OH$	$N\text{—}(CH_2CH_3)_3$
Dimethylethanolamine	Triethylamine

Synergists usually are tertiary amines [23–28], but amides and ureas [29,30] have also been used. Other compounds [31,32] such as alcohols and ethers

TABLE 3.3. Examples of Hydrogen Abstraction-Type Free Radical Generating Photoinitiators.

COMPOUND	APPEARANCE	ABSORPTION, nm
Benzophenone	White solid	250–350
Trimethylbenzophenone with methyl benzophenone	Clear liquid	250–334
4-Methylbenzophenone	White solid	—
Bis-(4,4′-dimethylamino)benzophenone*	Liquid	360 (main)
Benzil	Solid	—
Xanthone	Solid	—
Thioxanthone	Pale cream solid	~320–340
Isopropylthioxanthone	Pale yellow solid	258–382
2-Chlorothioxanthone	Pale yellow solid	360 (main)
9,10-Phenanthrenequinone	White solid	420 (main)
9,10-Anthraquinone	Solid	—

* Also known as Michler's ketone; it does not require an added amine synergist because of the presence of a hydrogen group on a carbon in the alpha position to a nitrogen group.

TABLE 3.4. Examples of Some Common Low Molecular Weight Synergists.

SPECIFIC COMPOUNDS
2-Ethylhexyl-p-dimethylaminobenzoate
2-Ethyl-p-(N,N-dimethylamino)benzoate
2-n-Butoxyethyl-4-dimethylaminobenzoate
Dimethylethanolamine
Methyldiethanolamine
N,N-Dimethyl-p-toluidine
Triethanolamine
Triethylamine
Important General Types
Amines
Disubstituted amides
Tetrasubstituted ureas

that contain a hydrogen atom on a carbon atom positioned alpha to the hydroxyl or ether oxygen, as well as thiols that have a hydrogen attached to a sulfur atom (R-S-H), will function as synergists. However, radiation excited compounds will react most readily with amine compounds containing hydrogen in the alpha position to a nitrogen [33], and the electron-rich environment associated with the amino radical is a very efficient initiator for electron-poor ethylenically unsaturated monomers such as the acrylates [27,34–35]. A long list of the photoinitiators used in combination with a variety of amine synergists and the resultant increase in cure speed can be found in the literature [35]. Primary and secondary amines are not usually used as synergists because they have a tendency to undergo Michael addition to acrylate monomers, which are widely used in free-radical radiation curing, and causing subsequent gelation during storage.

In a general sense, synergists are low molecular weight, monoamine functional compounds used in conjunction with aryl ketone photoinitiators such as benzophenone. When low molecular weight synergists are used, they may have an adverse effect on coatings as described in Table 3.5. Certain hydrogen-abstraction photoinitiators will function without an added synergist because they contain an internal synergist. Two of these are mentioned in Table 3.4. One of these type photoinitiators claims to do away with many of the adverse effects caused by synergists [36]. Polymeric/oligomeric amine

TABLE 3.5. Potential Adverse Effects When Low Molecular Weight Synergists are Used.

Decreased Storage Stability
Emulsification of Certain Inks
Loss of Mechanical Properties-Embrittlement
Unpleasant Odors
Yellowing

synergists [37] also negate many of the deficiencies of low molecular weight synergists.

Synergists function by interacting with the radiation-excited photoinitiator by an electron-transfer mechanism to form a complex that is termed an "exiplex," as described in the following reaction scheme. The exiplex then undergoes an internal, intermolecular

$$C_6H_5-\overset{\overset{\displaystyle O}{\|}}{C}-C_6H_5 \xrightarrow{h\nu} \left[C_6H_5-\overset{\overset{\displaystyle O}{\|}}{C}-C_6H_5\right]^*$$

Benzophenone — Radiation-excited Benzophenone

$$\left[C_6H_5-\overset{\overset{\displaystyle O}{\|}}{C}-C_6H_5\right]^* + CH_3CH_2-\overset{\overset{\displaystyle CH_2CH_3}{|}}{N}-CH_2CH_3 \longrightarrow \left[C_6H_5-\overset{\overset{\displaystyle O\,\delta^-}{|}}{C}-C_6H_5 \quad \overset{\delta^+}{}\underset{\underset{\displaystyle CH_2CH_3}{|}}{N}-(CH_2CH_3)_2\right]$$

Radiation-excited Benzophenone — Triethylamine — EXIPLEX

$$\longrightarrow C_6H_5-\overset{\overset{\displaystyle OH}{|}}{\dot{C}}-C_6H_5 + \overset{\overset{\displaystyle N-(CH_2CH_3)_2}{|}}{\dot{C}H}-CH_3$$

Non-initiating Free Radical — Initiating Free Radical

hydrogen transfer. As a result, a hydrogen atom on the triethylamine portion of the exiplex is freed and is accepted by the benzophenone portion of the exiplex. The altered benzophenone is a free radical (a ketyl radical), but it is not capable of initiating and propagating polymerization at the rapid rate required for radiation curing. It decays to an inert species that does not enter into the polymerization reaction, or it acts as a chain terminator [38,39]. The altered amine is also a free radical, and it is capable of initiating very rapid polymerization of acrylates and related ethylenically unsaturated compounds. Studies have shown that the amino radical is markedly more effective in initiating polymerization than the ketyl radical [40].

During the polymerization termination step, synergist molecules become incorporated into the polymeric network through recombination of free radicals. Usually, this incorporation has been more or less ignored or looked on as a deficiency because synergists were low molecular weight and their use had adverse effects (Table 3.5). In spite of the adverse effects, synergists were used because of their rate-enhancing characteristics, and they were not associated with property improvements. The use of polymeric/oligomeric amine synergists [37] and their derivatives eliminated the disadvantages of conventional synergists and improved mechanical properties of the final cured coating. Besides improving flexibility, gloss, toughness, and other functional and decorative properties, these synergists

- Reduce the amount of acrylates in formulation.
- Eliminate the need for nitrogen or other inert blanketing.
- Decrease shrinkage.
- Improve adhesion.

These synergists were of the general form [37]:

$$R\left[\text{OLIGOMER—CHR}'''\text{—}\underset{}{\overset{\displaystyle R'\\|}{N}}\text{—}R''\right]_x$$

where *R* is a chemical residue of the compound used to make the oligomer, *x* is an integer related to the functionality of *R*, and *R′* and *R″* are the same or different and are hydrogen or alkyl depending on whether a tertiary, secondary, or primary amine is involved, and *R‴* is hydrogen, alkyl, or aryl. For example, if a poly(propylene oxide) triol is the oligomer, *R* is the residue of glycerol and *x* is three. These polymeric- oligomeric synergists can be used to modify coating characteristics from smooth to textured [41].

Novel hexa-arylbiimidazoles have been synthesized recently [42]. Although the radical formed when these compounds are irradiated is a poor photoinitiator for ethylenically unsaturated compounds, when it combines with a hydrogen donor co-initiator such as N-phenylglycine-2-mercaptobenzoazole, and similar compounds, the system has very high reactivity.

Photosensitizers

When polymerization is induced by a free radical photoinitiator, the process involves consumption of the photoinitiator. That is, it photolyzes or degrades under the action of radiation into other compounds. Photosensitizers, *S*, initiate polymerization by means of energy transfer. If a photosensitizer in its ground state, S^o, is irradiated with radiation of the proper wavelength, in one or more steps it will be raised to an excited, higher energy state, S^*. The excited photosensitizer then transfers energy to another compound, a photoinitiator that will photolyze and produce free radicals that will initiate polymerization. In this process, the photosensitizer falls back to some lower energy state or to its ground state. In some literature, authors or inventors will mistakenly describe compounds as photosensitizers, when in fact they are photoinitiators.

Thus, photosensitization is a process that is used to transfer energy from one compound, the photosensitizer, to produce an excited state in another molecule, the photoinitiator, that is activated at a wavelength other than that directly impinging on the compound that will effect polymerization. For example, if one attempts to use either quinoline-8-sulfonyl chloride or thioxanthone to photoinitiate the polymerization of methyl methacrylate at a wavelength greater than 360 nm, there is basically little or no reaction [25]. At such wavelengths, the quinoline absorbs very little radiation, and the thioxanthone is ineffective without a synergist (i.e., a hydrogen donor molecule). However, if a mixture of the two compounds is used, the thioxanthone absorbs radiation and transfers it to the quinoline, which undergoes photolysis, and free radicals capable of carrying out the polymerization are generated. The thioxanthone functions as a sensitizer for the quinone. Detailed studies indicate that sensitization of α-amino ketones or α-hydroxy ketones with an ethoxylated thioxanthone takes place by at least two pathways, photo reduction of the thioxanthone and triplet-triplet energy transfer [43].

Oxygen Inhibition

Oxygen, which is present in air environments, is an efficient quenchant for radiation-excited or activated molecules, and it leads to inhibition of polymerization with formulations involving fragmentation-type photoinitiators. Molecular oxygen may be considered to be a diradical. When it is in a singlet stage, it quickly undergoes a change to molecular oxygen's electronic ground state. The diradical does not have sufficient energy to initiate polymerization, and thus, may be thought of as stable. However, it does react willingly with existing free radicals, and thus, has a definite effect on systems that polymerize by a free radical mechanism [44,45]. This is the mechanism that keeps acrylate compounds stable during storage, but that acts as a polymerization inhibitor during radiation-induced, free radical polymerization. A result of oxygen inhibition can be a thin layer of unpolymerized molecules at the surface of an exposed film. The thickness of this unpolymerized or inhibited layer has been found to be linearly proportional to the inverse of exposure time, radiation intensity, and photoinitiator concentration [46].

Free radicals, FR·, have options other than reacting with monomer or oligomeric unsaturated compounds and initiating polymerization. For example, they can recombine and form compounds, FR—FR, that may again photolyze and generate new free radicals,

$$FR\cdot + FR\cdot \rightarrow FR{-}FR$$

and they can react with oxygen. The presence of oxygen can have two effects [4]. Oxygen can quench an excited state photoinitiator, $[PI]^*$, back to its ground state, $[PI]^o$, and form singlet oxygen, ${}^1O_2^*$,

$$[PI]^* + O_2 \rightarrow [PI]^o + {}^1O_2^*$$

which can retard the free radical polymerization of an ethylenically unsaturated monomer, *M*, by reacting with the unsaturation and forming peroxides and hydroperoxides.

$$\text{M}\cdot + \text{M} \rightarrow \text{M—M}\cdot$$

(Normal propagating monomer in absence of oxygen)

$$\text{M—M}\cdot\ (\text{or M}\cdot) + {}^1\text{O}_2{}^* \rightarrow \text{M—M—O—O}\cdot\ (\text{or M—O—O}\cdot)$$

(Free radical reaction with oxygen leads to termination)

$$\text{M—M—O—O}\cdot\ (\text{or M—O—O}\cdot) + \text{M} \rightarrow \text{No Reaction}$$

These reactions reduce the performance characteristics of a coating [2] by molecular weight reduction and incomplete cure that results in a tacky or slippery surface [47]. Even small concentrations of oxygen ($\sim 10^{-3}$ molar) can have marked effects and prevent polymerization because of the large rate constant that is associated with oxygen quenching [27]. Most studies are concerned with oxygen present near the surface because it results in an uncured, sticky or tacky top layer [48,49]. A study by Krongauz and coworkers [50] has centered on the effect of dissolved oxygen and its effect in thick layers, as will be found in adhesives and sealants. Such oxygen can be removed, but relatively long nitrogen purge times of more than 10 min. are needed to remove oxygen from the bulk of a film. Oxygen quenching is important when fragmentation-type photoinitiators are used. Hydrogen-abstraction photoinitiators and cationic photoinitiators are not affected by oxygen.

Many ways to prevent or minimize the effect of oxygen inhibition have been devised. These include:

- Curing in an inert atmosphere [51,54]
- Addition of oxygen scavengers [52]
- Addition of waxes [55,56]
- Use of shielding films [57]
- Use of dye sensitizers [58,59]
- Use of surface-active initiators [60,61]
- Use of high photoinitiator concentration [27]
- Increasing ultraviolet radiation intensity [27]

Curing in an inert environment such as nitrogen is an effective way to eliminate oxygen, however, it adds a cost to the coating operation. Techniques for "knifing" nitrogen onto a coating surface prior to ultraviolet radiation exposure have been developed [53].

Oxygen scavengers are an efficient way of minimizing or effectively eliminating the effects of oxygen. Systems based on hydrogen-abstraction photoinitiators cure well in an air environment because the amine radicals formed preferentially react with oxygen forming a peroxy radical which will act as transfer agents [62].

$$(\text{CH}_3\text{CH}_2)_2\text{—N—}\dot{\text{C}}\text{HCH}_3 + \text{O}_2 \rightarrow (\text{CH}_3\text{CH}_2)_2\text{—N—}\underset{\substack{|\\ \text{O—O}\cdot}}{\text{CHCH}_3}$$

These peroxy radicals will abstract hydrogen from other α-carbon-containing amine molecules, forming hydroperoxides and new amine radicals capable of initiating rapid polymerization.

$$(CH_3CH_2)_2\text{—N—}\underset{\substack{|\\ \text{O—O}\cdot}}{CHCH_3} + (CH_3CH_2)_3\text{—N} \rightarrow$$

$$(CH_3CH_2)_2\text{—N—}\underset{\substack{|\\ \text{O—OH}}}{CH_2CH_3} + (CH_3CH_2)_2\text{—N—}\dot{C}HCH_3$$

This "cyclic" process has been reported to repeat itself up to twelve times per initial amine radical [21,52]. It has also been said that air curing can be improved by addition of compounds containing urethane [63], amide [64], and triazine [65] moieties in their structure. When systems are to be cured in air, this insensitivity of hydrogen-abstraction photoinitiators to oxygen quenching represents an important advantage for amine co-initiation.

When inert gases, waxes, shielding films, or surface-active photoinitiators are used, oxygen is prevented from reaching the surface to be radiation cured, and it is necessary only for any dissolved oxygen to be depleted before polymerization takes place. The latter is readily accomplished, and under such conditions cure is rapidly effected. Shielding films are transparent to ultraviolet radiation, and they prohibit oxygen from reaching the surface. In one instance [66], the film being cured was used as the shielding film. The film was first exposed to the radiation and left with a partially uncured, tacky surface. The film was then taken through a 180-degree rotation using a roller system, and what was the tacky surface ended at the bottom. It was then passed through the radiation source again, and with the uncured portion of the film at the bottom where it was covered with the radiation transparent cured portion, any oxygen present was quickly depleted and complete cure was effected.

Waxes are expected to rise to the coating's surface and exclude oxygen from the reactive solution below. When surface-active photoinitiators are photolyzed, they release a component that will rise to the surface and exclude oxygen, or they can concentrate at the surface, form a barrier, and then photolyze under the barrier to initiate polymerization. When the degree of cure and rate of polymerization in an inert atmosphere were compared to that when a wax barrier coating was used, no quantitative difference was apparent [67]. The effect of an inerting atmosphere when various photoinitiators, different commercial radiation sources, and pigment were used has been studied [68]. The degree of cure was measured by postbaking the ultraviolet or electron-beam irradiated specimens and determining the amount of weight loss. Compounds such as N-vinyl pyrrolidone and N-vinyl caprolactam improved the cure of 1,6-hexanediol diacrylate in air [69].

It should be pointed out that oxygen inhibition will affect property results obtained with photo differential scanning calorimetry [70]. Experiments must be properly designed if potentially misleading results are to be avoided. The results also pointed out that oxygen inhibition is directly related to sample viscosity. The difficulty in comparing results obtained with different viscosity materials cured in air is discussed [70].

Visible Radiation Photoinitiators

The use of formulations that cure with light or visible radiation has a number of advantages. For example, light allows greater depths of cure at the same intensity [71], cure through clear, colored, and translucent substrates [72], and relief from repeated exposure to other forms of radiation as in dentistry [71]. Recently, fluorone photoinitiators that are useful throughout the visible spectra have been described [72]. These may be generalized as:

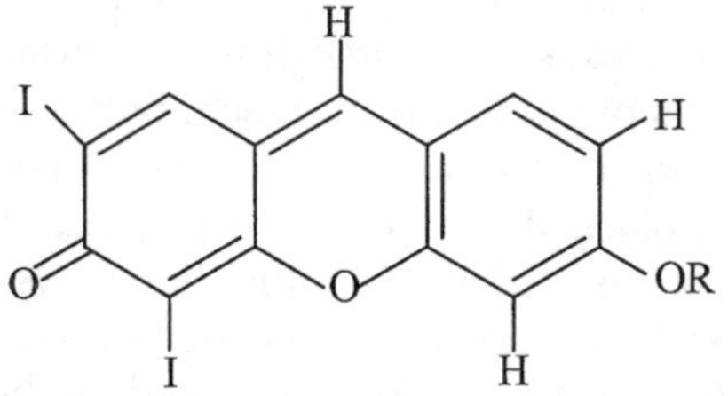

When *R* is methyl, ethyl, or butyl, the maximum absorption is at 470 nm, and at 472 nm if it is octyl. If *R* is butyl, the compound is 5,7-diiodo-3-butoxy-6-fluorone [73]. By changing the hydrogen group on the center ring to CN and altering the other hydrogen atoms and alkoxy group on the right-hand ring, absorptions in the 630–638 nm and 580–590 nm can be obtained. Further alteration of the groups on this right-hand ring results in maximum absorptions at 534–538 nm. These compounds are useful in the polymerization of acrylates. These compounds have very high absorptivity and only small quantities, about 0.01–0.15 %, are needed to achieve good depth of cure. Coinitiators for these photoinitiators are compounds such as tetramethylammonium triphenylbutyl borate and other proprietary compounds [74]. The coinitiators improve cure response and improve bleaching characteristics of compounds such as 5,7-diiodo-3-butoxy-6-fluorone, so essentially colorless cured films are obtained.

Other light-activated photoinitiators that have been used are D,L-camphorquinone [75,76], which absorbs at 460 nm; benzil, fluorenone, rose bengal, and uranyl nitrate [77]; 1-hydroxy-1-cyclohexylphenyl ketone, 2-hydroxy-2,2-dimethyl acetophenone, and mixtures of these compounds with azobis (isobutyronitrile) [78]. Still other photoinitiators are activated in the upper regions of the visible spectra (~650 nm) and the near infrared region (~1200–1500 nm). Typical of these compounds might be N,N-di-n-butyl-N′,N′-bis(di-n-butylaminophenyl)-p-benzoquinone diimmonium hexafluoroantimonate, whose preparation has been detailed [79].

The halogenated fluoresceins have also been used as light photoinitiators for curing unsaturated polyester composite systems [80]. The compounds are used in combination with amines and oxime compounds. These fluoresceins have the following structural formula:

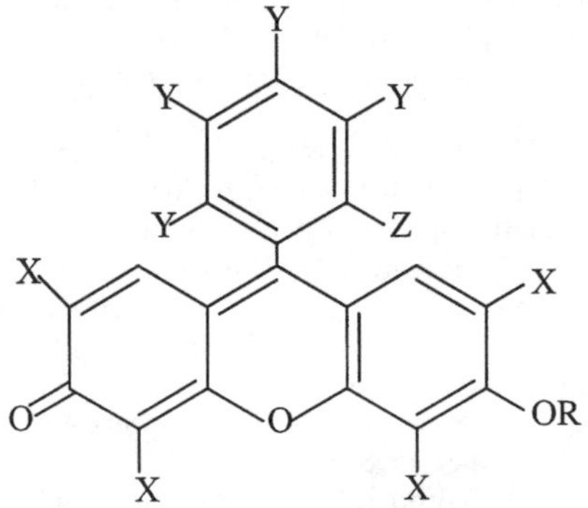

Generalized fluorescein

In this structure, *X* and *Y* are either hydrogen or halogen, *Z* is hydrogen or ester, and *R* is alkyl. The amine additive is an N,N-disubstituted benzylamine, and the oxime is a benzoyl substituted oxime carbonate ester:

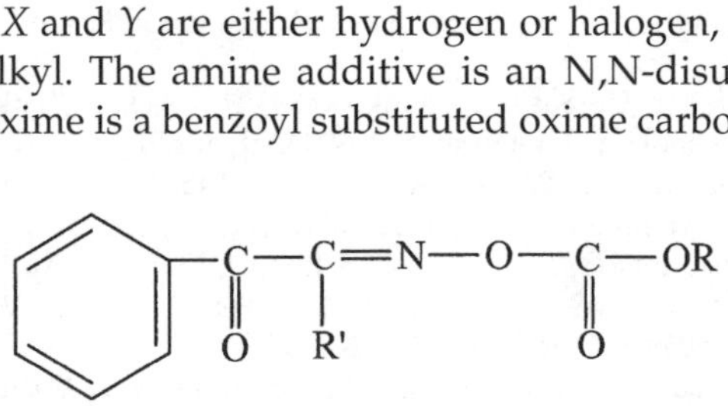

In this structural formula, *R* is alkyl, aralkyl, or aryl, and *R′* is alkyl or aryl. Although these oximes have no appreciable light absorption, low concentrations (0.1–0.5%) had a significant effect on both cure rate and final degree of cure. Formulated products based on these photoinitiator systems could be handled for reasonable periods of time under normal lighting conditions, but they cured rapidly when the light intensity was significantly increased. Other studies of oximes in photocuring have been carried out [81].

One of these fluoresceins, rose bengal, has been studied in combination with an iron arene complex, cumene hydroperoxide, or methyldiethanolamine [82]. The photochemical and physical characteristics were investigated with laser flash and steady-state photolysis techniques. The results indicated that the rose bengal and iron arene absorb light with a resultant semioxidized rose bengal and a new iron arene complex.

References

[1] Billmeyer, F. W., Jr., *Textbook of Polymer Science*, 3rd ed., Interscience Publishers, Division of John Wiley and Sons, New York, 1984.
[2] Roffey, C. G., *Photopolymerization of Surface Coatings*, John Wiley & Sons, New York, 1982.
[3] Bamford, C. H. and Norrish, R. G. W., *Journal of the Chemical Society*, 1935, p. 1504.
[4] Hartley, G. H. and Guillet, J. E., *Macromolecules*, Vol. 1, No. 65, 1968.
[5] Trozzolo, A. M. and Winslow, F. H., *Macromolecules*, Vol. 1, No. 98, 1968.
[6] Ciamician, G. and Silber, P., *Berichte*, Vol. 34, 1901, p. 1530.
[7] Patent: Christ, R. E., E. I. duPont de Nemours & Company, "Cementing Process," U. S. 2,367,670 (1945).
[8] Patent: Renfrew, M. M., E. I. duPont de Nemours & Company, U. S. 2,335,133 (1943).
[9] Patent: Agre, C. L., E. I. duPont de Nemours & Company, "Process of Photopolymerization," U. S. 2,367,661 (1945).
[10] Patent: Renfrew, M. M., E. I. duPont de Nemours & Company, U. S. 2,448,828 (1946).
[11] Heine, H. G. and Traenchner, H. J., *Progress in Organic Coatings*, Vol. 3, 1975, p.115.
[12] Osborn, C. L. and Sandner, M. R., *American Chemical Society Division of Organic Coatings and Plastics Chemistry Preprints*, Vol. 34, No. 1, 1974, p. 660.
[13] Ciba Specialty Chemicals, "Photoinitiators for UV Curing, Key Products Selection Guide," Publication Number A-805M49 (1998).
[14] Beck, E., Keil, E., Lokai, M., Schroder, J., and Sass, K., "Acylphosphine Oxides as Photoinitiators for Pigmented Coatings," *Modern Paint and Coatings*, Vol. 88, No. 3, March 1998, p. 36.
[15] Decker, C., Zahouily, K., Decker, D., and Nguyen, T., "Performance of Acylphosphine Oxide Photoinitiators in UV-Curable Systems," *Proceedings of RadTech 2000*, Baltimore, MD, 9–12 April, 2000, p. 595.
[16] Reich, W., Glotfelter, C., Sass, K., Bankowsky, H., Beck, E., Lokai, M., and Noe, R., "Acylphosphine Oxides as Photoinitiators Not Only for Pigmented Coatings," *Proceedings of RadTech 2000*, Baltimore, MD, 9–12 April, 2000, p. 545.
[17] Luchterhandt, T., Weinmann, W., and Gubbenberger, R., "Acyl Phosphine Oxides in Dental Materials," *Proceedings of RadTech 2000*, Baltimore, MD, 9–12 April, 2000, p. 560.
[18] Sumiyoshi, T., Schnable, W., Henne, A., and Lechtken, P., *Polymer*, Vol. 26, 1985, p. 141.
[19] Schnabel, W., *Radiation Curing*, Vol. 1, No. 26, 1986.
[20] Baxter, J. B., Davidson, R. S., Hagemann, H. J., and Overeem, T., *Makromolecular Chemie*, Vol. 189, 1988, p. 2769.
[21] Wilczak, W. A., "Sensitization of Phosphine Oxide Photoinitiators with Optical Brighteners in Certain Graphic Arts Applications," *Proceedings of RadTech 2000*, Baltimore, MD, 9–12 April, 2000, p. 570.
[22] Burglin, M., Kohler, M., Dietliker, K., Wolf, J. P., Gatlik, I., Gescheidt, G., Neshchadin, D., and Rist, G., "Curing of Outdoor Water-Borne Coatings—A Novel Application of Bisacylphosphine Oxide Photoinitiators," *Proceedings of RadTech 2000*, Baltimore, MD, 9–12 April, 2000, p. 577.
[23] Sandner, M. R., Osborn, C. L., and Trecker, D. J., *Journal of Polymer Science, Polymer Chemistry Edition*, Vol. 10, 1972, p. 3173.
[24] Osborn, C. L. and Stevens, J. J., *American Paint Journal*, January 24, 1977, p. 32.
[25] Pappas, S. P. and McGinniss, V. D., "Photoinitiation of Radical Polymerization," in *UV Curing: Science and Technology*, S. Peter Pappas, Ed., Technology Marketing Corporation, Stamford, CT, 1978.

[26] Green, P. N., "Photoinitiators, Types and Properties," *Polymers Paint and Colour Journal*, Vol. 175, No. 4141, April 17, 1985, p. 246.
[27] Vesley, G. F., *Journal of Radiation Curing*, Vol. 13, No. 1, 1986, p. 4.
[28] Christensen, J. E., Wooten, W. L., and Whitman, P. J., *Experimental Photoinitiators IV: 2-Ethylhexyl-p-N,N-dimethylaminobenzoate as Coinitiator with Benzophenone*, The Upjohn Company, North Haven, CT, April 1988.
[29] Jacobine, A. F. and Scanio, C. J. V., *Journal of Radiation Curing*, Vol. 10, No. 26, July 1983.
[30] Patents: Jacobine, A. F., U. S. 4,425,208; 4,446,247; and 4,518,473 (1984).
[31] Patent: Chang, Catherine, The-Lin, E. I. duPont de Nemours and Company, "Photopolymerizable Compositions Containing Aminophenyl Ketones," U. S. 3,661,588 (1972).
[32] Patent: Moore, E. M., E. I. duPont de Nemours and Company, "Photopolymerizable Products," U. S. 3,495,987 (1970).
[33] Cohen, S. G., Parola, A., and Parsons, G. H., Jr., *Chemical Reviews*, Vol. 73, No. 141, 1973.
[34] Kinstle, J. F. and Watson, S. L., Jr., *Journal of Radiation Curing*, Vol. 3, No. 1, 1976, p. 2.
[35] Patent: Osborn, C. L., Trecker, D. J., Union Carbide Corporation, "Photopolymerization Process Using Combinations of Organic Carbonyls and Amines," U. S. 3,759,807 (1973).
[36] Nagarajan, R. and Cui, H., "Novel Photoinitiator Blends—II," *Proceedings of RadTech 2000*, Baltimore, MD, 9–12 April, 2000, p. 366.
[37] Gerkin, R. M., Beckett, A. D., and Koleske, J. V., "Nonacrylate Formulating Compounds for Free Radical-Initiated, Radiation-Cure Acrylate Systems," *Proceedings of RadTech '90-North America*, Chicago, IL, Vol. 2, 25–29 March, 1990, p. 177.
[38] Kuhlmann, R. and Schnabel, W., *Polymer*, Vol. 17, 1976, p. 419.
[39] Hutchison, J., Lambert, M. C., and Ledwith, A., *Polymer*, Vol. 14, 1973, p. 250.
[40] Kinstle, J. F. and Watson, S. L., Jr., *Journal of Radiation Curing*, Vol. 2, No. 2, 1975, p. 7.
[41] Patent: Beckett, A. D., Koleske, J. V., Gerkin, R. M., U. S. 5,212,271 (1993).
[42] Kaji, M. and Hidaka, T., "Mechanistic Analysis of Molecular Design of Heterocyclic Photoinitiator System," *Proceedings of RadTech Europe '99*, Berlin, Germany, 8–10 November, 1999, p. 393.
[43] Dietliker, K, Rembold, M. W., Rist, G., Rutsch, W., and Sitek, F., "Sensitization of Photoinitiators by Triplet Sensitizers," *Proceedings of Radcure Europe '87*, Munich, West Germany, 4–7 May, 1987, pp. 3–37.
[44] Hanrahan, M. J., "Overcoming the Effects of Oxygen on Cured and Uncured UV Formulations," *Journal of Radiation Curing/Radiation Curing*, Vol. 19, No. 40, Spring 1992.
[45] Hanrahan, M. J., "Oxygen Inhibition: Causes and Solutions," *RADTECH Report*, Vol. 4, No. 2, March/April 1990, p. 14.
[46] Wight, F. R. and Numez, I. M., "Oxygen Inhibition of Acrylate Photopolymerization," *Journal of Radiation Curing*, Vol. 16, No. 1/2, January/April 1989, p. 3.
[47] Ivanov, V. B. and Shlyapintokh, V. Ya., *Journal of Polymer Science*, Polymer Chemistry Edition, Vol. 16, 1979, p. 899.
[48] Walia, J. and Skinner, D., "Understanding the Process and Benefits of Nitrogen Inerting for UV Curable Coatings," *End User Conference, RadTech 2000*, Baltimore, MD, 9–12 April, 2000, p. 230.
[49] Menzel, K., Bankowsky, H. H., Enenkel, P., and Lokai, M., "Latest Investigations in UV-Inert Technology," *End User Conference, RadTech 2000*, Baltimore, MD, 9–12 April, 2000, p. 242.
[50] Krongauz, V. V., Chawla, C. P., and Woodman, R. K., "Oxygen and Coating Thickness Effects on Photocuring Kinetics," *RadTech 2000*, Technical Proceedings, Baltimore, MD, 9–12 April, 2000, p. 260.
[51] Patent: Miller, L. S., U. S. 3,520,714 (1970).
[52] Bartholomew, R. F. and Davidson, R. S., *Journal of the Chemical Society*, 1971, p. 2342.
[53] Union Carbide Corporation, Publication L-4310B, *Linde™ Photocure Systems: PS-700 Series*, 1984.
[54] Morgan, C. R. and Kyle, D. R., *Radiation Curing*, Vol. 10, No. 4, 1983, p. 4.

[55] Miranda, T. J. and Huemmer, J., *Paint Technology*, Vol. 41, 1969, p. 118.
[56] Patent: Bolon, D. A., Lucas, G. M., Jaffe, M. S., General Electric Company, Column 4 of U. S. 3,989,644 (1976).
[57] Gardon, J. L. and Prane, J. W., *Nonpolluting Coatings and Coating Processes*, Plenum Press, New York, 1973, p. 272.
[58] Decker, C., *Makromolecular Chemie*, Vol. 180, 1979, p. 2027.
[59] Decker, C., Faure, J., Fizet, M., and Rychia, L., *Photographic Science and Engineering*, Vol. 23, 1979, p. 137.
[60] Patents: Hageman, J. and coworkers, European 0/037,152; 0/037,604 (1980).
[61] Hult, A. and Ranby, B. G., *Polymer Preprints*, American Chemical Society, Division of Polymer Chem., 1984.
[62] Davidson, R. S., in *Radiation Curing in Polymer Science and Technology*, Vol. III, J. P. Fouassier and J. F. Rabek, Eds., New York, 1993, p. 153.
[63] Dowbenko, R., Friedlander, C., Gruber, G., Prucnal, P., and Wismer, M., *Progress in Organic Coatings*, Vol. 11, 1983, p. 71.
[64] Patent: Gruber, G. W. and coworkers, British 2,002,374 (1979).
[65] Patent: Hall, R. P., SCM Corporation, "Curing by Actinic Radiation," U. S. 3,899,611 (1975).
[66] Patent: Hall, R. P., SCM Corporation, "Process for Curing Air-Inhibited Resins by Radiation," U. S. 3,644,161 (1972).
[67] Bolon, D. A. and Webb, K. K., "Barrier Coats Versus Inert Atmospheres. The Elimination of Oxygen Inhibition in Free-Radical Polymerizations," *Journal of Applied Polymer Science*, Vol. 22, 1978, p. 2543.
[68] Krajewski, J. J., Packer, E. S., and Yamazaki, T., "Effect of Photosensitizer System, Inerting Atmosphere, and Radiation Source on Degree of Conversion in Radiation Cured Films," *Journal of Coatings Technology*, Vol. 48, No. 623, December 1976, p. 43.
[69] Miller, C. W., Jonsson, S., Hoyle, C. E. and coworkers, and Ng, L. T., "Analysis of the Reduction of Oxygen Inhibition By N-Vinylamides in Free Radical Photocuring of Acrylic Formulations," *Proceedings of RadTech 2000*, Baltimore, MD, 9–12 April, 2000, p. 754.
[70] Herlihy, S., "The Influence of UV Light Absorption and Oxygen Inhibition on Photo DSC Results," *Proceedings, RadTech Europe '99*, Berlin, Germany, 8–10 November, 1999, p. 489.
[71] Rogers, S. C., "Visible Light Curing Provides Low Shrinkage and Good Depth," *Adhesives Age*, Vol. 31, No. 4, April 1988, p. 20.
[72] Marino, T. L., Martin, D., and Neckers, D. C., "Chemistry and Properties of Novel Fluorone Visible Light Photoinitiators," *Adhesives Age*, Vol. 38, No. 4, April 1995, p. 22.
[73] Neckers, D. C. and Shi, J., "Fluorone and Pyronin Y Derivatives," Spectra Group Ltd., U. S. Patent 5,451,343 (1995); Spectra Group Ltd., "H-Nu 470: A UV/Visible Photoinitiator," Information Sheets (2001).
[74] Neckers, D. C. and Shi, J., "Photooxidizable Initiator Composition and Photosensitive Materials Contain the Same," Spectra Group Ltd., U. S. Patent 5,395,862 (1995); Spectra Group Ltd., "New Coinitiators," Information Sheets (2001).
[75] Jakubiak, Julita, Linden, Lars-Ake, and Rabek, Jan F., "A Reappraisal of Camphorquinone-amines Photoinitiating Systems for Polymerization of Multifunctional Monomers," *Proceedings of RadTech Europe '99*, Berlin Germany, 8–10 November, 1999, p. 417.
[76] Angiolini, L., Caretti, D., and Salatelli, E., "Novel Polymeric Photoinitiators Absorbing Visible Light," *Proceedings of RadTech Europe '99*, Berlin, Germany, 8–10 November, 1999, p. 387.
[77] Patent: Dart, E. C., Perry, R., Nemcek, J., Imperial Chemical Industries Ltd., "Photocurable Resin Impregnated Fabric for Forming Rigid Orthopedic Devices and Method," U. S. 3,874,376 (1975).
[78] Patent: Buck, C. J., Johnson & Johnson Products, Inc., "Visible Light Cured Orthopedic Polymer Casts," U. S. 4,512,340 (1985).
[79] Patent: Cohen, M. S., Epolin, Inc., "Quinone Diimmonium Salts and Their Use to Cure Epoxies," U. S. 5,686,639 (1997).

[80] Alexander, I. J. and Scott, R. J., "Curing Polyester Resins with Visible Light," *British Polymer Journal*, Vol. 15, No. 1, 1983, p. 30.

[81] Tsunooka, M., Tachi, H., Asakino, K., and Suyama, K., "Oxime Derivatives as Photobase Generators: Their Use for Novel Photo-Crosslinking Systems," *Proceedings of RadTech Europe '99*, Berlin, Germany, 8–10 November, 1999, p. 405.

[82] Burget, D., Grotzinger, C., Louerat, F., Jacques, P., and Fouassier, J. P., "Photochemical and Photophysical Behavior of a Visible Photo-initiating System: Xanthenic Dye/Iron Arene Complex/Hydroperoxide/Amine," *Proceedings of RadTech Europe '99*, Berlin, Germany, 8–10 November, 1999, p. 399.

4

Cationic Photoinitiators and Initiation Mechanism

Introduction

CATIONIC PHOTOINITIATORS are onium compounds that photolyze to form Lewis or Bronsted acids when excited with ultraviolet radiation. In the photolysis process, free radicals are also formed, and therefore, both epoxides and ethylenically unsaturated compounds can be polymerized in a dual-cure process. Although photoinitiators in use today generate Bronsted or protonic acids from iodonium or sulfonium metallic salts, it is important from a historical standpoint to discuss first the Lewis-acid generating or diazonium type photoinitiators.

Watt [1] has published a history of the cationic curing of epoxides over the period of 1965 to 1981. These studies dealt with the use of the photolysis products from onium salts to cure principally cycloaliphatic epoxides into coatings. We say principally cycloaliphatic epoxides because, as will be discussed in a later chapter, a significant amount of these epoxides must be present to ensure cure at a commercially effective rate. The earliest of these studies were concerned with the photographic process rather than the cure of industrial coatings. The photographic area was abandoned when silver salts were found to be much better for this process. In the sense of coatings, the earliest studies and commercial products involved diazonium salts as photoinitiators and later, including currently, iodonium and sulfonium salts.

Onium Salts, General

The most familiar onium compound is ammonium hydroxide. In this compound, water interacts with ammonia in the following reaction.

$$\underset{\text{Ammonia}}{NH_3} + \underset{\text{Water}}{H_2O} \rightarrow \underset{\text{Ammonium ion}}{NH_4^+} + \underset{\text{Hydroxyl ion}}{OH^-}$$

Ammonium hydroxide is formed when a proton from water adds to ammonia to form the ammonium ion, NH_4^+, leaving a hydroxyl group from water as the counter ion. Another familiar reaction is the addition of a proton to a water molecule to form the oxonium ion H_3O^+. Other compounds containing an element with an unshared pair of electrons such as sulfur and phosphorous can form onium salts.

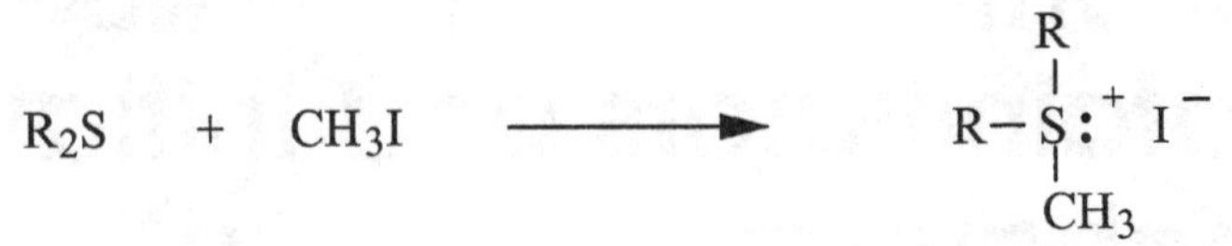

$$R_2S \quad + \quad CH_3I \longrightarrow R-\overset{R}{\underset{CH_3}{\overset{|}{\underset{|}{S}}}}:^{+}\ I^{-}$$

Sulfide Alkyl Halide Sulfonium Salt

Diazonium Salts, Lewis Acids

Early studies of diazonium salt photodecomposition were carried out in solution [2] and focused on the relationship between thermal and photochemical breakdown of the salts [3]. North American Aviation scientists were the first to use onium salts to photocure cycloaliphatic epoxides [4].[1] Aryldiazonium fluoroborates, which yielded boron

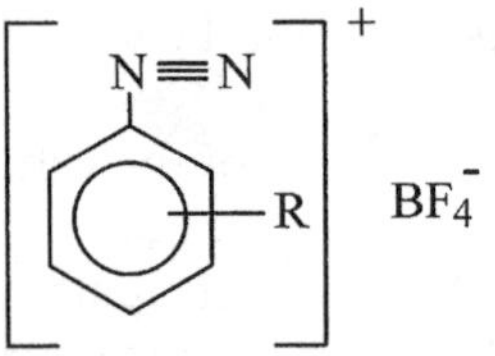

$$\left[R-C_6H_4-N{\equiv}N \right]^{+} \quad BF_4^{-}$$

trifluoride on photolysis, were used as photoinitiators for sealants used in the manufacture of printed-circuit assemblies. Other anions with low nucleophilicity, AsF_6^-, $FeCl_4^-$, PF_6^-, $SbCl_6^-$, SbF_6^- in combination with diazonium aryl groups were investigated, and they generated AsF_5, $FeCl_3$, PF_5, and so on, when irradiated [7–11]. Other diazonium salts can liberate trifluoromethane [12] or perchloric acid [13] on irradiation. These compounds will also very rapidly polymerize cycloaliphatic epoxides.

A few years later, in the early 1970s, Schlesinger and Watt working at American Can Company found that diazonium hexafluorophosphates and antimonates were superior to the corresponding borates in effecting photocure [7–9, 14]. At this company, a photocure technology known as "Americure" was developed.

Photolysis of the diazonium salts can be described by the following re-

[1] The first use of diazonium salts to initiate the cure of cyclic ethers was also carried out in 1965 [5]. The investigators initiated the cationic polymerization of tetrahydrofuran to high molecular weight poly(tetramethylene oxide) by the thermal decomposition of a compound that was later identified as p-chlorobenzenediazonium hexafluorophosphate [6].

action in which the diazonium salt degrades into gaseous nitrogen, an aryl fluoride, and the Lewis acid photoinitiator phosphorous pentafluoride.

$$A{-}N{=}N^{+}PF_6^{-} \xrightarrow[\text{RADIATION}]{\text{ULTRAVIOLET}} N_2\uparrow + A{-}F + PF_5$$

Aryl hexafluorodiazonium phosphorous salt — Nitrogen — Aryl fluoride — Phosphorous Pentafluoride (A Lewis Acid)

The Lewis acid will interact with an epoxide environment, preferably cycloaliphatic epoxide, and initiate the polymerization of the epoxide into a highly crosslinked film. The following is a simplified description of the polymerization process with a monofunctional epoxide.

PF_5 + O → $\bar{O}PF_5$ +

EPOXIDE

$\bar{O}PF_5$ O O +

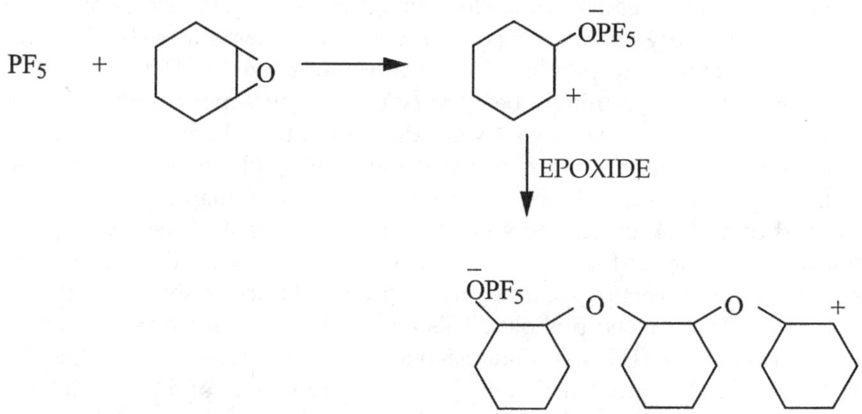

Cross-linking is easier to visualize if one considers multifunctional epoxide monomers rather than the monofunctional epoxide. Nitrogen or other inert atmosphere was not required for good surface and through cure with these cationic systems.

Americure technology was considered a major breakthrough in the effort to develop high-solids, low-energy-curable, environmentally friendly coating systems that were of definite interest at this time. The technology was introduced into the marketplace, but the coating industry really wasn't ready for a technology that required new equipment and learning. Even with this reluctance in the marketplace, a variety of coating and ink applications were investigated, including metal containers, flexible plastics and paperboard, and rigid plastic containers. Specific items that were successfully coated included food cartons, the exterior of two-piece metal beverage cans, oil containers, plastic bottles and cups, disposable paper cups, plates, bowls, and so on. The Americure technology did achieve commercial use as overprint varnishes for metal cans, and there was some success in the flexible packaging area.

The radiation-absorbing portion of the diazonium salt structure could be easily modified, and a wide variety of compounds with tailored adsorption

TABLE 4.1. Summary of Americure Diazonium Salt Technology.

TECHNOLOGY	REFERENCES
Basic chemistry	[7, 15–17]
End uses	[18–23]
Epoxide blends	[14, 24–26]
Photoresists	[23]
Pigmented systems	[27–29]
Solvents	[30]
Stabilizers	[31–34]
Thermal properties	[35]
Ultraviolet radiation dosimetry	[36]

characteristics have been made [8,9]. These compounds will absorb in the ultraviolet and blue region of the electromagnetic spectrum. Schlesinger and coworkers broadly patented this technology and related aspects. Their inventions and the corresponding patents are summarized in Table 4.1.

Regardless of seemingly being very useful, the Americure technology had two major deficiencies that were deterrents to full commercial acceptance. First, the gaseous nitrogen that formed during photolysis was trapped in the coating as tiny bubbles if the coating was more than a few ten-thousandths of an inch in thickness (about 5 to 10 microns). If the coating was thicker, when the gas formed it could not escape from the coating because the extremely rapid reaction solidified the coating and trapped the unsightly gas bubbles and/or caused pinholes. This limited the Americure technology to very thin coatings. The second deficiency involved the instability of the diazonium salts. Formulated systems, which were blends of a cycloaliphatic epoxide, other epoxides, diazonium salt, and additives, were unstable without exposure to ultraviolet radiation. The degree of instability was found to be related to epoxide structure. For example, allyl glycidyl ethers were found to react rapidly with heat evolution when mixed with p-chlorobenzenediazonium hexafluorophosphate. The heat evolution enhanced the reaction rate and shortened stability time. In addition, the particular aryl diazonium salt and the substituents on the aryl group play a role in storage stability. The systems could be stabilized by the use of Lewis bases, but these quenched the cationic reaction that was supposed to take place when the formulation was exposed to ultraviolet radiation. Some compounds, such as sulfoxides, ureas, amides, and nitriles, will impart limited storage improvement without interfering with the radiation-induced polymerization [37–40]. This short pot life was more pronounced if the systems were heated during use, transport, or storage. Recently, it has been found that in certain diazomicro films and printing plates the diazonium compounds are thermally stable for several months or years [41]. The study pointed out that the onium salt and monomer used caused electron transfer reactions started by radical induced cation formation. The monomers used were 4-alkyloxydiazonium salts with various anions, ethylene glycol divinylether, and a cycloaliphatic epoxide with the following structure:

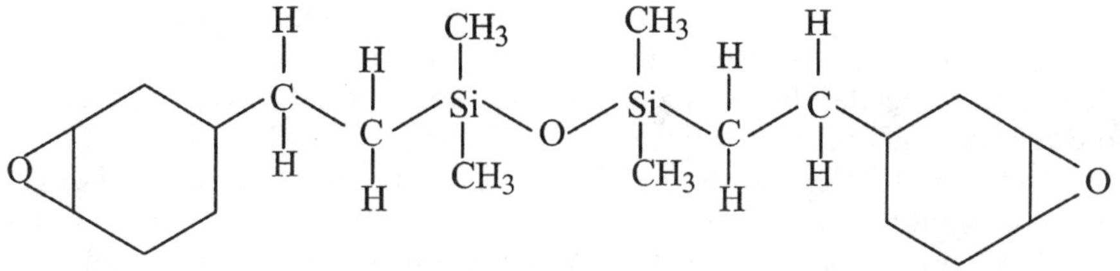

A cycloaliphatic epoxide

What might be called a third deficiency was the fact that the diazonium salts were inherently moisture sensitive.

Copolymers of cyclic monomers and epoxides were prepared by photolysis of the diazonium salts [42]. The cyclic monomers investigated were β-propiolactone and ε-caprolactone.

In summary, the diazonium-initiated systems were commercially used, but market acceptance and growth were not attained because of the stated deficiencies and the advent of iodonium and sulfonium salt technology that involved proton or Bronsted acid initiation. However, before leaving the diazonium salts, it certainly should be pointed out that the efforts at North American Aviation and American Can represented pioneering efforts and fostered the realization of others that the benefits from such technology were increased production, decreased costs, and a potential way to meet environmental regulations. The efforts of these investigators certainly spurred the early scientific and later developmental and commercial efforts that others dedicated to iodonium and sulfonium salt chemistry.

Iodonium and Sulfonium Salts, Bronsted Acids

Early studies of iodonium and sulfonium salt preparation, as well as diazonium salt preparation, date back to the end of the 19th century. At that time, and in the early 20th century, investigators of these compounds, as was in vogue at the time, were mainly concerned with preparation and proving structures rather than searching for commercial utility. However, as early as 1903, Büchner published studies dealing with the photolysis and thermal decomposition of diphenyliodonium halides [43]. In the late 1920s, triphenylselenonium chloride was synthesized in benzene by reacting aluminum chloride with diphenylselenium dichloride [44]. This salt was then used to prepare a number of salts by a metathesis reaction. A novel process for preparing various sulfonium salts involved the use of aluminum chloride in benzene, to which sulfur monochloride and optionally other ingredients were added [45].

Although onium salt technology development began earlier [46, 47], it was not until the 1970s that these salts were independently commercialized by General Electric and 3M Companies [45–51]. Academic research and developmental studies at North Dakota State University contributed a great deal to an understanding of the technology and certainly were important to ultimate commercialization [56,57].

In the mid 1970s, the potential for and beginning of commercial development could be seen. For example, Aloye [58] compared ultraviolet-radiation (UV) cured coatings based on epoxides, on acrylates, and on polythiolene and cured on metal substrates. Deviny [59] of 3M Company described coating concentrates (formulated products) that were useful for preparation of ultraviolet-radiation curable coatings. Tarwid and Kester [60] reported on the variations in adhesion and softening point that could be obtained in epoxide-based coatings applied to metal and cured with onium salts.

Cationic photoinitiators are aryl sulfonium metallic salts and aryl iodonium metallic salts. The aryl sulfonium salts are often described as triarylsulfonium salts, but actually, these photoinitiators are a complex mixture of aryl sulfonium salts with the predominant species having the following structures:

$$S \quad S^{+}(MF_6)^{-} \quad \text{and} \quad (MF_6)^{-+}S \quad S \quad S^{+}(MF_6)^{-}$$

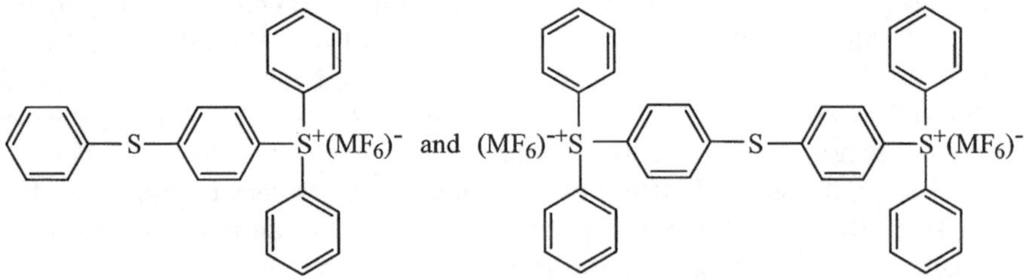

wherein the letter *M* indicates a metal that is usually phosphorous or antimony, though boron and arsenic have been used. When the arylsulfonium salts are prepared, there is some triarylsulfonium salt [61,62] with the following structure,

$$S^{+}\ (MF_6)^{-}$$

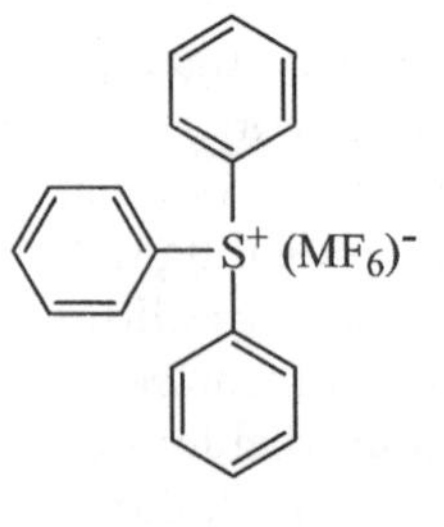

present in the complex mixture. However, as described by Watt [1], when this component is isolated and investigated, it has a markedly slower cure rate with conventional ultraviolet radiation lamps than the complete reaction

TABLE 4.2. Currently or Formerly Commercial Arylsulfoniumhexafluorometallic Salt Photoinitiators.

General form: $A_NS_XMF_6$
(A = Aryl, S = Sulfur, M = metal, F= Fluorine, Subscripts N and X refer to the fact that the photoinitiators can contain 1 or more aryl or sulfur group)

SUPPLIER	TRADENAME	METAL	COUNTRY
3M Company	FX-512	Phosphorus	U. S. A.
Asahi Denka Kogo KK	OPTOMER SP-150*	Phosphorus	Japan
Asahi Denka Kogo KK	OPTOMER SP-170*	Antimony	Japan
Degussa A. G.	Degacure KI-85	Phosphorus	Germany
General Electric Company	UVE-1016*	Phosphorus	U.S.A.
General Electric Company	UVE-1014*	Antimony	U.S.A.
Sartomer Company, Inc.	SarCat KI-85	Phosphorus	U.S.A.
Sartomer Company, Inc.	SarCat SR 1011	Phosphorus	U.S.A.
Sartomer Company, Inc.	SarCat SR 1010	Antimony	U.S.A.
UCB Chemicals-	Uvacure 1590	Phosphorus	U.S.A.
Union Carbide Corp.***	CYRACURE UVI-6992**	Phosphorus	U.S.A.
Union Carbide Corp.***	CYRACURE UVI-6976**	Antimony	U.S.A.

* Not believed to be currently commercially available.
** UVI-6990 and UVI-6974 can be found in the early literature as designations for UVI-6992 and UVI-6976, respectively.
*** A subsidiary of The Dow Chemical Company.

mixture that Watt described with the acronym MASH (Mixed Aromatic Sulfonium Hexafluorophosphates). Ultraviolet radiation absorption maximums and other characteristics of these and other cationic photoinitiators have been tabulated by Gaube [63].

The reason for this difference is that the triarylsulfonium salt has its main absorbance at the lower wavelengths of less than about 260 nm and has nil absorbance above 285 nm. In contrast, MASH has absorbance similar to the triarylsulfonium salt at the low wavelengths, but it maintains relatively high absorbance over the wavelength range of 260 nm to about 320 nm, where many ultraviolet radiation sources operate.

Aryl sulfonium hexafluorophosphate and hexafluoroantimonate photoinitiators are available from various suppliers, including Union Carbide Corporation[2], Sartomer Company, Inc., UCB Chemical, and others. Synthesis methods and Bronsted acid generation mechanisms for these compounds can be found in the literature [64,65].

Table 4.2 is a listing of some commercially available or formerly available arylsulfoniumhexafluorometallic salts. Listing formerly available salts is useful, since when reading early patents, papers, and company literature in this field, they are referred to by trade name. The salts are almost always marketed as pale yellow or amber liquids that are 50% solutions of the photoinitiator in propylene carbonate, though some early versions of cationic photoinitiators such as 3M Company's FX-512 used γ–butyrolactone as the solvent [66].

[2] Union Carbide Corporation is a subsidiary of The Dow Chemical Company.

The propylene carbonate solvent used for the photoinitiators has been said to have an effect on solvent migration in some cured inks [67] and to have an effect on volatile organic content. The latter difficulty may have been related to the fact that the recommended photoinitiator content was quite high in the formulations. One study investigated removing a portion of the solvent and replacing it with a cycloaliphatic epoxide [68]. The final mixture contained 40% photoinitiator, 40% cycloaliphatic diepoxide, and 20% propylene carbonate. The new photoinitiator blend and the usual 50/50 mixture were then tested in formulations suitable for rigid and flexible packaging applications. Black, blue, red, and yellow flexographic inks were prepared using either 12.5% of the new photoinitiator blend or 10% of the original blend, i.e., all blends contained the same concentration of the arylsulfonium salt photoinitiator. The films all cured well with the new initiator blend imparting a slightly faster cure rate, and the cured films had excellent adhesion to thin-film, plastic substrates. Clear coatings formulated to contain 2% actual cationic photoinitiator were tested on aluminum and tin-free steel substrates. Overall, the results indicated that adhesion, cure rate, flexibility, hardness, and solvent resistance were not affected when the new photoinitiator blend was used. The viscosities with the new initiator were slightly higher than with the old initiator, but this could be remedied with a small amount of reactive diluent.

The aryl iodonium salts are of the form

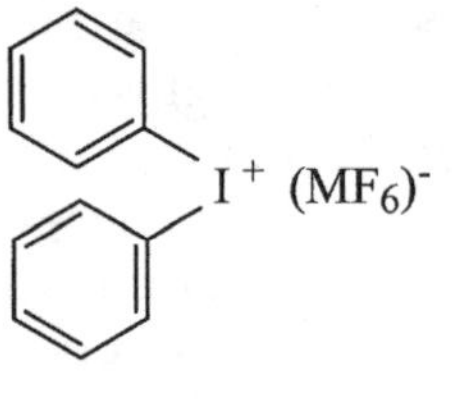

$$(C_6H_5)_2I^+ \ (MF_6)^-$$

and apparently only the diaryl- or diphenyliodonium hexafluoroantimonate compound [45,69,70] is commercially available [71]. Simple synthesis and characterization of alkoxy-substituted diaryliodonium salts are available [72]. The compounds have a low order of toxicity.

The iodonium salts are marketed as off-white powders. Table 4.3 contains a listing of some commercially available or formerly available diaryliodoniumhexafluoroantimonate photoinitiators. Iodonium salt photoini-

TABLE 4.3. Currently or Formerly Commercial Diaryliodoniumhexafluoroantimonate Photoinitiators.

General form: A_2ISbF_6
(A = Aryl, I = Iodine, Sb = Antimony, F= Fluorine)

COMPANY	TRADEMARK	METAL	COUNTRY
3M Company	FC-509	Antimony	U.S.A.
Sartomer Company, Inc.	SarCat SR 1012	Antimony	U. S. A.

tiators are often used with a photosensitizer such as isopropylthioxanthone, which was once marketed by 3M Company as FC-510. Sensitizers in general are treated in a later section of this chapter.

Since some iodonium salt photoinitiators can be immiscible or partially immiscible in epoxy silicones [73], modified iodonium salts were designed and developed for improved miscibility in organofunctional polysiloxanes [74]. These photoinitiators, bis(dodecyl phenyl)iodonium hexafluoroantimonate,

$$CH_3-(CH_2)_6-CH_2-O-C_6H_4-\overset{+}{I}-C_6H_5\quad SbF_6^-$$

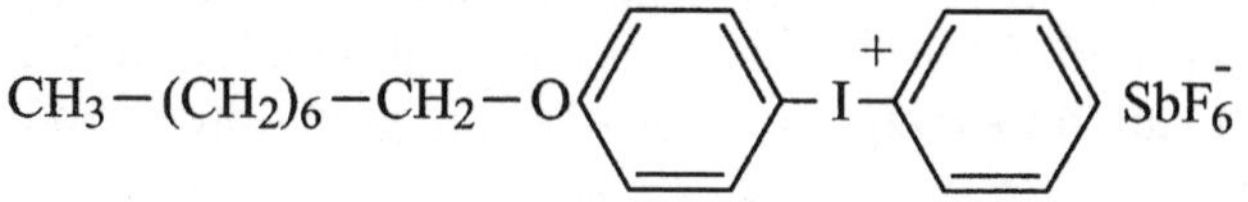

and an octyl diphenyliodonium hexafluroantimonate,

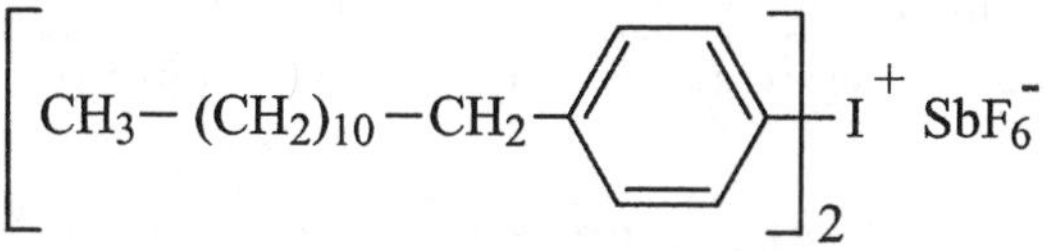

$$\left[CH_3-(CH_2)_{10}-CH_2-C_6H_4\right]_2-\overset{+}{I}\quad SbF_6^-$$

and related compounds were important to the development of ultraviolet radiation curable release coatings[3] and other silicone coating compositions. The dodecyl compound is a complex mixture, due to the nature of the starting material, and the octyl compound is a low melting, crystalline iodonium salt.

A newer cationic iodonium photoinitiator [75–77] that has higher conversion (95.3% vs. 75.5%) and a markedly shorter induction time (1.4 s vs. 30.3 s) than diaryliodonium hexafluoroantimonate in photo-DSC heat generation has the following structure:

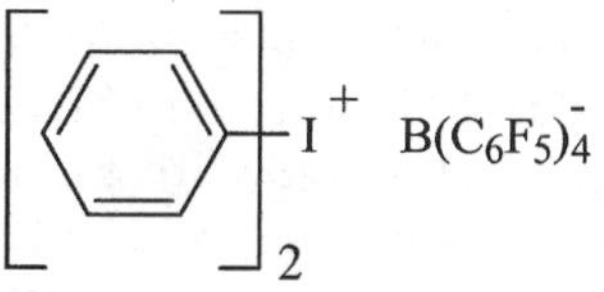

$$\left[C_6H_5\right]_2-\overset{+}{I}\quad B(C_6F_5)_4^-$$

Diaryliodonium tetrakis(pentafluorophenyl) borate

This photoinitiator is a crystalline salt with a novel anion. The anion is highly resonance stabilized, bulky, and very weakly coordinated with the cation. These features result in the generation of a very strong acid for initiation of cycloaliphatic epoxides. In addition, the molecule is designed to have good solubility in nonpolar liquids, such as the epoxy silicones used in release coatings. Studies indicate the iodonium borate yields the fastest cure rate of

[3] See Chapter 10, "Adhesives," for a discussion of release coatings.

any photoinitiator used in ultraviolet radiation initiated reaction of epoxy silicones.

Methods for preparing diaryliodonium, diphenyliodonium, alkylaryliodonium salts can be found in the literature [78–80], as well as catalysts and inhibitors for reactions of diphenyliodonium ions with anions [81]. One patented process [80] involves a reaction between aryl iodonium tosylate and an active hydrogen aliphatic organic compound, such as dimedone. This produces an intermediate arylalkyliodonium tosylate salt that is subsequently reacted with a polyfluorometal or metalloid salt of an alkali or alkaline earth metal to produce the arylalkyliodonium multifluorometal or metalloid salt, an example of which is dimedonylphenyliodonium hexafluoroarsenate.

The photochemical reaction when aryliodonium salts are irradiated with ultraviolet radiation has been postulated to be a homolytic cleavage of one of the aryl bonds followed by hydrogen abstraction from a suitable donor (solvent or other molecule with an active hydrogen) and loss of a proton to yield the Bronsted acid HX.

$$\underset{\text{Iodonium salt}}{Ar_2I^+PF_6^-} \xrightarrow{h\nu} \underset{\text{Excited State}}{[Ar_2I^+PF_6^-]^*} \rightarrow \underset{\text{Free radicals}}{ArI\cdot^+ + Ar\cdot} + \underset{\text{Anion}}{PF_6^-}$$

$$ArI\cdot^+ + \underset{\text{Active hydrogen source}}{R\text{–}H} \rightarrow ArI^+H + R\cdot$$

$$ArI^+H \rightarrow ArI + \underset{\text{Bronsted Acid}}{H^+PF_6^-}$$

The anion X may be PF_6, as in the example above, or it may be SbF_6, BF_4, and so on. The protonic acid formed is very powerful, and it will first protonate an oxirane group, and ring opening polymerization will take place.

The design and synthesis of efficient Bronsted acids that will undergo photolysis have been considered [82]. The study indicated that a heteroatom carbon that would undergo cleavage with radiation and form a singlet cation-radical/radical pair was needed. Molecular orbits were considered and photo acids were prepared that provided a proton with a high quantum yield and without producing free radical species.

Organometallic Compounds

Organometallic compounds are also used as cationic photoinitiators. Included in this category are the iron arene salts [83,84] that may be described by the following general structural formula wherein *M* represents the metals boron, phosphorus, arsenic, and antimony and the subscript to fluorine is four in the case of boron:

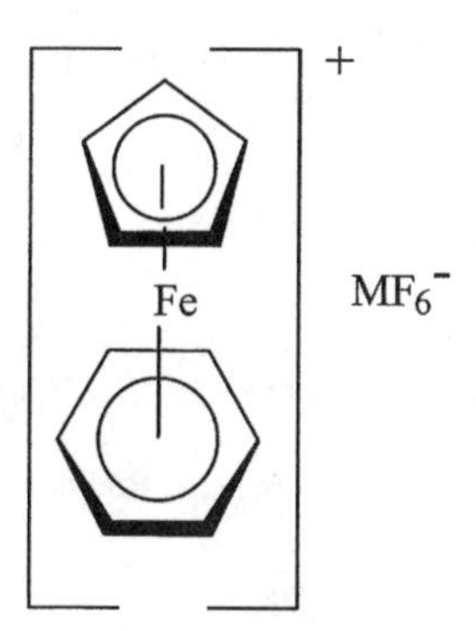

Methods for preparation of these iron salts from ferrocene can be found in the literature [51]. Such ferrocenium complexes are thermally stable and radiation sensitive. The salts photolyze at the higher end of the ultraviolet region and in the visible region. For example, cyclopentadienyl cumyl iron (II), a highly colored compound, photolyzes at wavelengths of about 320–540 nm [85].

Under the action of visible and/or ultraviolet radiation, these compounds photolyze and form a Lewis acid that may be typified by:

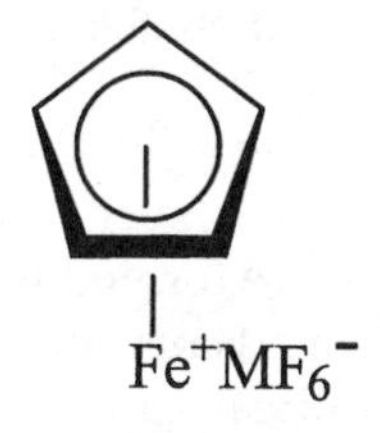

A Lewis acid

and an uncharged arene ligand.

Other organometallic compounds that have been used as cationic photoinitiators for the polymerization of epoxides include zirconocene dichloride [87] and manganese decacarbonyl [88], which initiate polymerization of epichlorohydrin; organohalogen compounds in combination with organometal aluminum complexes for epoxides [89, 90]; as well as other compounds [91–95].

Photolysis of the aryl sulfonium salts can be described in the series of reactions described in Figure 4.1 and in various references [96–100]. In these reactions, ultraviolet radiation first interacts with the photoinitiator salt in the presence of an active hydrogen source and raises it to a higher energy or excited state. The active hydrogen source is not deliberately added, since it is usually present in various ways as from the solvent in which the photoinitiator is dissolved or some other ingredient in the formulation. The excited salt then undergoes spontaneous homolytic cleavage to form free radicals, a

EXCITATION

$$A_N S_X^+ MF_6^- \xrightarrow[\text{ACTIVE HYDROGEN}]{\text{ULTRAVIOLET RADIATION}} \left[A_N S_X^+ MF_6^-\right]^*$$

Sulfonium Salt — Excited State

HOMOLYTIC CLEAVAGE

$$\left[A_N S_X^+ MF_6^-\right]^* \longrightarrow A\cdot + AS\cdot + A_{(N-2)}S_{(X-1)}^+ + MF_6^-$$

Excited Salt — Free Radicals — Radical Cation — Anionic Species

INTERACTION WITH ACTIVE HYDROGEN

$$A_{(N-2)}S_{(X-1)}^+ + R{-}H \longrightarrow A_{(N-2)}S_{(X-1)}H^+ + R\cdot$$

Active H Source

DEPROTONATION AND CATION FORMATION

$$A_{(N-2)}S_{(X-1)}H^+ \longrightarrow A_{(N-2)}S_{(X-1)} \quad H^+ + MF_6^-$$

Cation — Counter ion

FIG. 4.1—Photolysis of An Onium Salt Photoinitiator—Mechanism of Cation Formation.

radical cation, and an anionic species such as phosphorous or antimony hexafluoride. The free radicals that are formed will readily initiate the polymerization of ethylenically unsaturated compounds, such as the acrylates, if present in a formulation. The radical cation then interacts with the active hydrogen source, accepts a proton from it, and another free radical is formed. Almost instantaneously, the new short-lived compound forms an inactive species, a long-lived cation that is balanced by the anionic hexafluorometallic compound. The latter two groups form the Bronsted or protonic acid that is the actual initiator of the polymerization. The cation readily protonates an epoxide group, such as that found in cycloaliphatic epoxides, and ring opening takes place in the polymerization initiation step. In this ring-opening reaction, a very active hydroxyl group that is capable of reacting with other epoxide groups is formed, and the propagation phase of the polymerization is underway. The final product is a highly cross-linked polymeric mass that can range in character from hard and brittle to tough and elastic, depending on the formulating ingredients.

Photosensitization

In general, the diaryliodonium salts discussed only absorb light in the ultraviolet region of the electromagnetic spectrum, and only sparingly in the near ultraviolet region, and not at all in the visible region (400–800 nm). However, these salts can be sensitized to the near ultraviolet and visible regions by compounds known as "photosensitizers," or merely "sensitizers." Some sensitizers will function such that free radicals will be generated when the photoinitiator/sensitizer combination is exposed to visible radiation [101,102], and others function such that cations will be generated from the combination when it is exposed to visible radiation [97,103,104].

Illustrative of the sensitizers that will promote free radical generation in the visible region are diphenylmethane, xanthene, acridine, thiazol, aminoketone, and others. Some specific compounds are anthracene, 9-methylanthracene, 2-ethyl-9,10-dimethoxyanthracene, perylene, and tetraphenylbenzidine. Examples of the sensitizers that produce cations from irradiation in the visible region are Acridine Orange, C.I.[4] 46005; Acridine Yellow, C. I. 46025; Phosphine R, C. I. 46045; Benzoflavin, C. I. 46065; Setoflavin T, C. I. 49005; isopropylthioxanthone, and others. Each of the sensitizers tends to have a characteristic response in the visible and near ultraviolet regions. Thus, if the compounds are mixed and used together, it should be possible to broaden the radiation response and/or increase cure rate [101].

References

[1] Watt, W. R., "UV Curing of Epoxides by Cationic Polymerization," *Radiation Curing*, Vol. 13, No. 4, 1986, p. 7.

[2] Lee, W. E., Calvert, J. G., and Malmberg, E. W., "The Photodecomposition of Diazonium Salt Solutions," *Journal of the American Chemical Society*, Vol. 83, 1961, p. 1928.

[3] Lewis, E. S., Holliday, R. E., and Hartung, L. D., "Relation Between Thermal and Photochemical Hydrolysis of Diazonium Salts," *Journal of the American Chemical Society*, Vol. 91, No. 2, 1969, p. 430.

[4] Patent: Licari, J. J., Crepeau, W., Crepeau, P. C., North American Aviation, U. S. 3,205,157 (1965).

[5] Dreyfuss, M. P. and Dreyfuss, P., *Polymer*, Vol. 6, 1965, p. 93.

[6] Dreyfuss, M. P. and Dreyfuss, P., *Journal of Polymer Science*, A-1, Vol. 4, 1966, p. 2179.

[7] Patent: Schlesinger, S. I., American Can Company, U. S. 3,708,296 (1973).

[8] Schlesinger, S. I., *Polymer Engineering and Science*, Vol. 14, No. 7, 1974, p. 513.

[9] Schlesinger, S. I., *Journal of Photographic Science and Engineering*, Vol. 18, No. 4, 1974, p. 387.

[10] Patent: De Moira, P. P., Murphy, J. P., Ozalid Company, Ltd., "Photopolymerisable Compositions and Their Uses with Diazonium Salts as Photocatalysts," U. S. 3,930,856 (1976).

[11] Patent: Fischer, E., Farbwerke Hoechst Aktlengesellschraft, "Polymerization of Cyclic Ethers by Diazonium Salts of Perchloric or Perfluorocarboxylic Acids," U. S. 3,236,784 (1966).

[12] Patent: DiPippo, C. A., James River Graphics, Inc., "Light-Sensitive Diazonium Trifluoromethane Sulfonates," U. S. 4,482,489 (1984).

[13] Patent: Cripps, H. N., E. I. duPont de Nemours and Company, "Photopolymerizable Compositions and Process," U. S. 3,347,676 (1967).

[14] Patent: Watt, W. R., American Can Company, U. S. 3,794,576 (1974).

[4] C.I. = Color Index

[15] Patents: Schlesinger, S. I., American Can Company, U. S. 3,816,279 (1974); U. S. 3,835,003 (1974); U. S. 3,895,952 (1975); U. S. 3,949,143 (1976); U. S. 3,996,052 (1976); U. S. 4,113,497 (1978).
[16] Patent: Feinberg, J. H., American Can Company, U. S. 3,829,369 (1974).
[17] Patent: Roteman, J., American Can Company, "Epoxy Resin Photoresist with Iodoform and Bismuth Triphenyl," U. S. 3,895,954 (1975).
[18] Patent: Schlesinger, S. I., American Can Company, U. S. 3,826,650 (1974).
[19] Patent: Schlesinger, S. I., Boszak, R. J., Shuppert, L. V., Cowling, R. J., American Can Company, U. S. 3,890,149 (1975).
[20] Patent: Watt, W. R., American Can Company, U. S. 4,105,806 (1978).
[21] Patent: Watt, W. R., Astolfi, E. G., Laufer, J. K., American Can Company, U. S. 4,113,895 (1978).
[22] Patents: Schlesinger, S. I., Boszak, R. J., American Can Company, U. S. 3,997,344 (1976); U. S. 4,054,732; U. S. 4,076,536 (1978); U. S. 4,100,321 (1978).
[23] Patents: Schlesinger, S. I., Cochran, V., American Can Company, U. S. 4,054,451 (1977); U. S. 4,054,452 (1977); U. S. 4,054,055 (1977); U. S. 4,054,635 (1977); U. S. 4,056,393 (1977); U. S. 4,071,671 (1978); U. S. 4,210,499 (1980).
[24] Patent: Watt, W. R., American Can Company, U. S. 3,936,557 (1976).
[25] Patent: Schlesinger, S. I., American Can Company, U. S. 4,080,274 (1978).
[26] Patent: Tortorello, A. J., American Can Company, U. S. 4,289,595 (1981).
[27] Patent: Tarwid, W. A., American Can Company, U. S. 4,054,498 (1977).
[28] Patent: Karoly, G., Gardon, J. L., American Can Company, U. S. 4,107,353 (1978).
[29] Patent: Schlesinger, S. I., American Can Company, U. S. 4,287,228 (1981).
[30] Patent: Feinberg, J. H., American Can Company, U. S. 3,960,684 (1976).
[31] Patent: Feinberg, J. H., American Can Company, U. S. 3,711,390 (1973); U. S. 3,711,391 (1973); U. S. 3,816,281 (1974); U. S. 3,817,845 (1974); U. S. 3,817,850 (1974).
[32] Patents: Watt, W. R., American Can Company, U. S. 3,721,616 (1973); U. S. 3,721,617 (1973); U. S. 3,816,278 (1974); U. S. 3,816,280 (1974).
[33] Patent: Schlesinger, S. I., American Can Company, U. S. 3,9951,769 (1976).
[34] Patent: Jacobs, S. S., American Can Company, U. S. 4,000,115 (1976).
[35] Patent: Guarnery, J. M., Watt, W. R., American Can Company, U. S. 3,775,122 (1973).
[36] Patent: Schlesinger, S. I., American Can Company, U. S. 3,775,122 (1973).
[37] Patent: Watt, W. R., American Can Company, U. S. 3,721,616 (1973).
[38] Patent: Watt, W. R., American Can Company, U. S. 3,721,617 (1973).
[39] Patent: Feinberg, J. H., American Can Company, U. S. 3,771,390 (1973).
[40] Patent: Feinberg, J. H., American Can Company, U. S. 3,771,391 (1973).
[41] Deubzer, B., Herzig, Ch., Morke, W., Muller, U., and Utterodt, A., "The Truth About Diazonium Salts in Photosensitive Vinylether and Epoxy Formulations," *Conference Proceedings of RadTech Europe '99*, Berlin, Germany, 8–10 November, 1999, p. 821.
[42] Patent: Schlesinger, S. I., American Can Company, "Photopolymerization of Lactone-Epoxide Mixtures with Aromatic Diazonium Salts as Photocatalyst," U. S. 4,080,274 (1978).
[43] Büchner, E. H., Koninklijke Akad. V. Weltenschappen le Amsterdam, *Proceedings of Sec. Science*, Vol. 8, 1903, p. 646 (In English).
[44] Leicester, H. M. and Bergstrom, F. W., "Salts of Triphenylselenonium Hydroxide," *Journal of the American Chemical Society.*, Vol. 51, 1929, p. 3587.
[45] Patent: Pitt, H. M., Stauffer Chemical Company, "Process for Making Sulfonium Compounds," U. S. 2,807,648 (1957).
[46] Knapczyk, J. W. and McEwen, W. E., "Reactions of Triarylsulfonium Salts with Bases," *Journal of the American Chemical Society*, Vol. 91, No. 1, 1969, p. 145.
[47] Knapczyk, J. W. and McEwen, W. E., "Photolysis of Triarylsulfonium Salts in Alcohol," *Journal of Organic Chemistry*, Vol. 35, No. 8, 1970, p. 2539.
[48] Crivello, J. V. and Lam, J. H. W., Symposium No. 56, *Journal of Polymer Science*, 1976, p. 383.

[49] Crivello, J. V. and Lam, J. H. W., *Proceedings of Organic Coatings and Plastics Chemistry*, American Chemical Society, Vol. 39, 1978, p. 31.
[50] Patent: Smith, G. H., 3M Company, Belgium 828,841 (1975).
[51] Patent: Smith, G. H., Olofson, P. M., U. S. 4,173,476 (1979).
[52] Pappas, S. P. and Jilek, J. H., "Photoinitiated Cationic Polymerization by Photosensitization of Onium Salts," *Photographic Science and Engineering*, Vol. 23, No. 3, 1979, p. 140.
[53] Jilek, J. H., "Photosensitized Cationic Polymerization," PhD Thesis, North Dakota State University, 1980, University Microfilms Int. # 7915508.
[54] Pappas, S. P., Jilek, J. H., and Gatechair, L. R., "Cationic UV Curing," *Proceedings 6 of VIth International Conference in Organic Coatings Science and Technology*, Athens, Greece, 14–18 July, 1980, p. 587.
[55] Pappas, S. P., Gatechair, L. R., and Jilek, J. H., "Photoinitiation of Cationic Polymerization. III. Photosensitization of Diphenyliodonium and Triphenylsulfonium Salts," *Journal of Polymer Science: Polymer Chemistry Edition*, Vol. 22, 1984, p. 77.
[56] Pappas, S. P., Jilek, J. H., and Gatechair, L. R., "Photogeneration of Cationic Initiators by Donor-Acceptor Interactions-7. Silver (I), Copper (II), and Onium Salts as Acceptors," *Journal of Imaging Science*, Vol. 30, No. 5, 1980, p. 218.
[57] Pappas, S. P., Pappas, B. C., Gatechair, L. R., and Jilek, J. H., "Photoinitiation of Cationic Polymerization. IV. Direct and Sensitized Photolysis of Aryl Iodonium and Sulfonium Salts," *Polymer Photochemistry*, Vol. 5, No. 1, 1984, p. 1.
[58] Aloye, J. A., Technical Paper FC-505, Society of Manufacturing Engineers, Dearborn, MI, 1976.
[59] Deviny, E. J., Technical Paper FC-503, Society of Manufacturing Engineers, Dearborn, MI, 1976.
[60] Kester, D. E. and Tarwid, W. A., in "Container Coatings," ACS Symposium Series No. 78, R. C. Strand, Ed., 1976, pp. 38–47.
[61] Knapczyk, J. W. and McEwen, W. E., *Journal of the American Chemical Society*, Vol. 91, 1969, p. 1451.
[62] Crivello, J. V., *Journal of Radiation Curing*, Vol. 5, No. 1, January 1978, p. 2.
[63] Gaube, H. G., "UV Curing of Epoxy Coatings by Selected Cationic Initiators—A Comparative Study on Variables Affecting Reactivity and Cured Film Properties," *Proceedings of Radcure '86*, Baltimore, MD, 8–11 September, 1986, pp. 15–27.
[64] Hacker, N. P. and Dektar, J. L., "Cationic Photoinitiators: Generation from the Photodecomposition of Triphenylsulfonium and Triphenylselenonium Salts," *Proceedings of the ACS Division of Polymeric Materials: Science and Engineering*, Miami Beach, FL, Vol. 61, Fall 1989, p. 76.
[65] Crivello, J. V., "The Chemistry of Photoacid Generating Compounds," *Proceedings of the ACS Division of Polymeric Materials: Science and Engineering*, Miami Beach, FL, Vol. 61, Fall 1989, p. 62.
[66] Dougherty, J. A., Vara, F. J., and Anderson, L. R., "Vinyl Ethers for Cationic UV Curing," *Proceedings of RADCURE '86*, Baltimore, MD, 8–11 September 1986, p. 15-1.
[67] Carter, J. W. and Jupina, M. J., "Cationic UV Ink Migration and Safety Assessment," *Conference Proceedings, RadTech Europe '97*, Lyons, France, 1997, p. 123.
[68] Carter, W., Jupina, M., and Lamb, K., "Performance of Higher-Solids Cationic Photoinitiator Mixture," *Conference Proceeding, RadTech Europe '99*, Berlin, Germany, 1999, p. 515.
[69] Crivello, J. V. and Lam, J. H. Q., *Macromolecules*, Vol. 10, No. 6, 1977, p. 1307.
[70] Crivello, J. V., "Applications of Photoinitiated Cationic Polymerization in Coatings," *Journal of Coatings Technology*, Vol. 63, No. 793, February 1991, p. 35.
[71] Sartomer Company, Technical Solutions, Product Catalog, 1999.
[72] Crivello, J. V. and Lee, J. L., "Alkoxy-Substituted Diaryliodonium Salt Cationic Photoinitiators," *Proceedings of RadTech '90—North America*, Chicago, IL, Vol. 1, 25–29 March, 1990, p. 424.

[73] Eckberg, R. P., Riding, K. D., and Farley, D. E., "Novel Photocurable Organosilicon Compositions," *Proceedings of RadTech '90 North America*, Chicago, IL, 25–29 March, 1990, p. 358.

[74] Patent: Eckberg, R. P., LaRochelle, R. W., General Electric Company, "Ultraviolet Curable Epoxy Silicone Coating Compositions," U. S. 4,279,717 (1981).

[75] Kerr, S. R., III, "Next Generation UV Silicone Release Coatings," *Adhesive Age*, Vol. 39, No. 9, August 1996, p. 26.

[76] Patent: Castellanos, F., Cavezzan, J., Fouassier, J. P., Priou, C., Rhone-Poulenc Chimie, "Onium borates/borates of Organometallic Complexes and Cationic Initiation of Polymerization Therewith," U. S. 5,468,902 (1995).

[77] Patent: Cavezzan, J., Priou, C., Rhone-Poulenc Chimie, "Cationically Crosslinkable Polyorganosiloxanes and Antiadhesive Coating Produced Therefrom," U. S. 5,340,898 (1994).

[78] Beringer, F. M., Falk, R. A., Karniol, M., Lillien, I., Masullo, G., Mausner, M., and Sommer, E., "Diaryliodonium Salts. IX. The Synthesis of Substituted Diphenyliodonium Salts," *Journal of the American Chemical Society*, Vol. 81, 1959, p. 342.

[79] Caserio, M., Glusker, D. L., and Roberts, J. D., "Hydrolysis of Diaryliodonium Salts," *Journal of the American Chemical Society*, Vol. 81, 1959, p. 336.

[80] Patent: Crivello, J. A., Lee, J. L., General Electric Company, "Alkylaryliodonium Salts and Method for Making," U. S. 4,450,360 (1984).

[81] Beringer, F. M., Gindler, M., Rapoport, M., and Taylor, R. J., "Diaryliodonium Salts. X. Catalysts and Inhibitors in the Reactions of Diphenyliodonium Ions with Anions and Hydroxylic Solvents," *Journal of the American Chemical Society*, Vol. 81, 1959, p. 351.

[82] Saeva, F. D., "Design and Synthesis of Efficient Bronsted Photoacid Systems," *Proceedings of the ACS Division of Polymeric Materials: Science and Engineering*, Miami Beach, FL, Vol. 61, Fall 1989, p. 72.

[83] Meier, K. and Zweifel, H., Technical Paper FC 85-417, *Rad-Cure Europe*, Basle, Switzerland, 1985.

[84] Lohse, F., Meier, K., and Zweifel, H., *Proceedings of the 11th International Conference on Organic Coatings Science and Technology*, Athens, Greece, 1985, p. 175.

[85] Schumann, H., *Chemiker-Zeitung*, Vol. 108, 1984, p. 345.

[86] Meier, K. and Zweifel, H., *Journal of Radiation Curing*, Vol. 13, No. 4, 1986, p. 26.

[87] Kaeriyama, K., *Makromolecular Chemie*, Vol. 153, 1972, p. 229; *Journal of Polymer Science, Polymer Chemistry Edition*, Vol. 14, 1976, p. 1547.

[88] Strohmeier, V. W. and Barbeau, C., *Makromolecular Chemie*, Vol. 81, 1965, p. 86.

[89] Hayase, S., Onishi, Y., Suzuki, S., and Wada, M., *Macromolecular Chemie*, Vol. 18, 1985, p. 1799.

[90] Hayase, S., Onishi, Y., Suzuki, S., and Wada, M., *Macromolecular Chemie*, Vol. 19, 1986, p. 986.

[91] Patent: Green, G. E., Irving, E., Ciba-Geigy Corporation, "Photopolymerizable and Thermally Polymerizable Compositions," U. S. 4,299,938 (1981).

[92] Curtis, E., Irving, E., and Johnson, B. F. G., *Chemistry in Britain*, April 1986, p. 327.

[93] Patent: Anderson, W. S., Shell Oil Company, "Process for Light-induced Curing of Epoxy Resins in Presence of Cyclopentadienylmanganese Tricarbonyl Compounds," U. S. 3,709,861 (1973).

[94] Patent: Roteman, J., American Can Company, "Epoxy Resin Photoresist with Iodoform and Bismuth Triphenyl," U. S. 3,895,954 (1975).

[95] Patent: Cella, J. A., General Electric Company, "Method of Applying and Curing Epoxy Coating Compositions Using Dicarbonyl Chelate of Group IIIa-Va Element and UV Irradiation," U. S. 4,086,091 (1978).

[96] Crivello, J. V., "Recent Progress in Photoinitiated Cation Polymerization," *Organic Coatings and Applied Polymer Science Proceedings*, 185th National Meeting, American Chemical Society, Seattle, WA, Vol. 48, 20–25 March, 1983, p. 226.

[97] Crivello, J. V., "Photoinitiated Cationic Polymerization," Chapter 2 in *UV Curing: Science and Technology*, S. P. Pappas, Ed., Technology Marketing Corporation, 1978.

[98] Maycock, A. L. and Berchtold, G. A., "Photochemical Reaction of Phenacyl- and Benzylsulfonium Salts," *Journal of Organic Chemistry*, Vol. 35, No. 8, 1970, p. 2532.

[99] McKean, D. R., Schaedeli, U., Kasai, P. H., and MacDonald, S. A., "Polymeric Matrix Effects on the Efficiency of Acid Generation from Triphenylsulfonium Salts," *Proceedings of the ACS Division of Polymeric Materials: Science and Engineering*, Miami Beach, FL, Vol. 61, Fall 1989, p. 81.

[100] Davidson, R. S. and Goodin, J. W., "Some Studies on the Photo-Initiated Cationic Polymerisation of Epoxides," *European Polymer Journal*, Vol. 18, 1982, p. 589.

[101] Patent: Smith, G. H., Minnesota Mining and Manufacturing Company, "Novel Photosensitive Systems Comprising Diaryliodonium Compounds and Their Use," U. S. 3,729,313 (1973).

[102] Patent: Smith, G. H., Minnesota Mining and Manufacturing Company, "Photocopolymerizable Compositions Based on Epoxy and Hydroxyl-Containing Materials," U. S. 4,256,828 (1981).

[103] Patent: Crivello, J. V., Schroeter, S. H., General Electric Company, "Photocurable Compositions and Methods," U. S. 4,026,705 (1977).

[104] Patent: Crivello, J. V., General Electric Company, "Photocurable Compositions Containing Group Via Aromatic Onium Salts," U. S. 4,058,401 (1977).

5

Free Radical Initiated Polymerization Systems

Free Radical Systems

FREE RADICAL RADIATION CURING of coatings began in the late 1950s at Ford Motor Company. Dr. William Burlant of this company was interested in ionizing radiation and the effect it had on organic molecules [1]. He knew that ionizing radiation was capable of initiating very rapid reactions and was interested in polymerizations that could be initiated with such high-energy radiation. In the beginning, the efforts were not really directed towards coatings. However, before the studies had gone very far, Burlant realized that radiation-cured coatings might be an assembly line application for the technology, and his group held a special demonstration for Henry Ford and Ford's senior engineers[1]. The process being revealed had the benefits of rapid cure rate, no appreciable substrate or cured-film temperature rise, and no volatile solvents. In the 1950s, the latter factor was of importance mainly because of energy costs and secondarily because of environmental concerns, since at this time it took about a half hour at elevated temperatures to remove solvent and to effect cross-linking of automobile coatings.

As might be expected, there were difficulties with the reliability of early and, often prototype, electron beam equipment, but by 1960 these problems were solved and coating systems were ready for commercialization. Ford Motor Company commercialized the electron beam process, but it was not used to coat complex-shaped car bodies. Rather, the process was devoted to painting interior plastic parts that were used in automobiles. For his pioneering efforts in this area, RadTech International—North America bestowed on Burlant the title "Father of Electron Beam Curing."

[1] When the electron-beam curing process was demonstrated to Henry Ford and senior staff, Burlant was requested to stop the curing line. The viewers were awed at what they saw and wanted to walk into and see the curing system. They wanted to be certain the demonstration wasn't a sham and that there wasn't a person someplace inside the curing device who was switching cured panels for uncured panels. In the 1950s, it wasn't thought possible that a coating could be cured in a fraction of a second or literally thousands of times faster than the half-hour times then needed for commercial cure lines [1].

Efforts that dealt with the use of electron beam cured coatings on plywood were carried out by Morganstern [2] at Radiation Dynamics, Inc., in the early 1960s. The investigations were done under a 1962 contract the company had with U. S. Plywood. This company was interested in decreasing cure time of coatings on wood substrates, and thereby, increasing productivity. A definite requirement was that the coating would have to cure under ambient conditions, and for this reason, the project was not successful, since some kind of inerting system is required with electron beam curing.

In the early days of radiation curing, there were visionaries who had ideas and foresight, and relatively rudimentary equipment. However, there were very few reactive ingredients available for use in formulating radiation curable systems. An important advance in the ingredient area was made in 1969 when Miranda and Huemmer [3] described the cure characteristics of certain ethylenically unsaturated monomers, their thinning or viscosity reduction power, and the effect an inert atmosphere had on their cure characteristics. The study also described methods for the preparation of acrylic, polyester, and allyl-modified polyester materials that could be used in radiation curing, and the curing characteristics of these compounds. This early study provided investigators with raw materials for potential use, and it also led to many new ideas for other products. Also included in this 1969 study is an early history of the radiation cross-linking of polymer molecules that began with the work of Charlesby [4] when he reported on the cross-linking of polyethylene in 1952. This work, in combination with that of Renfrew [5] in the early 1940s, may be looked on as the real beginning of the commercial radiation curing industry.

Many early studies involved the use of systems based on blends of monomeric compounds such as styrene, ethyl acrylate, and methyl methacrylate in combination with unsaturated polyesters. Results of these investigations suggested that mixtures of the polyesters with styrene were the most effectively cured combinations [6]. Two other types of free radical systems are of importance—the polyene/thiol systems and the acrylate systems. The latter are by far the most important free radical curable systems in use today.

Unsaturated Polyester/Styrene Systems

Unsaturated polyesters have been known for about 125 years. However, in the first few decades of the 20th century, all polyesters were lumped into the general classification of "alkyds," and it wasn't until the late 1930s that an effort was made to clarify and categorize the curing or "drying" nature of unsaturated polyesters [7–9]. These studies indicated that unsaturated polyesters could be converted to cross-linked structures by means of oxygen, heat, and radiation. In a general sense, the unsaturated polyesters may be described as linear, soluble, condensation products of dicarboxylic acids or their corresponding anhydrides and glycols, dihydric alcohols, or polyols. While unsaturation can be introduced into the polyester by means of an unsaturated alcohol [10], it is usually introduced through the use of fumaric acid, maleic acid, or maleic anhydride [11,12] during polyester preparation.

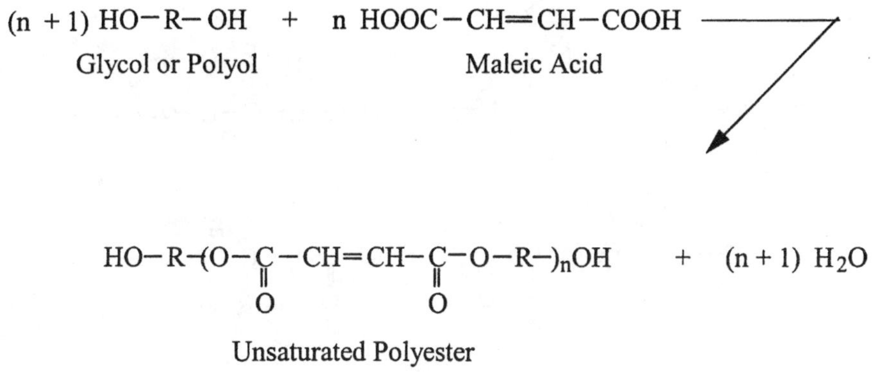

$$(n+1)\ \text{HO–R–OH} + n\ \text{HOOC–CH=CH–COOH} \longrightarrow$$

Glycol or Polyol — Maleic Acid

$$\text{HO–R}\!\left(\text{O–}\underset{\underset{\text{O}}{\|}}{\text{C}}\text{–CH=CH–}\underset{\underset{\text{O}}{\|}}{\text{C}}\text{–O–R–}\right)_n\text{OH} + (n+1)\ H_2O$$

Unsaturated Polyester

In practice, maleic anhydride is used to facilitate polyester formation by reducing the number of condensation reactions required by a factor of two. The unsaturation may also be terminal [10,13] or pendant. Alternatives to maleic acid or anhydride include fumaric, norbornyl, cinnamates, allyl, acrylic, and similar acids and anhydrides. In addition, combinations of saturated carboxylic acids (i.e., phthalic, isophthalic, succinic, adipic, azelaic, etc.) and unsaturated carboxylic acids can be copolymerized with the dihydric alcohols. Use of saturated acids or anhydrides in the polyester provides a means of controlling and altering properties of the final cured coating. Monofunctional carboxylic acids are used and function as molecular chain stoppers, and thus, provide a means of controlling molecular weight.

Although the unsaturated polyesters can be cured alone with radiation through reaction of the reactive double bonds in the molecules, i.e., without the addition of diluents such as styrene, usually their high molecular weight and attendant high viscosity require the use of a low molecular weight functional reactive diluent that will react with the polyester and become an integral part of the final coating [14]. After evaluating a number of factors, it was found that styrene was the best reactive diluent for this purpose [6,14]. Styrene is mixed with the unsaturated polyester to form either a liquid or a pastelike curable system, and a photoinitiator is included if the systems are to be cured with ultraviolet radiation. The styrene units act as cross-linking agents, as shown below. One or more polymerized styrene units act as the cross-link points in the matrix. The final polymerized article can also contain some homopolymerized styrene as well as some cross-links formed by intermolecular reaction of the unsaturation in the polyester molecules.

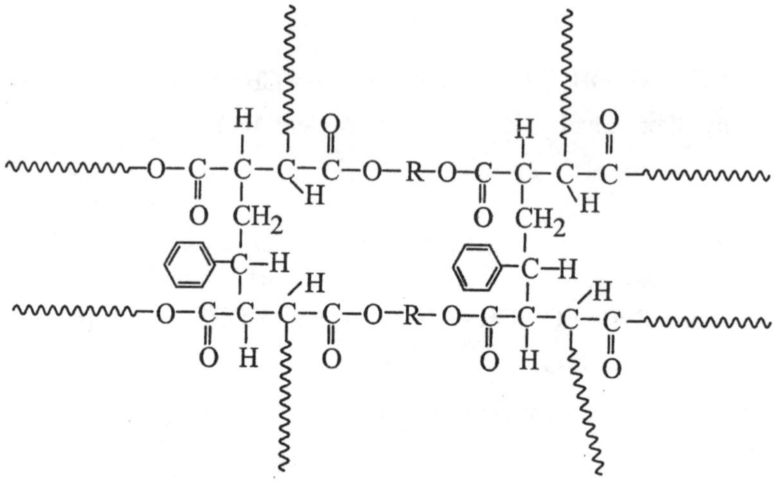

An unsaturated polyester crosslinked with styrene units.
∿∿∿∿ indicates other units of the polyester chain
which may or may not be crosslinked with styrene.

These unsaturated polyester/styrene mixtures were the earliest radiation-cure systems that reached commercialization. As early as 1984, they were well established in the wood industry [15]. They are particularly used in the wood industry because of their ability to cure at basically ambient temperatures and, thus, decrease volatility that caused both a pollution problem and a loss of reactive material. It should be noted that any residual or unreacted styrene would leave a noticeable and lingering odor.

These systems were not used in the radiation-cure area until 1955 when McCloskey and Bond reported on their cure with ultraviolet radiation [16]. They demonstrated that unsaturated polyesters will polymerize with this radiation if a compound that degrades into free radicals, i.e., a photoinitiator, was added. If the photoinitiator was not added, no reaction took place after the same irradiation time. Their work showed that the cure rates were much more rapid than could be obtained by thermal cure of the same systems, but using a peroxide initiator. Burlant and coworkers [17–19] investigated the electron beam or ionizing radiation curing of styrene and polyesters, silicone-modified binders [20], and later acrylates [21]. This work coupled with that of European investigators provided the technology to commercialize the systems as fillers and coatings for open-grain wood products, such as particleboard and fiberboard. The main attributes were rapid cure at low temperatures, which minimized volatile losses and, thus, gave better filling of the pores [22]. In addition, the boards could be sanded immediately after curing. Even though cure of these systems was considered rapid at the time, by today's standards the cure would be considered slow. Orthopedic casts were also made from fiberglass impregnated with unsaturated polyester/styrene/photoinitiator and radiation cured [23].

Wicks pointed out that surface cure was inhibited by the oxygen present in air, which resulted in a tacky surface [24], and that an inert atmosphere was important to a more complete cure. Other studies showed that, in addition to the styrene that volatilized during cure, relatively large amounts of unreacted styrene were present in some of the cured systems [25]. When waxes were incorporated in the formulation, volatility was decreased and the wax reduced oxygen inhibition of cure at the surface [26]. A wide variety of waxes were investigated, and the best was hydrocarbon paraffin wax. Effect of dose rate on gel formation and the relation to unreacted styrene has also been investigated [27]. The retarding effect on cure, due to the addition of titanium dioxide and of various polyols, has been studied, and Laws and coworkers [28] found the best results were obtained with unsaturated diethylene glycol-fumarate polyesters containing chlorendic acid as a saturated co-reactant.

Schwalm and coworkers [29] investigated the use of triethylene glycol divinyl ether, which has a viscosity at 25°C of 2.6 mPa·s, as a reactive diluent for an

$$CH_2{=}CH{-}O{-}(CH_2{-}CH_2{-}O)_2{-}CH_2{-}CH_2{-}O{-}CH{=}CH_2$$

Triethylene glycol divinyl ether

unsaturated polyester of unknown composition, but based on ethylene glycol and fumaric acid. It was found that the vinyl ether was an effective diluent and could copolymerize in an alternating copolymerization effectively when a hydroxyalkyl phenyl ketone free-radical photoinitiator was used. Effectively complete conversion of the vinyl ether was obtained if an aryl sulfonium salt cationic photoinitiator was included in the formulation.

Polyene/Thiol Compositions.

In the late 1960s and early 1970s, W. R. Grace & Co. investigators were investigating new radiation curable systems that came to be known as the "Polyene-Polythiol compositions" [30]. In an overall sense, these ultraviolet radiation- and electron beam-curable compositions comprised a particular polyene component, a polythiol component, and a photoinitiator [31–35]. The latter compound was not needed if electron beams were used, however, ultraviolet radiation of 300–400 nanometers was recommended for cure.

There were various ways to make the polyenes, as for example, they could be prepared by reacting allyl isocyanate with a polyether, polyester, or polylactone polyol that had a hydroxyl functionality of two or greater.

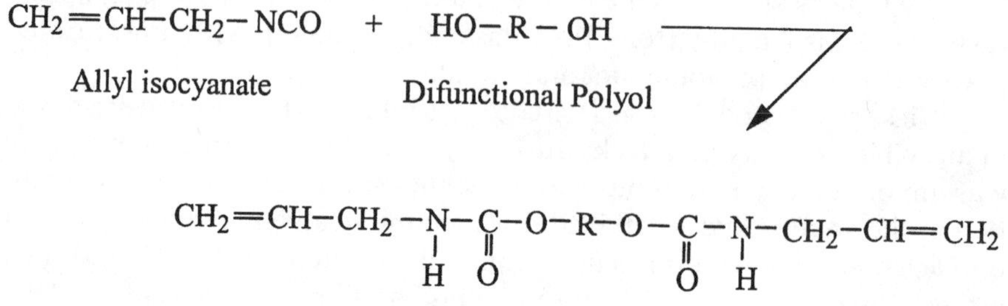

$$CH_2{=}CH{-}CH_2{-}NCO \quad + \quad HO{-}R{-}OH \longrightarrow$$

Allyl isocyanate — Difunctional Polyol

$$CH_2{=}CH{-}CH_2{-}\underset{H}{N}{-}\overset{O}{\overset{\|}{C}}{-}O{-}R{-}O{-}\overset{O}{\overset{\|}{C}}{-}\underset{H}{N}{-}CH_2{-}CH{=}CH_2$$

A difunctional Polyene

If a trifunctional polyol had been used, a trifunctional polyene would have resulted; if a tetrafunctional polyol had been used, a tetrafunctional polyene would have resulted; and so on. Final molecular weight of this component usually ranged from about 200 to 10 000. Urethane reaction catalysts such as stannous octanoate, dibutyltindilaurate, and the like were used to enhance reaction rate between hydroxyl and isocyanate groups.

The polythiol component was of the general form

$$R{-}(SH)_n$$

wherein *R* could be any polyvalent organic moiety that did not have reactive carbon-to-carbon unsaturation and *n* was an integer that was two or greater. Polythiols are the mercaptate esters formed by reaction of thiol-containing acids and polyols. Examples are compounds such as ethylene glycol bis(thioglycolate), trimethylolpropane tris(β-mercaptopropionate), pentaerythritol tetrakis(thioglycolate), pentaerythritol tetrakis(mercaptopropionate), and the like. The latter compound has the following structure:

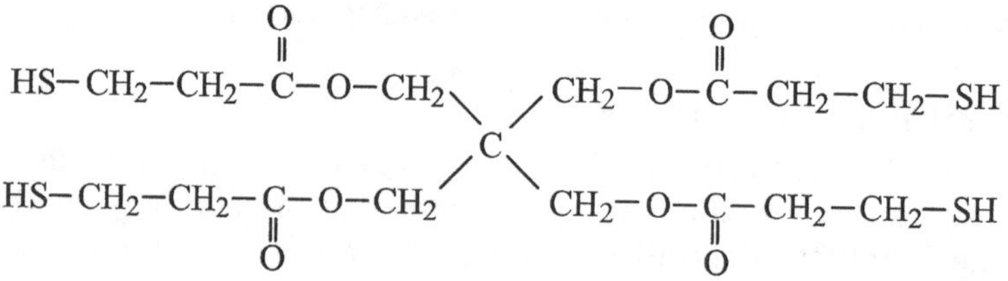

$$C\left(CH_2{-}O{-}\overset{O}{\overset{\|}{C}}{-}CH_2{-}CH_2{-}SH\right)_4$$

Pentaerythritol tetrakis (mercaptopropionate)

It was preferred that the polythiol compounds have substantial molecular weight to minimize the odor problem that is associated with such compounds. The molecular weight of the pentaerythritol derivate shown above is 488. Reaction in the final coating also decreased the odor problem, but even this often did not remove all traces of the sulfurous odor.

Suitable photoinitiators were compounds such as benzophenone, benzanthrone, acetophenone, thioxanthanone, fluorine-9-one, and the like. Initiation with a compound such as benzophenone took place in the usual hydrogen-transfer manner, except the mercaptate ester takes the place of an amine with an alpha-carbon atom [36], and a thiyl radical is formed.

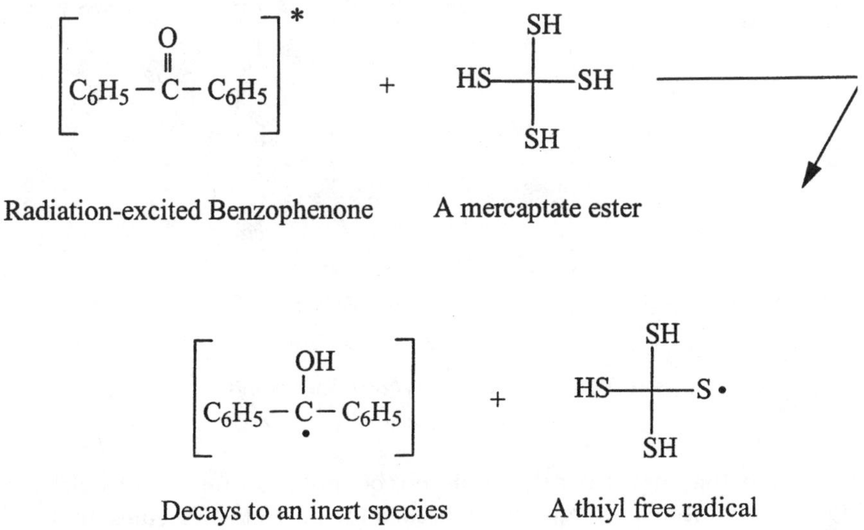

$$\left[C_6H_5-\overset{\overset{\large O}{\|}}{C}-C_6H_5 \right]^* \quad + \quad HS-\underset{\underset{\large SH}{|}}{\overset{\overset{\large SH}{|}}{C}}-SH \longrightarrow$$

Radiation-excited Benzophenone   A mercaptate ester

$$\left[C_6H_5-\underset{\bullet}{\overset{\overset{\large OH}{|}}{C}}-C_6H_5 \right] \quad + \quad HS-\underset{\underset{\large SH}{|}}{\overset{\overset{\large SH}{|}}{C}}-S\bullet$$

Decays to an inert species   A thiyl free radical

The thiyl radical is capable of initiating polymerization of the polyene unsaturation. Compounds such as triphenyl phosphine were found to increase reaction rate and improve mechanical properties. If a tetrafunctional mercaptate ester is simply represented as indicated above and a difunctional polyene as:

$$CH_2{=}CH-\sim\sim\sim-CH{=}CH_2$$

the reaction product of the two compounds will be a highly cross-linked mass that may be described as:

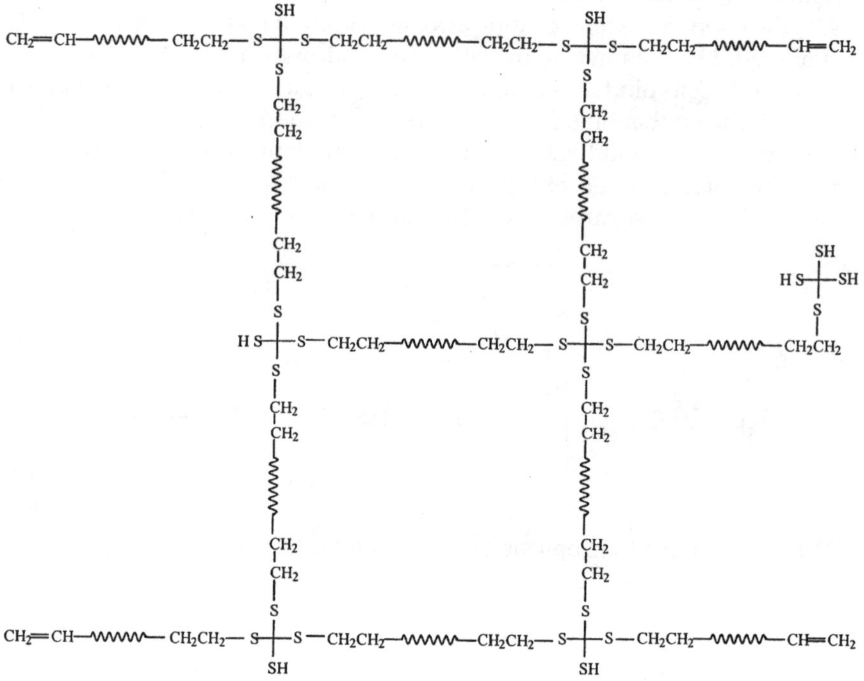

```
                                 SH                                              SH
CH2=CH—wwww— CH2CH2— S—|—S—CH2CH2—wwww—CH2CH2— S—|—S—CH2CH2—wwww—CH=CH2
                         S                                               S
                         CH2                                             CH2
                         CH2                                             CH2
                         wwww                                            wwww
                         CH2                                             CH2                   SH
                         CH2                                             CH2              HS—|—SH
                         S                                               S                     S
                    HS—|—S—CH2CH2—wwww—CH2CH2— S—|—S—CH2CH2—wwww—CH2CH2
                         S                                               S
                         CH2                                             CH2
                         CH2                                             CH2
                         wwww                                            wwww
                         CH2                                             CH2
                         CH2                                             CH2
                         S                                               S
CH2=CH—wwww— CH2CH2— S—|—S—CH2CH2—wwww—CH2CH2— S—|—S—CH2CH2—wwww—CH=CH2
                         SH                                              SH
```

A cross-linked polyene/thiol polymer

Of course, the final structure would not be uniform due to molecular flexibility and molecular weight differences in the polyene molecules. In this example, cross-linking is readily achieved due to the high functionality of the mercaptate ester used. If only difunctional compounds are used, it is possible to control molecular weight by the addition of monomeric compounds such as n-butyl mercaptan or n-dodecyl mercaptan. Because these compounds have high chain transfer constants due to the weakness of the sulfur hydrogen bond, they can be used in low concentrations.

The cured compositions were suggested for use as screen printing inks, elastomeric sealants, electronic component encapsulants, and decorative and functional coatings such as conformal coatings, dielectric coatings, high gloss overprint varnishes, and arc-resistant coatings.

The effect of decreasing the amount of thiol in the formulations by adding acrylates has been investigated [37]. This led to coating systems that had the attributes of the polyene-thiol products, and it alleviated some of the difficulties with odor that were encountered with them. Others combined silicone-modified unsaturated polyesters with the polyene-thiol systems to form radiation curable coating systems that had good adhesion to polycarbonate [38].

Another group of investigators [39] investigated the use of cyclic olefins such as norbornene derivatives as replacements for the polyenes that had olefinic end groups. The ene-polymers of this study were prepared by reacting a polyol or a polyamine with 2-ene-5-carboxylic acid chloride or by reacting cyclopentadiene with acrylic derivatives to yield the norbornyl functional oligomers as shown below.

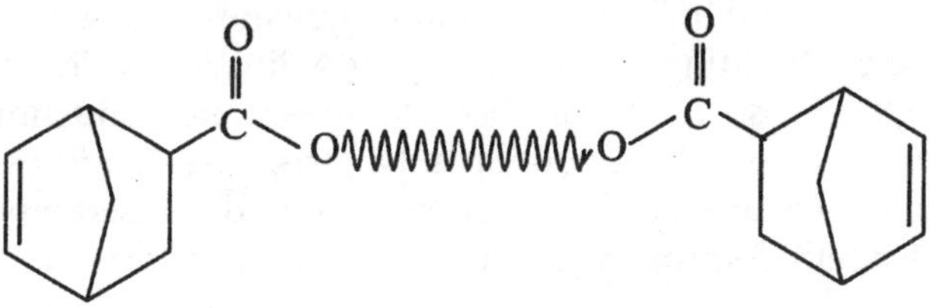

Norbornyl functional oligomer

These oligomers reacted very rapidly with multifunctional thiol cross-linking agents when exposed to ultraviolet radiation.

Acrylate Compositions

Acrylates are the backbone materials of the free radical radiation-curing industry. These ethylenically unsaturated compounds have a structure that is particularly susceptible to initiation with a variety of free radical-generating photoinitiators. The cure efficiency of various photoinitiators has been studied [40]. The generalized acrylates structure is given by the following,

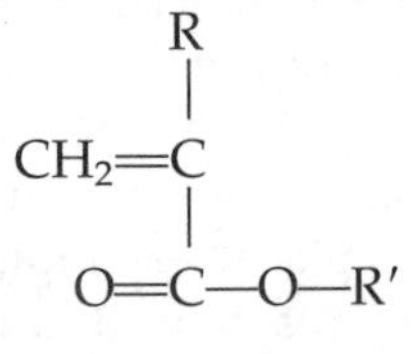

$$CH_2{=}C(R){-}C({=}O){-}O{-}R'$$

wherein *R* is hydrogen for an acrylate and is methyl for a methacrylate. As described in an earlier chapter, acrylates polymerize more rapidly than methacrylates. Quite often the term "acrylates" is taken to mean both acrylates and methacrylates, which is satisfactory if it is stated, and if the reaction differences are kept in mind when dealing with radiation curable systems. The *R′* group can be any one of a number of groups, as will be described later. Included in the ethylenically unsaturated compounds that can be copolymerized with the acrylates are the vinyl ethers and N-vinyl pyrrolidone.

When acrylates are formulated for radiation curing, various structural types are used, and these include:

- Monofunctional acrylates (as shown above)
- Difunctional and higher functional acrylates
- Oligomeric acrylates that are from low to high molecular weight
- Low viscosity to high viscosity acrylates
- Reactive diluents such as vinyl ethers and N-vinyl pyrrolidone

Monofunctional acrylates are used to decrease viscosity. In an overall sense, if monofunctional acrylates are used alone, the final coating properties are poor. Difunctional and higher functionality acrylates are used to aid in rapid molecular weight building and to provide cross-linking, which improves mechanical properties, solvent resistance, stain resistance, and similar properties that cross-linking usually produces. Di-, tri-, tetra-, and hexa-functional acrylates are known and used. One of the most widely used multifunctional acrylates is trimethylolpropane and its alkoxylated adducts in the triacrylate form [41].

The oligomeric acrylates are of two main types[2] that are known as acrylated epoxides (often termed "epoxy acrylates," which is incorrect since there is no epoxy functionality in the molecules) and urethane acrylates. Usually, the acrylated epoxides and urethane acrylates are difunctional in nature. These oligomeric materials are usually high viscosity compounds and are sold diluted with mono- or difunctional acrylates for ease of use. Urethane acrylates are of two basic types that depend on whether an aliphatic or an aromatic isocyanate was used for their preparation. These types can be divided into subclasses that depend on the nature of the polyol used in their manufacture.

As was discussed earlier, radiation polymerization involves the same steps involved in any polymerization. First there is "initiation" that is followed by "propagation" of a growing molecular chain, and then "termination" that occurs when the growing molecular chain is ended or terminated. When radiation strikes a system formulated with a photoinitiator and acrylates, a huge number of free radicals are formed almost instantaneously. Thus, a large number of active sites are formed at the same time and the propagation or polymerization takes place very rapidly. Even though heat is released very rapidly during this reaction, it is usually contained and controlled very easily by the very nature of the polymerizing system. The film cured is very thin, on the order of a thousandth of an in. (0.0254 mm), and the substrate acts as a heat sink that rapidly takes up the energy release accompanying polymerization.

[2] Although these are the main oligomeric acrylates in terms of volume usage, there are other important acrylated oligomers used in industry. These include acrylated acrylates, oils, polyesters, polybutadiene, silicones, and amine-modified polyethers. Some of these will be discussed later in the chapter.

With either free radical or cationic initiation, the polymerization takes place very rapidly. However, free radicals are short-lived, and thus, the polymerization period is very short—basically, it occurs just while the system is in the radiation zone. If monomers or oligomers are not polymerized during this short period, they will remain unchanged in the final matrix. For example, in the study cited above involving unsaturated polyesters and vinyl ethers [29], when an acrylated epoxide containing 3% 1-benzoylcyclohexanol was subjected to an irradiation dosage of 1800 mJ/cm^2, only about 55% of the acrylate double bonds were consumed. Cations are long-lived, and although a tack-free, strong, tough coating can result after ultraviolet radiation exposure, if there are any remaining epoxide groups in the polymerized mass, they can react as a function of time in the absence of further irradiation. The final step in the reaction is termination, which can take place by combination of the propagating species and an initiating free radical, combination of a propagating species with an initiated species, and combination of one propagating species with another propagating species. These mechanism steps are summarized in Figure 5.1 for the polymerization of a monofunctional acrylate. Reactions similar to these occur when multifunctional acrylates and/or oligomeric acrylates are polymerized. The main difference is that instead of a linear polymeric molecule, a highly cross-linked network will result.

It should be pointed out that acrylates can be skin and eye irritants, they can be toxic, and certain acrylates can be human sensitizers. Although attempts have been made to decrease these factors by increasing molecular weight, it is important that those who use these acrylates—monofunctional as well as higher functionality—and other chemicals obtain and read the Material Safety Data Sheets, which will contain specific hazard and precautionary information, prior to using these materials. If there are questions about the materials to be used, the best rule is to ask the supplier for information.

Monofunctional Acrylates. Monofunctional acrylates are one of the main three formulating ingredients in acrylate-based free-radical curable systems. Monofunctional acrylates are low viscosity compounds that function as reactive diluents. They provide viscosity reduction for the formulation that aids in developing a formulation with good application characteristics. Overall, acrylates have good weatherability, and certain ones can impart special characteristics to a final coating. Acrylates and methacrylates contain a polymerization inhibitor, often hydroquinone, monomethyl ether of hydroquinone, or mixtures of these inhibitors, methoxymethyl hydroquinone, benzoquinone, p-tert-butyl catechol, and similar compounds that prevent polymerization during storage, and thus, enhance shelf life in the presence of air [42]. Usually, a specific shelf life for optimum performance will be given on or with products obtained. Several commercially available

INITIATION

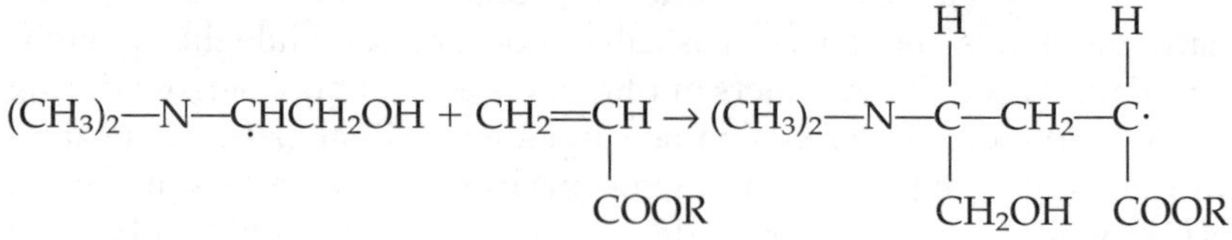

$$(CH_3)_2\text{—N—}\dot{C}HCH_2OH + CH_2\text{=}\underset{\displaystyle COOR}{\underset{|}{C}H} \rightarrow (CH_3)_2\text{—N—}\overset{\displaystyle H}{\overset{|}{\underset{\displaystyle CH_2OH}{\underset{|}{C}}}}\text{—}CH_2\text{—}\overset{\displaystyle H}{\overset{|}{\underset{\displaystyle COOR}{\underset{|}{C}}}}\cdot$$

Free Radical Species, FR· Acrylate Monomer II·, Initiated Species

PROPAGATION

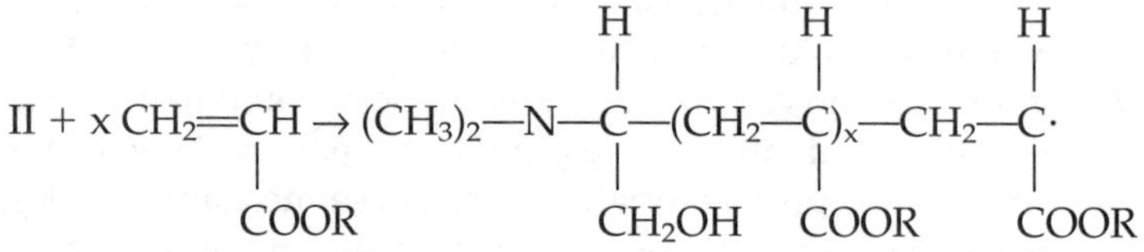

$$II + x\,CH_2\text{=}\underset{\displaystyle COOR}{\underset{|}{C}H} \rightarrow (CH_3)_2\text{—N—}\overset{\displaystyle H}{\overset{|}{\underset{\displaystyle CH_2OH}{\underset{|}{C}}}}\text{—}(CH_2\text{—}\overset{\displaystyle H}{\overset{|}{\underset{\displaystyle COOR}{\underset{|}{C}}}})_x\text{—}CH_2\text{—}\overset{\displaystyle H}{\overset{|}{\underset{\displaystyle COOR}{\underset{|}{C}}}}\cdot$$

Monomer III·, Propagating Species

TERMINATION

Combination of propagating species, III·, and an initiating free radical, FR·

$$FR\cdot + III\cdot \rightarrow FR\text{—}III$$

Combination of a propagating species, III·, with an initiated species, ·II

$$III\cdot + II\cdot \rightarrow III\text{—}II$$

Combination of one propagating species, III·, with another propagating species, III·

$$III\cdot + III'. \rightarrow III\text{—}III'$$

FIG. 5.1—Depiction of the three steps in the free-radical polymerization of an acrylate

monofunctional acrylates, their properties, and features are given in Table 5.1.

2-Hydroxyethyl acrylate is not listed in Table 5.1, because there are several health hazards associated with it. It is usually used under very controlled conditions and usually not in radiation-cure formulations. However, it had desirable features such as low viscosity and attendant good formulation viscosity reduction, and because of the terminal hydroxyl group it contained,

TABLE 5.1. Monofunctional Acrylates and Characteristic Properties (Adapted from [48]).

COMPOUND	MOLECULAR WEIGHT	VISCOSITY, mPa·s, 25°C	LOW SKIN IRRITATION?	FEATURES
2(2-Ethoxyethoxy)ethyl acrylate	188	6	No	Slightly water dispersible
2-Phenoxyethyl acrylate	192	12	Yes	Low volatility, adhesion promoter
4-Hydroxybutyl acrylate	144	...	No	Imparts excellent scratch resistance
4-Hydroxymethyl-cyclohexyl acrylate	174	...	No	Imparts excellent acid resistance
Alkoxylated nonyl phenol acrylate	...	129	No	Low viscosity and Odor
Caprolactone acrylate	344	80	Yes	Hydroxyl end group, low volatility & odor
Isobornyl acrylate	208	9	No	Increases Tg of final coating, excellent diluent
Isodecyl acrylate	212	5	No	Long hydrophobic side chain
Isooctyl acrylate	184	5	No	Good diluent
Methoxypolyethylene glycol (350) mono-methacrylate	494	19	No	Water soluble
Propylene glycol monomethacrylate	405	35	Yes	Hydroxyl end group, low volatility and odor
Stearyl acrylate	324	Solid	No	Long hydrophobic chain, may act as a bubble breaker

Data for this table was taken from literature of various raw material suppliers.

the monomer imparted good adhesion. To obtain some of these characteristics and add others such as flexibility, toughness, and pigment or filler wetting, the caprolactone acrylate listed in the table was developed [43]. The caprolactone acrylate is the reaction product of hydroxyethyl acrylate and ε-caprolactone:

$$CH_2{=}CH{-}C({=}O){-}O{-}CH_2CH_2OH + 2\ (\epsilon\text{-caprolactone}) \longrightarrow CH_2{=}CH(CO)O{-}(CH_2)_2O{-}\!\!\left(C({=}O){-}(CH_2)_5{-}O\right)_2\!H$$

2-Hydroxyethyl acrylate ϵ-Caprolactone Caprolactone acrylate

Addition of caprolactone to the acrylate increased molecular weight, decreased viscosity, and made the compound less of an irritant. The long, flexible tail added to the molecule imparted flexibility and toughness to formulations containing it. The caprolactone portion also provided the good pigment or filler wetting that is attributed to polycaprolactone polyols. These compounds were also used to form urethane acrylates [44], monoisocyanate capped adducts [45], and epoxide terminated adducts [46].

Ethylene oxide and propylene oxide acrylate adducts can be made by condensing acrylic acid with the appropriate alkylene oxide polyol, methoxy alkylene oxide polyol, or glycol. Again, increasing molecular weight improves handling characteristics. In the case of ethylene oxide-containing products, water solubility or sensitivity will be imparted, depending on the length of the ethylene oxide chain added and the quantity of the compound used in the formulation.

Other nonacrylate reactive diluents for acrylate systems [47] include N-vinyl-2-pyrrolidone, which has a low vapor pressure (<0.10 mm Hg at 20°C) and a viscosity of

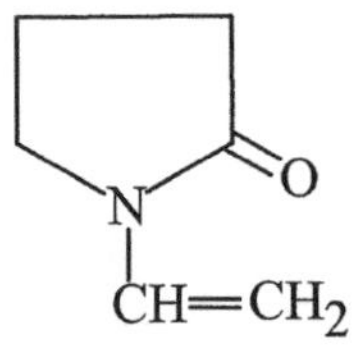

N-vinyl-2-pyrrolidone

two mPa·s at 25°C. It is used in small quantities, but it has a strong viscosity reducing power and it increases cure speed, adhesion, hardness, and strength. The compound is a toxic irritant, and proper care should be taken if it is used. Another similar reactive diluent is N-vinyl-2-caprolactam, which is a crystalline solid at room

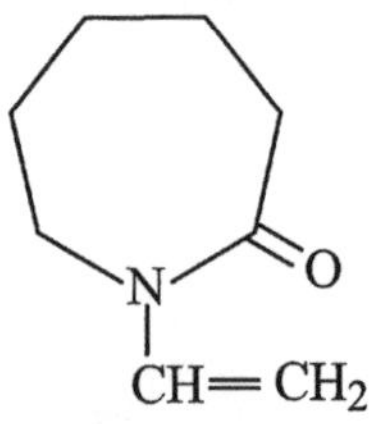

N-vinyl-2-caprolactam

temperature with a 35°C melting point and a viscosity of 3.4 mPa·s at 40°C. Assuming this reactive diluent is soluble in the formulation, it will impart similar diluent properties to the system as N-vinyl-2-pyrrolidone plus improved water resistance. The compound does have a strong, undesirable odor.

Vinyl ethers such as diethylene glycol monovinylether are also useful reactive diluents. These are low viscosity, low odor additives that can add to the complete cure of acrylates. Although vinyl ethers do not homopolymerize under free radical conditions, they will randomly copolymerize with acrylates under these conditions. In one instance of copolymerization [29], it has been shown that the addition of a vinyl ether at a 10% level to an acrylate formulation resulted in a more complete cure. Without the vinyl ether, only 55% of the acrylate function disappeared after irradiation. When irradiated with the vinyl ether, essentially all of the acrylate functionality was consumed.

Polyfunctional Acrylates. Polyfunctional components are used in radiation-curable formulations to provide cross-linking, and thus, rapid molecular weight build. Without a means of attaining high molecular weight or a cross-linked network, good mechanical and chemical properties would not develop, because of the huge number of initiation sites that are formed when the system is irradiated. Many difunctional monomers are used to obtain a good balance of cross-link density, fast reactivity, and low viscosity. Some of the difunctional acrylates that are commercially available are listed in Table 5.2, and a more detailed list can be obtained from suppliers' product litera-

TABLE 5.2. Difunctional Acrylates and Characteristic Properties (Adapted from [48]).

COMPOUND	MOLECULAR WEIGHT	VISCOSITY, mPa·s, 25°C	LOW SKIN IRRITATION?	FEATURES
1,3-Butylene glycol diacrylate	198	9	No	High stain resistance
1,4-Butanediol diacrylate	198	8	No	Very good solvency
1,6-Hexanediol diacrylate	226	9	No	Fast cure response and low volatility
Alkoxylated cyclohexane dimethanol diacrylate	...	70	No	Imparts toughness
Alkoxylated hexanediol diacrylate	...	46	No	Fast cure response
Diethylene glycol diacrylate	214	12	No	Low volatility and viscosity
Ethoxylated bisphenol A diacrylate	468	1600	Yes	Very low volatility, thermal resistance
Neopentyl glycol diacrylate	212	10	No	Low volatility and viscosity
Polyethylene glycol (200) diacrylate	770	25	No	Low volatility and viscosity
Polyethylene glycol (400) diacrylate	508	57	Yes	Water soluble
Triethylene glycol diacrylate	258	15	No	Low volatility and viscosity
Tripropylene glycol diacrylate	300	15	No	Low volatility and viscosity

ture [48–50] or from the Internet. Most oligomeric acrylates are difunctional, and they will be discussed under the appropriate section.

The effect of a chain transfer agent, dodecanethiol, on the photopolymerization of difunctional acrylates caused a decrease in the maximum rate of polymerization and a slight decrease in conversion [51]. The monomers investigated were 1,6-hexanediol diacrylate and poly(ethylene glycol 600) diacrylate and dimethacrylate. The kinetics of the systems indicated that the termination process is important in cross-linked systems.

Representative tri-, tetra-, and penta-functional acrylates are given in Table 5.3. Notice that skin irritation potential appears to decrease as the molecular weight of the acrylate increases, with or without prior reaction with ethylene oxide or propylene oxide. Reaction with the alkylene oxides provides flexibility to the final coatings, and the ethylene oxide derivatives are either water dispersible or water soluble, depending on the degree of ethoxylation. These highly functional compounds are important formulating tools and will impart very rapid reaction and accompanying molecular weight build and cross-linking.

Oligomeric Acrylates—Epoxy, Urethane, Oils.

Acrylated Epoxides or Epoxy Acrylates. Acrylated epoxides, or epoxy acrylates as they are sometimes termed, are usually the reaction products of a

TABLE 5.3. Tri- and Tetrafunctional Acrylate and Characteristic Properties (Adapted from [48]).

COMPOUND	MOLECULAR WEIGHT	VISCOSITY, mPa·s, 25°C	LOW SKIN IRRITATION?	FEATURES
		Trifunctional		
Ethoxylated trimethylol-propane triacrylate	500–1200	60–225	Yes	Water dispersible or soluble, flexibility
Pentaerythritol triacrylate	298	520	No	Hydroxyl functional
Propoxylated glycerol triacrylate	400–600	95	Yes	Fast surface cure
Propoxylated trimethylol-propane triacrylate	470–650	90–125	Yes	Flexibility
Trimethylolpropane triacrylate	296	106	No	Fast cure, low volatility
		Tetra- and higher functionality		
Dipentaerythritol pentaacrylate	525	13,600	Yes	Pentaacrylate for very high cross-link density
Di-trimethylolpropane Tetraacrylate	482	600	Yes	High cross-link density
Ethoxylated pentaeryth-ritol tetraacrylate	528	150	Yes	High cross-link density
Pentaerythritol Tetraacrylate	352	342*	No	High cross-link density

* At 38°C.

diglycidyl ether of bisphenol A and glacial acrylic acid. The starting diglycidyl ether is prepared by the reaction of bisphenol A and epichlorohydrin in the presence of a base such as sodium hydroxide [52].

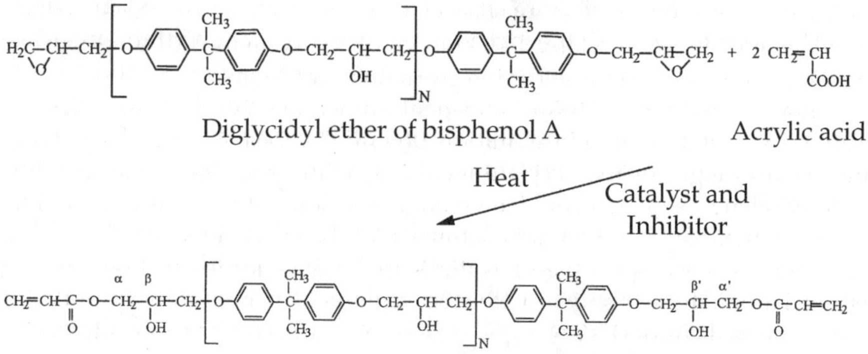

Diglycidyl ether of bisphenol A + 2 $CH_2{=}CH{-}COOH$ (Acrylic acid)

Heat, Catalyst and Inhibitor

An acrylated epoxide

Note α, β ,α', and β' positions (see text)

In commercial epoxides of this type, the value of *N* can range from about zero to 12 or so, but usually ranges from about zero to four in acrylated epoxides. Most desirably the value of *N* is kept small, ideally zero, so viscosity will be at a minimum. However, the starting epoxides have a distribution of molecular weights that depend on the number-average molecular weight of the diepoxide. This factor, coupled with the strong hydrogen bonding between the generated hydroxyl groups and some reaction between these hydroxyl groups and unreacted glycidyl groups that results in increased molecular weight, causes neat acrylated epoxides to have very high viscosity on the order of 5000 to 10 000 or more poise (5×10^5 to 10^6 mPa·s) at room temperature. As a result, these oligomers are often diluted with 20–30% of a low viscosity monomeric acrylate when marketed. If used in a neat form, the viscosity is usually reported at a temperature greater than room temperature, and they are warmed during formulation to facilitate handling. A study concerned with the blending of an acrylated epoxide with monofunctional, difunctional, and trifunctional acrylates and the accompanying viscosity decreases and cured property characteristics can be found in the literature [53]. Other studies have been concerned with the effect of electron-beam dosage on cured properties of acrylic oligomers [54] and with monomer structure on film properties [55].

These compounds have been known in the patent literature since the late1950s [52, 56–66]. In the 1970s [67], they began to be used in the radiation curing field because of their reactivity under free radical initiation and of the chemical and physical property enhancement they gave the final coating. In addition to the diglycidyl ethers of bisphenol A, a wide variety of epoxides for reaction with carboxylic acids is mentioned in the early literature. These include 4,4′-bis(2,3-epoxypropoxyl)benzene, di(2,3-epoxybutyl)adipate, di(4,5-epoxyoctadecyl)malonate, 3,4-epoxyhexyl-3,4-epoxypentanoate, glycerol tri(2,3-epoxycyclohexanoate), and many others. Catalysts for the reaction include tertiary amines, quaternary ammonium salts, and organo-substituted phosphines, which are used in amounts of about 0.05–3% by weight of the reactants [56] with the inclusion of a polymerization inhibitor such as quinone, 2,5-dimethyl-p-benzoquinone, 1,4-naphthaquinone, and the like [58]. Later studies indicated that colorless products with good storage stability could be prepared without the use of a polymerization inhibitor if a mixture of triphenyl stilbine and triphenyl phosphine were used as the catalyst system [59].

When acrylic acid reacts with the diglycidyl ether of bisphenol A, it has an option of adding to either the alpha or the beta position of the epoxide group. These positions are shown in the above reaction scheme. As drawn above, the structure is the α,α′-diacrylate compound, which is the predominant species formed. Nuclear magnetic resonance studies by Jackson and coworkers [57] indicated that about 15% of the β-isomers are produced—α, β′-, β, β′-, α, β-diacrylate esters. These same workers found there was little difference in reactivity between the α,α′-diacrylates and fractions containing predominantly the β-acrylate isomers.

As mentioned before, and as is apparent for the structure given above, the compounds formed are acrylated epoxides or acrylate esters of epoxides, rather than the popularly used name "epoxy acrylates." However, it is possible to make epoxy acrylates that would have the following species in the product mixture:

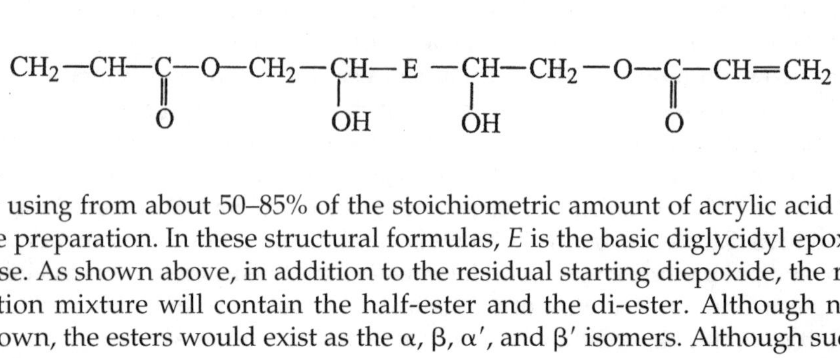

```
CH2—CH—E—CH—CH2
  \ /      \ /
   O        O

CH2—CH—E—CH—CH2—O—C—CH=CH2
  \ /      |         ||
   O       OH        O

CH2—CH—C—O—CH2—CH—E —CH—CH2—O—C—CH=CH2
       ||         |     |          ||
       O          OH    OH         O
```

by using from about 50–85% of the stoichiometric amount of acrylic acid in the preparation. In these structural formulas, *E* is the basic diglycidyl epoxy base. As shown above, in addition to the residual starting diepoxide, the reaction mixture will contain the half-ester and the di-ester. Although not shown, the esters would exist as the α, β, α′, and β′ isomers. Although such a product may appear attractive because one could have an easy synthesis, relatively low viscosity, and the potential for a dual radiation cure followed by a thermal postcure, the half esters have poor shelf life [67], due to residual catalyst and unreacted epoxide. Rapid viscosity increase and even gelation can be expected from such systems. However, a proprietary, undisclosed method of stabilizing the half-esters has been noted [67], wherein the viscosity increase can be kept to a minimum. For example, when an essentially 100% half ester that had an initial viscosity of 3500 mPa·s at 60°C was stabilized and stored at 60°C for about 85 days, the viscosity only increased to 4900 mPa·s at 60°C.

Even though the early work indicated shelf life difficulties with the partially acrylated epoxides, work in the area has continued as investigators have sought methods to decrease the viscosity of acrylated epoxides and to investigate dual-cure systems that could be cured with a combination of radiation and thermal energy. When about 0.5 to 0.85 equivalents of acrylic acid were reacted with glycidyl ethers of bisphenol A, and the resultant compositions were exposed to ultraviolet radiation, tack-free, heat-curable compositions with excellent shelf life were obtained [68]. The resultant film was a partially cured B-stage material that could be completely cured when heated. Others investigated urethane acrylates with varying amounts of acrylate and epoxide functionality to determine the effect on molecular weight and cured properties [69]. They found that to achieve their ultimate properties, the oligomers with higher epoxide functionality tended to be less sensitive to the photoinitiator used and more dependent on the postbake conditions. Conversely, the oligomers with higher acrylate functionality were dependent on photoinitiator selection and to a lesser degree on postbake conditions to achieve ultimate performance characteristics.

Other studies have centered on viscosity reduction. One method [70] of reducing viscosity involved reacting the diepoxide with an excess of a monofunctional aliphatic alcohol or aliphatic ether alcohol, removing the excess alcohol by distillation, and then reacting the resulting product with acrylic acid. Another method involved capping the secondary hydroxyl groups that were formed when the diepoxide was reacted with acrylic acid [71]. This method resulted in a marked loss of physical properties. This was confirmed [72] when a 1 000 000 mPa·s room temperature viscosity urethane acrylate was similarly capped with a six-carbon compound—the ambient viscosity decreased to 85 000 mPa·s; however, cure response was decreased by a factor of two. Another study [73] found that the addition of a small amount of methanol (1.81%) and lithium bromide (0.19%) to an acrylated epoxide (98%) with a neat viscosity of 1 200 000 mPa·s decreased the viscosity to 150 000 mPa·s at 25°C. Addition of these compounds had no effect on the radiation-cured properties of the oligomer when it was formulated with trimethylolpropane triacrylate and a photoinitiator. Addition of methanol alone was only effective in reducing viscosity, but the reduction was only to 275 000 mPa·s at 25°C.

TABLE 5.4. Characteristic Properties of Some Acrylated Epoxides and Acrylated Epoxide Blends (Note temperature at which viscosity was determined).

DESIGNATION	DESCRIPTION	VISCOSITY, mPa·s @ (°C)	REF.
Actilane® 320	Epoxy acrylate	150,000–200,000 @ (25)	50
Actilane® 320HD20	Actilane 320 with 20% 1,6-hexanediol diacrylate	5,000–15,000 @ (25)	50
Actilane® 340	Epoxy novolac acrylate at 65% solids in butoxyethyl acetate	11,000–18,000 @ (25)	50
EBECRYL® 600	Bisphenol A epoxy diacrylate	3,650 @ (60)	49
EBECRYL® 604	EBECRYL 600 with 10% 1,6-hexanediol diacrylate	8,800 @ (25)	49
EBECRYL® 605	EBECRYL 600 with 25% tripropylene glycol diacrylate	7,900 @ (25)	49
Sartomer CN 104	Epoxy acrylate	18,900 @ (49)	48
Sartomer CN 104 A80	CN 104 with tripropylene glycol diacrylate*	36,100 @ (25)	48
Sartomer CN 104 B80	CN 104 with 1,6-hexanediol diacrylate*	13,200 @ (25)	48
Sartomer CN 112 C60	Epoxy novolac acrylate with trimethylolpropane triacrylate*	57,900 @ (25)	48

* No amount specified for the diluent.

Acrylated epoxides have also been made by reacting hydroxyethyl acrylate or hydroxypropyl acrylate with the diglycidyl ethers of bisphenol A [74].

Because of the high viscosity of the acrylated epoxides, they are often diluted with reactive diluents or diluents as described in Table 5.4 for some commercially available products. Many others are available from suppliers. One study [75] described the viscosity and cured tensile properties of a blend containing 75% of a bisphenol A acrylated epoxide and 25% of various mono- and difunctional compounds, such as isobornyl acrylate (~20,000 mPa·s), vinyl pyrrolidone (~1500 mPa·s), hexanediol diacrylate (5000 mPa·s), tripropylene glycol diacrylate (~7000 mPa·s), propoxylated neopentyl glycol diacrylate (~20,000 mPa·s), and others. These mixtures were blended with 4 parts per 100 of a di-methoxy benzoin ether as the photoinitiator and cured with four passes under a 300 watt/in. (~120 watt/cm) medium pressure mercury vapor lamp. As might be predicted, the cured formulations were characterized by high modulus, low elongation, and good tensile strength.

Urethane Acrylates. Early urethane acrylates were prepared by reacting hydroxy acrylates such as 2-hydroxyethyl acrylate, 2-hydroxypropyl acrylate, 2-hydroxybutyl acrylate, and others with a diisocyanate or other polyfunctional isocyanate [76]. When such products were recovered, they often were crystalline solids and were diluted with a reactive solvent, such as styrene, to make syrupy liquids.

Later studies added polyols and basic polyurethane technology to the preparation of these oligomers by adding polyols to the reaction mixture. These urethane acrylates are prepared [77] by reacting a polyol, which is almost always difunctional to prevent gelling, with a diisocyanate to form an isocyanate-prepolymer. The prepolymer is then end capped by reacting the isocyanato groups with 2-hydroxyethyl acrylate.

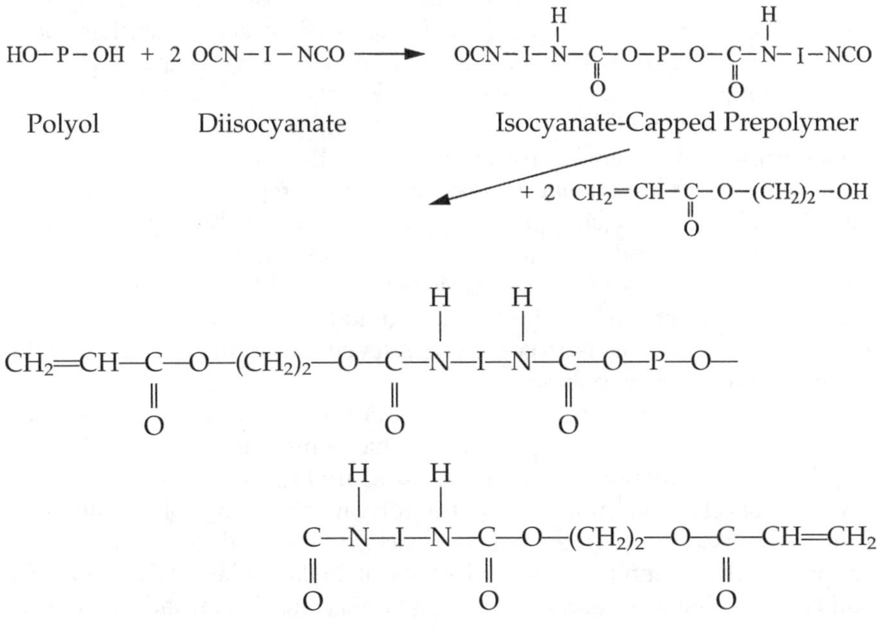

$$\text{HO-P-OH} + 2\ \text{OCN-I-NCO} \longrightarrow \text{OCN-I-}\overset{\text{H}}{\overset{|}{\text{N}}}\text{-}\underset{\underset{\text{O}}{\|}}{\text{C}}\text{-O-P-O-}\underset{\underset{\text{O}}{\|}}{\text{C}}\text{-}\overset{\text{H}}{\overset{|}{\text{N}}}\text{-I-NCO}$$

Polyol Diisocyanate Isocyanate-Capped Prepolymer

$$+\ 2\ CH_2{=}CH\text{-}\underset{\underset{O}{\|}}{C}\text{-O-}(CH_2)_2\text{-OH}$$

$$CH_2{=}CH\text{—}\underset{\underset{O}{\|}}{C}\text{—O—}(CH_2)_2\text{—O—}\underset{\underset{O}{\|}}{C}\text{—}\overset{H}{\overset{|}{N}}\text{—I—}\overset{H}{\overset{|}{N}}\text{—}\underset{\underset{O}{\|}}{C}\text{—O—P—O—}$$

$$\underset{\underset{O}{\|}}{C}\text{—}\overset{H}{\overset{|}{N}}\text{—I—}\overset{H}{\overset{|}{N}}\text{—}\underset{\underset{O}{\|}}{C}\text{—O—}(CH_2)_2\text{—O—}\underset{\underset{O}{\|}}{C}\text{—}CH{=}CH_2$$

A Urethane Acrylate

The above scheme is the idealized way one would like the reaction to progress. However, if the polyol is added to the isocyanate, the first polyol molecules to encounter the diisocyanate probably form the above described prepolymer. But, as more and more polyol is added, there is a greater opportunity for the prepolymer to react with new polyol molecules. The result is that the urethane acrylate being prepared is much higher in molecular weight and viscosity than would be expected from the above structural formula. If the diisocyanate were added to the polyol, at some point a similar situation would exist, though a different distribution of molecular weights would be obtained than was obtained in the first case. Thus, urethane acrylates have a much higher viscosity than would be expected from the described chemistry. For example, 73% solutions of a variety of urethane acrylates dissolved in 2-butoxyethyl acrylate had viscosities that ranged from about 2000 to 20 000 mPa·s at 23°C. Neat urethane acrylates have room temperature viscosities on the order of a million or more centipoises.

Reactant feeding techniques to minimize molecular weight have been devised, but, though these help keep the viscosity to a minimum, it remains high. For example, a diisocyanate was first reacted with 2-hydroxyethyl acrylate in an acrylate solvent to form a monoisocyanate-terminated adduct [78]. Then, a polyol was added to the adduct to form the urethane acrylate at 70% solids with 2-(N-methylcarbamoyloxy)ethyl acrylate as the solvent. In comparative experiments, this technique resulted in viscosity reductions of about 20–30%. For example, in one case, a conventionally prepared urethane acrylate had a viscosity of 31 000 mPa·s at 25°C and 70% solids in the acrylate solvent. When prepared using the same ingredients and the adduct technique, the urethane acrylate had a viscosity of 21 000 mPa·s at 25°C and 70% solids. This represented a significant reduction, but the viscosity remained quite high. When a tetrafunctional polyol was used to prepare the urethane acrylate by conventional techniques, an apparently intractable mass with high viscosity resulted, and a coating could not be prepared from it. Although it was not stated, presumably the mass was cross-linked. When the tetraol was used in the adduct process, the product could be formulated with photoinitiator to form a coating. No viscosity was specified for this product, and it is assumed that it was very high.

Another investigator used mixtures of diols and triols of polycaprolactone and polyoxypropylene polyols to achieve urethane acrylates that had higher acrylate functionality than those prepared from diols alone [79]. One urethane acrylate containing 21% of a polycaprolactone triol and 46% of a polyoxypropylene diol (the remainder was isophorone diisocyanate and the capping hydroxy acrylate) had a viscosity of 25 000 mPa·s at 25°C and 70% solids in 2-(N-methylcarbamoyloxy)ethyl acrylate. When radiation cured with 2,3-di-sec-butoxyacetophenone, the resulting film had a tensile strength of 1800 psi (126.6 kg/cm^2) and an elongation of 105%. Another urethane acrylate had lower viscosity, 6640 mPa·s, tensile strength, 400 psi (28.1 kg/cm^2), and elongation, 80%.

A simple way of making urethane acrylates is through the use of isocyanatomethyl methacrylate [80]. This monoisocyanate is added to a polyol,

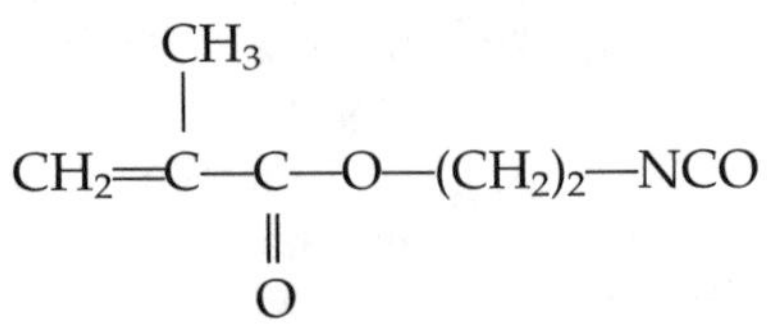

$$CH_2{=}C(CH_3){-}C({=}O){-}O{-}(CH_2)_2{-}NCO$$

Isocyanatoethyl methacrylate

and it directly end caps the molecule via an urethane linkage to form an acrylate functional molecule. In this case, there is no opportunity for high molecular weight oligomers to form, and triol, tetraols, hexols, and so on can be reacted without gelation.

$$2\ CH_2{=}C(CH_3){-}C({=}O){-}O{-}(CH_2)_2{-}NCO + HO{\sim\sim}P{\sim\sim}OH \xrightarrow{\text{Heat, Catalyst}}$$

Polyol

$$CH_2{=}C(CH_3){-}C({=}O){-}O{-}(CH_2)_2{-}N(H){-}C({=}O){-}O{\sim\sim}P{\sim\sim}O{-}C({=}O){-}N(H){-}(CH_2)_2{-}O{-}C({=}O){-}C(CH_3){=}CH_2$$

Urethane Acrylate

This isocyanate was available as a developmental monomer in the late 1970s [82] and early to mid 1980s [83] and was used to prepare various urethane acrylates [84] and other blocked isocyanate derivatives [85]. However, the monomer does not appear to be currently available. The nonavailability may be related to handling and physiological characteristics. One publication [83] indicates the monomer has an extremely severe effect on eyes and that contact could lead to permanent impairment of vision. Further, the compound has a severe effect on the skin, and inhalation tests show that short exposures could be harmful. Other information can be found in the EPA Chemical Profile [86].

There are many studies that discuss preparation of special urethane acrylates. For example, acrylate functionality can be introduced in side positions along a molecular chain [87]. It is possible to synthesize colored urethane acrylates by using compounds such as anthraquinone and azo dyes as chain extenders [88]. Water dispersible urethane acrylates have been made by employing compounds such as 2,2-bis(hydroxymethyl)propionic acid as an integral part of the urethane acrylate oligomeric chain [89]. Again high viscosity products resulted. These urethane acrylates are an innovation on earlier developed water-dilutable polyurethane technology [90].

In a similar manner as can be done with polyurethanes, a very large number of urethane acrylates can be prepared—polyols such as the polyesters [91–95], polycaprolactones [77–79,96], polyoxypropylenes [97,98], oxypropylene/oxyethylene copolymers [98], poly(tetramethylene oxides) [89,99], polycarbonates, polyhydrocarbons [100,101], hydrocarbon diols [102], as well as glycols, diols, and triols [91] can be used. If urethane acrylates are prepared from polyoxypropylene polyols, it is important to note that unsaturation can replace a portion of the hydroxyl end groups [103]. This unsaturation results in a lower number of available hydroxyl groups compared to what would be expected. When the large number of chemically different polyols is coupled with the fact that they are available in a variety of molecular weights, and as diols, triols, tetraols, hexols, and even higher functionality compounds, it is easy to see that a myriad of urethane acrylates can be made.

The property trends that exist between urethane acrylates (ester and ether based) cured with electron beam or ultraviolet radiation were studied

[104]. Significant differences were apparent in cure speed and in performance characteristics, such as adhesion and chemical resistance. Tensile, dynamic, and other mechanical properties of urethane acrylates and other multifunctional acrylates have been investigated [99, 105–112). In addition to mechanical properties, the electrical properties of cured urethane acrylates and acrylated epoxides were characterized [113]. Dielectric constant, volume resistivity, and surface resistivity were determined. The effect of water contamination on urethane acrylate tensile properties has been investigated [114]. Mechanical properties of cured polydimethylsiloxane-urea acrylates were studied [115].

Table 5.5 is a listing of some of the commercially available aliphatic urethane acrylates, and Table 5.6 is the same for aromatic urethane acrylates. Many others are available from listed and unlisted suppliers. As is the case with acrylated epoxides, urethane acrylates are usually marketed in a diluted form to improve their handling characteristics. The diluent almost always is a reactive compound, usually an acrylate, that will copolymerize into the final cured product. A study concerned with the blending of an urethane acrylate with monofunctional, difunctional, and trifunctional acrylates and the accompanying viscosity decreases and cured property characteristics can be found in the literature [53]. A summary of the various methods—monomer or inert solvent dilution, warming, water addition, or low viscosity adduct addition—has been detailed [116].

Hybrid urethane acrylates that have an acrylate group on one end of the molecule and an isocyanate group on the other end have been devised [117]. Oligomers such as these can be combined with multifunctional polyols, a

TABLE 5.5. Characteristic Properties of Some ALIPHATIC Urethane Acrylates and Urethane Acrylate Blends (Note temperature at which viscosity was determined).

DESIGNATION	DESCRIPTION	VISCOSITY, mPa·s @ (°C)	REF.
EBECRYL® 1290	Hexafunctional urethane acrylate with an acrylated polyol diluent*	2,000 @ (60)	49
EBECRYL® 230	Urethane diacrylate	44,000 @ (25)	49
EBECRYL® 244	EBECRYL 230 with 10% 1,6-hexanediol diacrylate	7,900 @ (60)	49
EBECRYL® 264	Trifunctional urethane acrylate with 10% 1,6-hexanediol diacrylate	45,000 @ (25)	49
Sartomer CN 934	Urethane acrylate	20,100 @ (40)	48
Sartomer CN 934 X50	CN 934 blended with ethoxylated bisphenol A diacrylate*	10,200 @ (25)	48
Sartomer CN 962	Urethane acrylate	58,250 @ (60)	48
Sartomer CN 963 B80	CN 962 blended with 1,6-hexanediol diacrylate*	1,100 @ (60)	48
Sartomer CN944 B85	CN 934 blended with 1,6-hexanediol diacrylate*	1,375 @ (60)	48

* No amount specified for the diluent.

TABLE 5.6. Characteristic Properties of Some AROMATIC Urethane Acrylates and Urethane Acrylate Blends (Note temperature at which viscosity was determined).

DESIGNATION	DESCRIPTION	VISCOSITY, mPa·s @ (°C)	REF.
Actilane® 130	Tetrafunctional urethane acrylate	90,000 to 120,000 @ (25)	50
Actilane® 167	Difunctional urethane acrylate	100,000 to 140,000 @ (25)	50
EBECRYL® 220	Hexafunctional urethane acrylate with an acrylated polyol diluent*	27,000 @ (25)	49
EBECRYL® 4827	Urethane diacrylate	4,100 @ (60)	49
EBECRYL® 4849	EBECRYL 4827 with 15% 1,6-hexanediol diacrylate	2,900 @ (60)	49
Sartomer CN 073 H85	Urethane acrylate with 2(2-ethoxyethoxy)ethyl acrylate*	6,050 @ (60)	48
Sartomer CN 972	Urethane acrylate	4,155 @ (60)	48
Sartomer CN 973	Urethane acrylate with tripropylene glycol diacrylate*	6,500 @ (60)	48

* No amount specified for the diluent.

photoinitiator, and a catalyst for the isocyanate/hydroxyl reaction and radiation cured to a tack-free state [118]. These partially cured materials in turn can be given a thermal treatment to form the final cured polyurethane coating.

A wide variety of diisocyanates are listed in the patent literature to make urethane acrylates. However, by and large only four diisocyanates are used. If the products are to be basically colorless and are expected to weather well, 4,4′-dicyclohexanemethane diisocyanate and isophorone diisocyanate are used. These isocyanates are free of aromatic groups, and thus, any color development caused by such groups is absent. If some color can be tolerated and if the products are to be used for interior coatings, toluene diisocyanate and 4,4′diphenylmethane diisocyanate are used.

During the preparation of urethane acrylates, catalytic amounts of urethane catalysts such as dibutyltin dilaurate, stannous octanoate, stannous laurate, dioctyltin diacetate, dibutyltin oxide, triethanolamine, triethylamine, and similar compounds are employed. Stabilizers such as hydroquinone and methoxymethyl hydroquinone that will prevent acrylate polymerization are often incorporated into the reaction mixture. Many of these catalysts and stabilizers are well known to those skilled in this art and need no further mention here.

Esterified Polyol Acrylates. Acrylate-capped polycaprolactone polyols that can be prepared by reaction of acrylic acid with a polyol have been described [77]. It was suggested that the reaction be carried out at the reflux temperature in the presence of an esterification catalyst. The ratio of acrylic acid to polyol hydroxyl content could be varied. If a one-to-one ratio of acid to hydroxyl was used, a fully acrylated polyol could be obtained. If less than the one-to-one ratio were used, a hydroxyl-containing product that could be used in a dual-cure mechanism might be employed. For example, if one mole of a difunctional polyol were reacted with one mole of acrylic acid, a product that has the following average structure could be obtained:

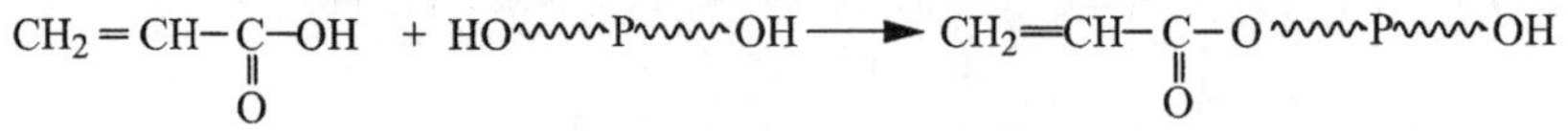

$$CH_2{=}CH{-}\underset{\underset{O}{\|}}{C}{-}OH \;+\; HO\sim\sim P\sim\sim OH \longrightarrow CH_2{=}CH{-}\underset{\underset{O}{\|}}{C}{-}O\sim\sim P\sim\sim OH$$

Acrylic Acid — Polyol — Mono-acrylate/mono-hydroxyl compound

Similar products could be synthesized from polyether or other polyols. If this compound is prepared from a caprolactone polyol, the product is somewhat similar to the caprolactone acrylate that was described earlier [43]. It differs in that the above average product will contain the mono-acrylate/mono-hydroxyl compound above, some diacrylate functional compound, and some residual unreacted polyol plus the caprolactone chains will contain the initiator used to make the polyol.

The thermal stability in terms of urethane acrylate structure was investigated [119]. It was found that urethane linkages yield a more stable cured film than urea linkages, that thermal stability is an inverse function of oligomer equivalent weight, that thermal stability is a function of the diisocyanate used in oligomer preparation in the following order:

Dicycloaliphatic > Aromatic > Cycloaliphatic

and that the polyol used affects thermal stability in the following order:

polycaprolactone > polycarbonate > poly(butylene oxide) > poly(ethylene oxide) > ether copolymers > poly(propylene oxide).

Acrylated Oils. Oligomeric, polyfunctional acrylates can be prepared from those vegetable oils known in the coating industry as drying oils. These oils, which include cottonseed oil, castor oil, linseed oil, oiticica oil, palm oil, safflower oil, soybean oil, sunflower oil, tall oil fatty acids, and tung oil, are a renewable resource of readily available unsaturated compounds that can be epoxidized and, subsequently, converted into acrylates.

Drying oils are unsaturated triglycerides of long chain fatty acids and have the following generalized structure [120]:

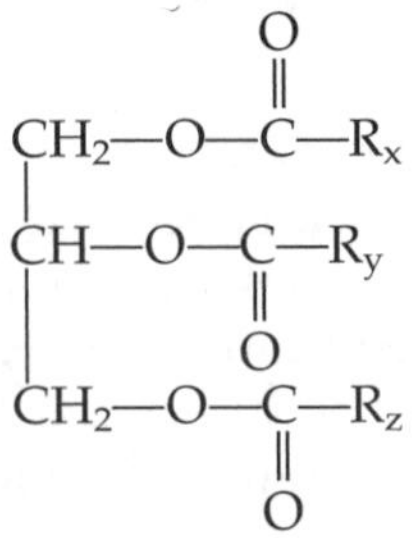

$$\begin{array}{l} CH_2{-}O{-}\overset{\overset{O}{\|}}{C}{-}R_x \\ | \\ CH{-}O{-}\underset{\underset{O}{\|}}{C}{-}R_y \\ | \\ CH_2{-}O{-}\underset{\underset{O}{\|}}{C}{-}R_z \end{array}$$

Triglyceride

in which R_x, R_y, and R_z are one or more of the hydrocarbon residue of a fatty acid such as linolenic acid:

$$HOOC(CH_2)_7CH{=}CHCH_2CH{=}CHCH_2CH{=}CHCH_2CH_3$$

Linolenic Acid

linoleic acid:

$$HOOC(CH_2)_7CH{=}CHCH_2CH{=}CH(CH_2)_4CH_3$$

Linoleic Acid

oleic acid, eleostearic acid, licanic acid, ricinoleic acid, and similar compounds. The main unsaturated acrylated oil derivatives investigated and commercially marketed for radiation curable coatings are soybean and linseed oil acrylates. The major fatty acid residues comprised in soybean oil are linoleic (51%), oleic (25%), and linolenic (9%), and in linseed oil are linoleic (16%), oleic (22%), and linolenic (52%). In Malaysia, there have been numerous studies carried out on the derivativization of palm oil to epoxides and acrylates [121–123]. Current research into the use of soybean oil and derivatives for radiation-cured coatings is being carried out at the Center for Photochemical Sciences at Bowling Green State University [124].

Epoxidized unsaturated vegetable, animal, and synthetic compounds have been known for a number of decades. They have been prepared by oxidation of the oils with hydrogen peroxide [125] or peracetic acid [126,127]. If we represent the oil as one of the possible arms of a triglyceride (linoleic acid residue in this instance), the epoxidation may proceed as follows:

$$—OOC(CH_2)_7CH{=}CHCH_2CH{=}CH(CH_2)_4CH_3 + 2\,H_2O_2 \longrightarrow$$

Residue of Linoleic Acid

$$—OOC(CH_2)_7\underset{\diagdown O \diagup}{CH—CH}CH_2\underset{\diagdown O \diagup}{CH—CH}(CH_2)_4CH_3 + 2\,H_2O$$

An Epoxidized Arm of a Triglyceride

One of the main uses of these epoxidized compounds is as oligomeric stabilizers, acid scavengers, and plasticizers for poly(vinyl chloride) and vinyl chloride copolymers [128]. Vernonia oil is a natural epoxidized vegetable oil [129].

While these epoxidized oils may be used directly in cationic radiation curing, for use in free radical radiation curing they must be reacted with acrylic acid or methacrylic acid to form an acrylate [130–132]. During the reaction step, a catalyst such as zinc dust or zinc hydrosulfite is added to promote reaction of the acid with the oxide, and a stabilizer or polymerization inhibitor such as hydroquinone or paramethoxyphenol is added to minimize reaction of the acrylate groups, which would result in a viscosity increase.

$$—OOC(CH_2)_7\underset{\diagdown O \diagup}{CH—CH}CH_2\underset{\diagdown O \diagup}{CH—CH}(CH_2)_4CH_3 + 2\ CH_2{=}CH{-}\overset{\displaystyle O}{\overset{\|}{C}}{-}OH \longrightarrow$$

$$—OOC(CH_2)_7\underset{OH}{CH}—\underset{O-C(=O)-CH=CH_2}{CH}CH_2\underset{OH}{CH}—\underset{O-C(=O)-CH=CH_2}{CH}(CH_2)_4CH_3$$

An Acrylated Oil

The hydroxyl group of acrylated soybean oil has been reacted with mono- and di-isocyanates to form urethane-containing compositions that had residual acrylate functionality [133]. The acrylated oil had good flow characteristics and imparted adhesion plus excellent pigment wetting. These same investigators reacted the acrylate group with amines such as diethanolamine to form derivatives of the acrylated oils. These compounds were formulated with other acrylates and radiation cured into coatings and inks [134–137].

Some of the trade names that epoxidized soybean oil and linseed oil were marketed under in the 1970s include UCAR® Actomer X-70, UCAR® Actomer X-80, and CRAYNOR 111. Today, such acrylates are available from various suppliers to the industry. One currently available product [49], EBECRYL® Resin 860, is an epoxidized soy bean oil acrylate with a functionality of three, an average molecular weight of 1200, and a viscosity of 26 900 mPa·s at 25°C. A similar product, Sartomer CN 111 [48], had a viscosity of 25 100 mPa·s at 25°C.

Shrinkage

When double bonds open and react as in polymerization, there is a volume reduction usually referred to as shrinkage. When such shrinkage occurs, as happens during the film forming process with rapidly curing radiation-cure systems, internal stresses result. The stresses are very high and have a marked effect on adhesion between the film and substrate. The volume shrinkage was measured [138] for certain mono-and poly-functional acrylates, as well as acrylate oligomers, by measuring the density before and after irradiation and calculating the volume change from the following expression:

$$\text{Volume Shrinkage} = \left[\frac{\text{Density after cure} - \text{Density before cure}}{\text{Density after cure}}\right] \times 100\%$$

The volume shrinkage ranged from about 5% to more than 13%, depending on the acrylate, and the particular values are tabulated in Table 5.7. Others

TABLE 5.7. Volume Shrinkage for Several Acrylates [138].

ACRYLATE	VOLUME SHRINKAGE, %
Hydroxypropyl acrylate	13.2
Poly(ethylene glycol) diacrylate*	12.0
Trimethylolpropane triacrylate	11.7
Pentaerythritol triacrylate	10.7
Urethane Acrylate*	8.7
Acrylated Epoxide*	5.4

* No molecular weight was specified for these materials.

have found that a typical formulation of acrylate monomers and oligomers will have volume shrinkage of 9 to 15% [139] and that values for common acrylic monomers range from 11–30% [140]. Linear shrinkage [141] of acrylated oligomers was measured [142], and volume shrinkage was calculated from these measurements to be 7–25% [139].

References

[1] Doyne, R. F., "Dr. William Burlant: A Pioneer in Electron Beam Curing," *Journal of Radiation Curing/Radiation Curing*, Vol. 19, Winter 1992, p. 6.
[2] Morganstern, K. H., "The Present Status of Electron Beam Curing of Coatings," in *Nonpolluting Coatings and Coating Processes*, J. L. Gardon and Joseph W. Prane, Eds., Plenum Press, NY, 1973.
[3] Miranda, T. J. and Huemmer, T. F., "Radiation Curing of Coatings," *Journal of Paint Technology*, Vol. 41, No. 529, 1969, p. 118.
[4] Charlesby, A., *Proceedings of the Royal Society (London)*, Vol. A215, 1952, p. 187.
[5] Patent: Renfrew, M. M., E. I. duPont de Nemours & Company, "Tooth Reconstruction," U. S. 2,335,133 (1943); "Photopolymerization," U. S. 2,448,828 (1946).
[6] Hoffman, A. S. and Smith, D. E., *Modern Plastics*, Vol. 43, No. 10, 1966, p. 111.
[7] Bevan, E. A., "Varnish Making," *Oil & Colour Chemists Association*, 1934, p. 34.
[8] Bradley, T. F., *Industrial and Engineering Chemistry*, Vol. 29, 1937, p. 440.
[9] Bradley, T. F., *Industrial and Engineering Chemistry*, Vol. 29, 1937, p. 579.
[10] Dannenberg, H., Bradley, T. F., and Evans, T. W., *Industrial and Engineering Chemistry*, Vol. 41, 1949, p. 1709.
[11] Vincent, H. L., *Industrial and Engineering Chemistry*, Vol. 29, 1937, p. 1267.
[12] Trivedi, B. C. and Culbertson, B. M., *Maleic Anhydride*, Plenum Press, NY, 1982.
[13] Patent: Thomas, P., Seine, B., Meyer, J., Pechiney-Saint-Gobain, U. S. 3,784,586 (1974).
[14] Rowland, C. S., *Interchemical Review*, Vol. 5, 1946–1947, p. 83.
[15] Garratt, P. G., "The Use of Unsaturated Polyester Resins in UV Curable Paint Formulations for Use in the Furniture Industry," *Conference Proceedings of RADCURE '84*, Atlanta, GA, 10–13 September,1984, pp. 3–13.
[16] McCloskey, C. M. and Bond, J., *Industrial and Engineering Chemistry*, Vol. 47, 1955, p. 2125.
[17] Burlant, W. J., Green, D., and Taylor, C., *Journal of Applied Polymer Science*, Vol. 1, No. 3, 1959, p. 296.
[18] Burlant, W. J. and Hinsch, J. E., *Journal of Polymer Science*, A, Vol. 3, 1965, p. 3587.
[19] Patent: Burlant, W. J., Ford Motor Company, "Process of Coating the Exterior Surface of Articles with a Polymerizable Material Subjected to High Energy Ionizing Irradiation," U. S. 3,247,012 (1966).
[20] Patent: Burlant, W. J., Tsou, I. H., Ford Motor Company, "Radiation Curable Cyclic Silicone-Modified Paint Binders," U. S. 3,437,512 (1969).

[21] Patent: Burlant, W. J., Taylor, C. R., Ford Motor Company, "Radiation-Curable Vinyl Resin Paints and Coated Articles," U. S. 3,528,844 (1970).
[22] Deniger, W. and Patheiger, M., *Industrie-Lackier-Betrieb*, Vol. 37, 1969, p. 85.
[23] Patent: Beightol, L. E., Solar Laboratories Inc., British 1,245,937 (1971).
[24] Wicks, Z. W., SME Technical Paper (Ser.) FC74-501 (1974).
[25] Klug, H. and Velic, M., *Farbe und Lack*, Vol. 77, No. 3, 1971, p. 231.
[26] Schroeter, S. H. and Moore, J. E., "The Ultraviolet Curing of Coatings," in *Nonpolluting Coatings and Coating Processes*, J. L. Gardon and Joseph W. Prane, Eds., Plenum Press, NY, 1973.
[27] Hoffman, A. S., Jameson, J. T., Salmon, W. A., Smith, D. E., and Trageser, D. A., *Industrial Engineering Chemistry: Product Research and Development*, Vol. 9, No. 2, 1970, p. 158.
[28] Laws, A., Lynn, S., and Hall, R., *Journal of Oil and Color Chemists Association*, Vol. 59, 1976, p. 193.
[29] Schwalm, R., Binder, H., Funhoff, D., Lokai, M., Schrof, W., and Weiguny, S., "Vinyl Ethers in UV Curing. Copolymers with Acrylates and Unsaturated Polyesters," *Conference Proceedings, RadTech Europe '99*, Berlin, Germany, 8–10 November, 1999, p. 103.
[30] Patent: Kehr, C. L., Wszolek, W. R., W. R. Grace & Company, "Photocurable Liquid Polyene-Polythiol Polymer Compositions," U. S. 3,661,744 (1972).
[31] Ketley, A. D., "UV-Curable Resins for Wire and Cable Coating," Society of Manufacturing Engineers Technical Paper FC75-331 (1975).
[32] Patent: Kehr, C. L., Wszolek, W. R., W. R. Grace & Company, "Water-Resistant Materials and Methods of Production and Use of Same," U. S. 3,676,195 (1972).
[33] Patent: Bush, R. W., Wood, L. L., W. R. Grace & Company, "Curable Solid Adducts of Polyene-Polythiol Compositions with Urea," U. S. 3,993,549 (1976).
[34] Morgan, C. R., Magnotta, F., and Ketley, A. D., *Journal of Polymer Science, Polymer Chemistry Edition*, Vol. 15, 1977, p. 627.
[35] Morgan, C. R. and Ketley, *Journal of Polymer Science, Polymer Letters Edition*, Vol. 16, 1978, p. 75.
[36] Morgan, C. R. and Ketley, A. D., *American Chemical Society Abstracts, 172nd Meeting Polymer Chemistry Division*, No. 119, 1976.
[37] Gush, D. P. and Ketley, A. D., *Chemical Coatings Conference II: Radiation Cured Coatings*, National Paint and Coatings Association, Washington, D. C., 10 May, 1978.
[38] Patent: Curry, H. L., Hall, W. L., General Electric Company, "Photocurable Polyene-Polythiol-Siloxane-Polyester Composition for Coating," U. S. 4,197,173 (1980).
[39] Jacobine, A. F., Glaser, D. M., and Nakos, S. T., "Photoinitiated Crosslinking of Norbornene Resins with Multifunctional Thiols," in *Radiation Curing of Polymeric Materials*, ACS Symposium Series 417, Charles E. Hoyle, and James F. Kinstle, Eds., 1990.
[40] Gatechair, L. R. and Wostratzky, D., "Photoinitiators: An Overview of Mechanisms and Applications," *Journal of Radiation Curing*, Vol. 10, No. 4, July 1983.
[41] Miller, H. C., "Trifunctional Acrylate Backbone Structure Physical Properties," *Proceedings, Volume 1, of RadTech '90-North America*, Chicago, IL, 25–29 March 1990, p. 61.
[42] Kensbock, E., "Chemical and Commercial Aspects of In-Can Stabilizers for UV-Ink and Coatings," *Proceedings of RadTech 2000*, Baltimore, MD, 9–12 April 2000, p. 610.
[43] Patent: Koleske, J. V., Domeier, L. A., Union Carbide Corporation, U.S. 4,683,287 (1987).
[44] Patent: Osborn, C. L., Koleske, J. V., Drake, K., Union Carbide Corporation, "Coating Compositions Prepared from Lactone-acrylate Adduct, Polyol, and Isocyanate," U. S. 4,618,635 (1986).
[45] Patent: Koleske, J. V., Smith, D. F., Weber, R. J., Jr., Union Carbide Corporation, "(N-Substituted Carbamoyloxy)alkanoyloxyalkyl Acrylate Polymers and Compositions Made Therefrom," U. S. 5,115,025 (1992).
[46] Patent: Stanley, J. P., Union Carbide Corporation, "Unsaturated Polylactone Acrylates and Derivatives Thereof," U. S. 5,557,007 (1996).

[47] Brochure, "V-Pyrol, V-Cap," International Specialty Products, 2302-081R or www.isp-corp.com/index2.html (2001).

[48] Sartomer Company, Inc., *Technical Solutions, Product Catalog,* 1999; Internet, www.sartomer.com/products.asp (2001).

[49] UCB Chemicals, Radcure Products, Internet, www.chemicals.ucb-group.com/b_units/b2indust/default.html (2001).

[50] Akzo Nobel, "Actilane® UV/EB Curing Chemicals for Coatings, Inks, and Adhesives," Brochure, Akcros Chemical Edition 3 (1/2000).

[51] Bowman, C. N., Lovell, L. G., Nie, J., Berchtold, K. A., and Hacioglu, B., "The Effects of Light Intensity and Photoinitiator Concentration on Cure Speed of Multi(meth)acrylates," *Proceedings of RadTech 2000,* Baltimore, MD, 9–12 April 2000, p. 507.

[52] Patent: Hall, M. W., Minnesota Mining & Manufacturing Company, "Acrylic Acid Derivatives of Epoxide Resins and Method of Curing Same," U. S. 2,824,851 (1958).

[53] Miller, H. C., "Urethane Acrylates: Expansion of Radiation Curable Epoxy Acrylate Coatings," *Proceedings, Volume 1, of RadTech '90-North America,* Chicago, IL, 25–29 March 1990, p. 201.

[54] Johnson, A. M. and Moorehead, T. E., "Effects of Electron Beam Dosage on Cured Homopolymer Properties of Acrylate Oligomers," *Proceedings, Volume 1, of RadTech '90-North America,* Chicago, IL, 25–29 March 1990, p. 71.

[55] Zwanenburg, R. C. W., "The Monomer, Its Structure and the Physical Properties of the Film—Is There a Correlation?" *Proceedings, Volume 1, of RadTech '90-North America,* Chicago, IL, 25–29 March 1990, p. 80.

[56] Patent: Newey, H. A., Shell Oil Company, "Epoxy-Containing Condensates of Polyepoxides and Acidic Materials, Their Preparation and Polymers," U. S. 2,970,983 (1961).

[57] Patent: Bowen, R., United States of America, Department of Commerce, "Dental Filling Material Comprising Vinyl Silane Treated Fused Silica and a Binder Consisting of the Reaction Product of Bisphenol and Glycidyl Acrylate," U. S. 3,066,112 (1962).

[58] Patent: Fekete, F., Keenan, P. J., Plant, W. J., H. H. Robertson Company, "Hydroxyl Polyether Polyesters Having Terminal Ethylenically Unsaturated Groups," U. S. 3,256,226 (1966).

[59] Patent: Doyle, T. E., Fekete, F., Keenan, P. J., Plant, W. J., H. H. Robertson Company, "Combination Catalyst Inhibitor for Beta-Hydroxy Carboxylic Acids," U. S. 3,317,465 (1967).

[60] Patent: May, C. A., Shell Oil Company, "Process for Preparing Polyesters and Resulting Products," U. S. 3,345,401 (1967).

[61] Patent: May, C. A., Shell Oil Company, "Reaction Products of Unsaturated Esters of Polyepoxides and Unsaturated Carboxylic Acids, and Polyisocyanates," U. S. 3,373,221 (1968).

[62] Patent: Newey, H. A., May, C. A., Shell Oil Company, "Process of Esterification of Polyepoxides with Ethylenically Unsaturated Monocarboxylic Acids," U. S. 3,377,406 (1968).

[63] Patent: May, C. A., Shell Oil Company, "Polyhydroxyl-Containing Unsaturated Polyesters and Their Preparation," U. S. 3,432,478 (1969).

[64] Patent: Dowd, R. T., Clark, T. D., Shell Oil Company, "Unsaturated Polyesters Esterified with Polycarboxylic Acid Anhydride and Containing Polyepoxide," U. S. 3,634,542 (1972).

[65] Patent: May, C. A., Shell Oil Company, "Unsaturated Polyesters from Epoxides and Ethylenically Unsaturated Monocarboxylic Acid Mixed with Solid Epoxide Resin," U. S. 3,637,618 (1972).

[66] Gamble, A. A., "Radiation Curable Offset Inks: A Technical and Marketing Overview," in Randell, D. R., *Radiation Curing of Polymers,* Special Publication No. 64 of the Royal Society of Chemistry, London, 1985.

[67] Jackson, R. J., Jones, P. D., and DeLaMare, H. E., "A Review of Epoxy-Derived Acrylates: Synthesis, Chemistry and Photocure," The Association for Finishing Processes of SME, AFP Technical Paper, FC76-494 (1976).

[68] Patent: De La Mare, H. E., Jones, P. D., Shell Oil Company, "Curable Epoxy-Vinyl Ester Compositions," U. S. 4,359,370 (1982).

[69] Christmas, B. K., Kemmerer, R. R., and Willard, F. K., "UV/Thermal Curing of Acrylate/Epoxide Functional Coatings," *Conference Papers RadTech '88—North America*, New Orleans, 24–28 April 1988, p. 274.

[70] Patent: Hicks, D. D., Interez, Inc., "Process for Preparing Low Viscosity UV Curable Polyacrylates," U. S. 4,631,300 (1986).

[71] Labana, S. S. and Aronoff, E. J., *Journal of Paint Technology*, Vol. 43, 1971, p. 77.

[72] Kemmerer, R. R., Christmas, B. K., and Scot, B. D., "A Novel Approach to Reducing the Viscosity of UV/EB Curable Oligomers," *Conference Papers RadTech '88—North America*, New Orleans, 24–28 April 1988, p. 290.

[73] Patent: Morris, W. J., Kemmerer, R. R., Christmas, B. K., Interez, Inc., "Method of Reducing Viscosity of Radiation Curable Acrylate Functional Resin," U. S. 4,687,806 (1987).

[74] Nagy, F. A., "Radiation Curable Epoxy/Acrylate-Hydroxyl Coating Composition," Mobil Oil Company, European Patent Application EP 82,603 (29 June 1983); *Chemical Abstracts* Vol. 99, 1983, 124202f.

[75] Miller, H. C., "UV/EB-Cure Urethane Acrylate and Epoxy Acrylate Coatings," *Modern Paint and Coatings*, Vol. 83, No. 13, December 1993, p. 40.

[76] Patent: Fekete, F., Keenan, P. J., Plant, W. J., H. H. Robertson Company, "Ethylenically Unsaturated Di- and Tetra-Urethane Monomers," U. S. 3,297,745 (1967).

[77] Patent: Smith, O. W., Weigel, J. E., Trecker, D. J., Union Carbide Corporation, "Radiation-Curable Acrylate-Capped Polycaprolactone Compositions," U. S. Reissue Re. 29,131 (1977).

[78] Patent: Watson, S. L., Jr., Union Carbide Corporation, "Procedure for Production of Lower Viscosity Radiation-Curable Acrylated Urethanes," U. S. 4,246,391 (1981).

[79] Patent: Hodakowski, L. E., Carder, C. H., Union Carbide Corporation, "Radiation Curable Acrylated Polyurethane," U. S. 4,131,602 (1978).

[80] Patent: Bortnick, N. M., Rohm & Haas Company, "Isocyanato Esters of Acrylic, Methacrylic, and Crotonic Acids," U. S. 2,718,516 (1955).

[81] Patent: Holtschmidt, H., Farbenfabriken Bayer A. G., "Production of Alkylisocyanate Esters of 2-Alkenoic Acids," U. S. 2,821,544 (1958).

[82] "Experimental Monomer XD-30153.00 Isocyanatoethyl Methacrylate," Brochure, Dow Chemical Company (12/07/78).

[83] "Developmental Monomer XAS-10743.00* Isocyanatoethyl Methacrylate," Brochure, Dow Chemical Company (undated, issued about 1980–1985).

[84] Speckhard, T. A., Hwang, K. K. S., Lin, S. B., Tsay, S. Y., Koshiba, M., Ding, Y. S., and Cooper, S. L., "Properties of UV Curable Polyurethane Acrylates: Effect of Reactive Diluent," *Journal of Applied Polymer Science*, Vol. 30, 1985, p. 647.

[85] Regulski, T. and Thomas, H. R., "Isocyanato Methacrylate II: The Blocked Isocyanate Derivatives, Preparation and Deblocking," *Division of Organic Coatings and Plastics Chemistry, ACS 185th National Meeting*, Seattle, Washington, 20–25 March 1983, pp. 998 and 1003 of Preprint papers.

[86] EPA Chemical Profile, CAS Registry Number 30674-80-7, "Methacryloyloxyethyl Isocyanate" (10/31/85, revised 11/30/87).

[87] Couvret, D., Brosse, J. C., Chevalier, S., and Senet, J. P., *European Polymer Journal*, Vol. 27, No. 2, 1991, p. 193.

[88] Khamis, M. A., Aboshosha, M., and Walsh, W. K., "Synthesis and Characterization of Radiation Curable, Colored, Acrylourethane Elastomers," *Radiation Curing*, Vol. 11, November 1984, p. 4.

[89] Patent: Park, K., Bryant, G. M., Union Carbide Corporation, "Acrylyl Capped Urethane Oligomers," U. S. 4,153,778 (1979).

[90] Patent: Milligan, C. L., Union Carbide Corporation, "Water-Dilutable Polyurethanes," U. S. 3,412,054 (1968).

[91] Patent: Marquardt, K., Wiest, H., Wacker Chemie GmbH., "Process for the Preparation of Polymers Containing Urethane Groups," U. S. 4,238,325 (1982).

[92] Patent: Hisamatsu, H., Takashashi, K., Takase, M., Dainippon Ink & Chemicals Inc., "Photopolymerizable Isocyanate-Containing Prepolymers," U. S. 3,891,523 (1975).

[93] Salim, M. S., "Urethane Acrylates for Radiation Cured Coatings," *Polymers Paint Colour Journal*, Vol. 177, No. 4203, November 1987, p. 762.

[94] Barclay, R., Jr., "Flexible Coatings from Ultraviolet-Cured Urethane Oligomer Formulations," The Association for Finishing Processes of SME, Technical Paper FC76-523 (1976).

[95] Barbeau, PH., Muzeau, E., and Hirsch-Askienazy, A., "The Mechanical Performance of Urethane Acrylate Oligomers," *Conference Proceedings, RadTech Europe '99*, Berlin, Germany, 8–10 November 1999, p. 123.

[96] Patent: Chang, W. H., PPG Industries, Inc., "Curable Lactone Derived Resins," U. S. 4,188,472 (1980).

[97] Morris, W. J., "Comparison of Acrylated Oligomers in Wood Finishes," *Journal of Coatings Technology*, Vol. 56, No. 715, 1984, p. 49.

[98] Patent: Schmidle, C. J., Thiokol Corporation, "Photocurable Compositions Based on Acrylate Polyester Urethanes," U. S. 4,377,679 (1983).

[99] Koshiba, M., Hwang, K. K., Hwang, S., Foley, S. K., Yarusso, D. J., and Cooper, S. L., "Properties of Ultra-violet Curable Polyurethane Acrylates," *Journal of Materials Science*, Vol. 17, No. 5, 1982, p. 1447.

[100] Dupont, M., Masse, M., and Schneider, J., "Rubber-Based Radiation Curable Pressure Sensitive Adhesives," *Conference Proceedings, RadTech Europe '99*, Berlin, Germany, 8–10 November 1999, p. 551.

[101] Patent: Ors, J. A., Small, R. D., Jr., AT&T Technologies, Inc., "Multilayer Circuit Board Fabrication Process and Polymer Insulator Used Therein," U. S. 4,628,022 (1986).

[102] Patent: Argyropoulos, J. N., Smith, O. W., Koleske, J. V., Union Carbide Corporation, "Polyurethane (Meth)acrylates and Processes for Preparing Same," U. S. 5,248,752 (1993).

[103] Fishback, T., Aviles, G., and Reichel, C., "Influence of Polyol Unsaturation on Urethane Properties," *Adhesives Age*, Vol. 39, June 1996, p. 20.

[104] Hutchinson, I., "Film Property Trends Between UV and EB Cured Urethane Acrylates," *Proceedings of RadTech 2000*, Baltimore, MD, 9–12 April 2000, p. 522.

[105] Oraby, W. and Walsh, W. K., "Elastomeric Electron Beam-Cured Coatings: Structure-Property Relationships. I. Oligomer Structure," and "II. Chain Transfer Agents," *Journal of Applied Polymer Science*, Vol. 23, 1979, pp. 3227 and 3243, respectively.

[106] Suzuki, Y., Fujimoto, T., Tsunoda, S., and Shibayama, K., "Dynamic Mechanical Properties of Interpenetrating Polymer Networks Formed by UV Curing Processes," *Journal of Macromolecular Science—Physics*, Vol. B17, No. 4, 1980, p. 787.

[107] Joseph, E., Wilkes, G., and Park, K., "Structure-Property Relationships of an Electron Beam Cured Model Urethane Prepolymer," *Journal of Applied Polymer Science*, Vol. 26, 1981, p. 3355.

[108] Wadhwa, L. H. and Walsh, W. K., "Morphology and Mechanical Properties of Radiation Polymerized Urethane Acrylates. I. Pure Oligomers," *Journal of Applied Polymer Science*, Vol. 27, 1982, p. 591.

[109] Priola, A., Renzi, F., and Cesca, S., "Structure-Property Relationships for Radiation Curable Coatings," *Journal of Coatings Technology*, Vol. 55, No. 703, August 1983, p. 63.

[110] Noren, G. K., Zimmerman, J. M., Krajewski, J. J., and Bishop, T. E., "Mechanical Properties of UV-Cured Coatings Containing Multifunctional Acrylates," *Proceedings of the ACS Division of Polymeric Materials: Science and Engineering*, Dallas, TX, Vol. 60, Spring 1989, p. 349.

[111] McConnell, J. A. and Willard, F. K., "Structure-Performance Relationships of Urethane Acrylates," *Proceedings of the ACS Division of Polymeric Materials: Science and Engineering*, Dallas, TX, Vol. 60, Spring 1989, p. 354.

[112] Zumbrum, M. A. and Ward, T. C., "Mechanical Property Changes Accompanying *in Situ* Photopolymerization of Thin Films," *Proceedings of the ACS Division of Polymeric Materials: Science and Engineering*, Dallas TX, Vol. 60, Spring 1989, p. 361.

[113] Van Landuyt, D. C. and Leyrer, S. P., "Physical and Electrical Properties of Acrylic Oligomers," *Radiation Curing*, Vol. 9, May 1982, p. 10.
[114] Kallendorf, C. J. and Woodruff, R. T., "The Effect of Water Contamination on the Performance of Urethane Acrylate Coatings," *Radiation Curing*, Vol. 14, No. 3, August 1987, p. 12.
[115] Yu, X., Nagarajan, M. R., Li, C., Speckhard, T. A., and Cooper, S. L., "Properties of Ultraviolet Cured Polydimethylsiloxane-Urea Acrylates," *Journal of Applied Polymer Science*, Vol. 30, 1985, p. 2115.
[116] Ashcroft, W. R. and Younger, J. R., "Trends in Low Viscosity Acrylate Resins for Radiation Cured Inks and Varnishes," *Proceedings RadTech '90—North America*, Chicago, IL, Vol. 1, 25–29 March 1990, p. 196.
[117] Demarteau et al., "Dual UV/Thermally Curable Formulations," *Proceedings–Radcure '84, AFP/SME*, September 1984, p. 1–1.
[118] Fischer, W., Meier-Westhues, U., and Hovestadt, W., "Dual-Cure, New Possibilities of Radiation Curing Coatings," *End User Conference, RadTech 2000*, Baltimore, MD, 9–12 April 2000, p. 38.
[119] Shama, S. A., "Structure-Stability Relationships of Urethane Acrylate Oligomers," *Proceedings, RadTech '90-North America*, Chicago, IL, Vol. 1, 25–29 March 1990, p. 242.
[120] Koleske, J. V., "Drying Oils," *Paint and Coating Testing Manual*, MNL 17, Joseph V. Koleske, Ed., ASTM International, West Conshohocken, PA, 1995, p. 26.
[121] Abadie, M. J. M. and Ionescu-Vasii, L., "Recent Advances in Cationic Photoinitiators–Application to the Photopolymerization of Epoxides and Vinyl Ethers," *Proceedings of RadTech Asia '95 Radiation Curing Conference*, Guilin, China, 20–24 November 1995, p. 53.
[122] Kumar, R. N., Kong, W. C., and Abubakar, A., *Journal of Applied Polymer Science*, Vol. 73, 1999, p. 1569.
[123] Basiron, Y. and Thiagarajan, T., "ASEAN: Oils and Fats Centre of the World," *Palm Oil Technical Bulletin*, Vol. 4, No. 6, 1998, p. 2.
[124] Howell, C., "UV Curing, Soybean Oil Research Continues," *Modern Paint & Coatings*, Vol. 91, No. 4, April 2001, p. 41.
[125] Patent: Swern, D., Findley, T. W., Secretary of Agriculture, United States of America, "Epoxidized Oils," U. S. 2,569,502 (1951).
[126] Patent: Phillips, B., Frostick, F., Union Carbide Corporation, U. S. 2,779,771 (1957).
[127] Patent: Wahlroos, A. W., Archer-Daniels-Midland Company, U. S. 2,813,878 (1957).
[128] Port, W. S., "Epoxy Compounds as Polymer Stabilizers and Plasticizers," Chapter 10 in *Epoxy Resins: Chemistry and Technology*, Clayton A. May and Yoshio Tanaka, Eds., Marcel Dekker, Inc., New York, 1973.
[129] Dirlikov, S. K., Frischinger, I., Islam, M. S., and Lepkowski, T. J., in *Biotechnology and Polymers*, C. G. Gebelein, Ed., Plenum Press, New York, 1991.
[130] Patent: Nevin, C. S., A. E. Staley Manufacturing Company, "Preparation of Polymerizable Vinylated Compounds," U. S. 3,125,592 (1964).
[131] Patent: Nevin, C. S., A. E. Staley Manufacturing Company, "Vicinal Acryloxy Hydroxy Long Chain Fatty Compounds and Polymers Thereof," U. S. 3,224,989 (1965).
[132] Patent: Nevin, C. S., A. E. Staley Manufacturing Company, "Vicinal Acryloxy Hydroxy Long Chain Fatty Compounds and Polymers Thereof," U. S. 3,256,225 (1966).
[133] Patent: Borden, G. W., Smith, O. W., Trecker, D. J., Union Carbide Corporation, "Acrylated Epoxidized Soybean Oil Amine Compositions and Method of Curing Same," U. S. 3,876,518 (1975); "Acrylated Epoxidized Soybean Oil Urethane Derivatives," U. S. 4,025,477 (1977).
[134] Patent: Trecker, D. J., Borden, G. W., Smith, O. W., Union Carbide Corporation, "Compositions of Acrylated Epoxidized Soybean Oil Amine Compounds Useful as Inks and Coatings," U. S. 3,931,071 (1976); "Acrylated Epoxidized Soybean Oil Amine Compositions and Method," U. S. 3,931,075 (1976).

[135] Patent: Trecker, D. J., Borden, G. W., Smith, O. W., Union Carbide Corporation, "Method for Curing Ink and Coating Compositions of Acrylated Epoxidized Soybean Oil Amine Compounds," U. S. 4,016,059 (1977).

[136] Patent: Borden, G. W., Smith, O. W., Trecker, D. J., Union Carbide Corporation, "Method for Coating or Printing Using Acrylated Epoxidized Soybean Oil Urethane Compositions," U. S. 4,224,369 (1980); "Ink and Coating Compositions and Method," U. S. 4,233,130 (1980).

[137] Patent: Hodakowski, L. E., Osborn, C. L., Union Carbide Corporation, "Radiation Polymerizable Cycloalkenyl Derivatives of Acrylated Epoxidized Fatty Oils or Fatty Acids," U. S. 4,100,046 (1978).

[138] van Neerbos, A., Hoefs, C. A. M., and Giezen, E. A., "UV Curing Resins for Tinplate Decoration," *3E-Activities, XVth FATIPEC Congress*, Amsterdam, The Netherlands, Vol. 1, 8–13 June 1980, p. I–319.

[139] Goodman, D. L. and Byrne, C. A., "Composite Curing with High Energy Electron Beams: Novel Materials and Processes," Technical Paper, *28th International SAMPE Technical Conference*, Seattle, WA, 4–7 November 1996.

[140] "Volume Shrinkage During Polymerization of Acrylic Monomers," 1964, Rohm and Haas Company, Washington Square, PA.

[141] Kim, C. S. and Smith, T. L., "An Improved Method for Measuring the Thermal Expansion Coefficient of Linear Expansion of Flexible Polymer Films," *Journal of Polymer Science: Part B: Polymer Physics*, Vol. 28, 1990, p. 2119.

[142] "UV/EB Curable Oligomers and Shrinkage Behavior," 1996, UCB Chemical Company, Radcure Business Unit, Smyrna, GA.

6

Cationic Initiated Polymerization Systems

PHOTOCURED SYSTEMS INVOLVING onium salt photoinitiators (Chapter 4), epoxides, and polyols have certain advantages over free radical cured systems involving acrylates of various types. One important advantage of the cationic systems is improved adhesion due to low shrinkage of less than 2% to less than 5% during the photopolymerization reaction. In addition, these systems are easy to handle because all products used are low viscosity liquids and the compounds are readily miscible. As a result, usually simple stirring is all that is needed to combine the formulating ingredients. It should be mentioned that some polyols that can be used as formulating ingredients are low melting solids that have low viscosity after melting. Some of the polycaprolactone and poly(tetramethylene oxide) polyols are in this category. The ingredients have low skin and eye irritation potential, though investigators should read manufacturers' Material Safety Data Sheets and other available information when working with these and with other compounds. The epoxide-based formulations do not require an inert atmosphere during cure, since they are not oxygen inhibited. These items and others are given in the comparison of cationic and free radical systems presented in Table 6.1. A number of these items as well as others will be dealt with in the following discussion.

Cationic Cycloaliphatic Epoxide Systems [1]

Cycloaliphatic epoxides form the basis for all or almost all cationic-initiated radiation-curable, coating formulations. The main epoxide used is the diepoxide 3,4-epoxycyclohexylmethyl-3,4-epoxycyclohexanecarboxylate:

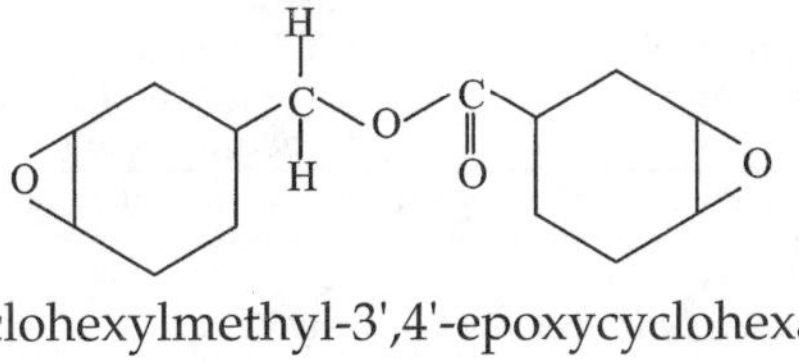

H H C O C O O O

3,4-Epoxycyclohexylmethyl-3',4'-epoxycyclohexanecarboxylate

TABLE 6.1. Comparison of Certain Features Involved in Cationic and in Free Radical Photocuring.

ATTRIBUTE	PHOTOINITIATOR TYPE	
	FREE RADICAL	CATIONIC
Dual free radical/cationic cure possible?	No	Yes
Formulated product shelf stability	Moderate	High
Handling characteristics	Often difficult*	Easy
Humidity effects	None	May slow reaction
Life of active species	Short	Long
Oxygen inhibited?	Yes/No, depends on type initiator	No
Photoinitiator physical state	Solid and Liquid	Solid and liquid
Reaction continues after radiation exposure?	No	Yes
System technology development		
Electron beam	High	Moderate
Ultraviolet radiation	High	High
Thermal conditions effect on rate of cure	Improves	High
Thermal post cure effect	May improve adhesion	Hastens any "dark" reaction
Volume Shrinkage Due to Photopolymerization	High	Low

* This refers to the very high viscosity and often-sticky character of the acrylated epoxides and urethane acrylates. This factor has been reduced somewhat by offering diluted products, but this approach in turn limits a formulator's options.

The diepoxide most often forms 30–40% or more of most formulated products. It is available from a number of sources under a variety of trade names as described in Table 6.2. The primary Irritation Index Draize Value for skin is 1.35 (0–8 scale) and for eye is 7.5 (0–110 scale) [2]. In general, this compound has a 25°C viscosity of about 350–450 mPa·s, though certain products such as CYRACURE™ UVR-6105 with a viscosity of 220–250 mPa·s and a Draize value for skin of 1.6 and for eye of 8.0 are stated to be the above diepoxide [2] and Uvacure™ 1500, 1501, and 1502 with viscosities of 275, 280, and 80 mPa·s[1][3].

[1] The structure for these compounds is not given. Uvacure™ 1500 is described as a cycloaliphatic epoxide that is the base compound for cationic curing, 1501 as a low viscosity cycloaliphatic epoxide that can decrease residual odor, and 1502 as a low viscosity cycloaliphatic epoxide that can be used alone or as a reactive diluent.

TABLE 6.2. Trade Names for the Diepoxide, 3,4-epoxycyclohexylmethyl-3,4-epoxycyclohexanecarboxylates.

COMPANY	TRADE NAME AND DESIGNATION*
CIBA-GEIGY Corporation*	ARALDITE™ CY-179
Daicel Chemical Industries	Celloxide™ 2021
Degussa AG	Degacure™ K126
Sartomer	SarCat™ K126
Union Carbide Corporation**	CYRACURE™ UVR-6105, UVR-6110
Union Carbide Corporation**	ERL-4221 and others
UCB Chemicals	Uvacure™ 1500, 1501

Each of these companies has related products, and their literature can be a rich source of information about particular products.

* Currently Vantico, Inc.

** A Subsidiary of The Dow Chemical Company.

Other cycloaliphatic epoxides include bis(3,4-epoxycyclohexylmethyl) adipate.

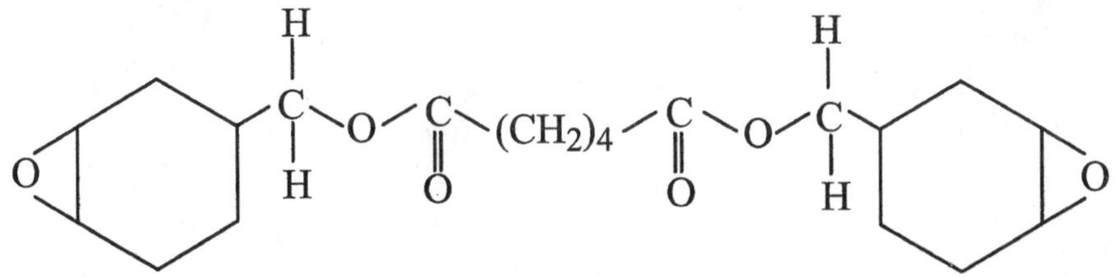

This epoxide is a compound that can impart flexibility to a cured product. It has the very low Draize values of 0.25 for skin and 4.0 for eye. This diepoxide has a viscosity of 550–750 mPa·s at 25° C, and it is or has been marketed by Union Carbide Corp.[2] as CYRACURE™ UVR-6128 and by CIBA-GEIGY[3] as ARALDITE™ CY-178.

Cycloaliphatic epoxides such as the above are prepared through the Diels-Alder reaction that involves the 1,4-addition of an alkene to a conjugated diene that might be exemplified by the reaction of 1,3-butadiene (the diene) with ethylene (the dienophile) to form cyclohexane.

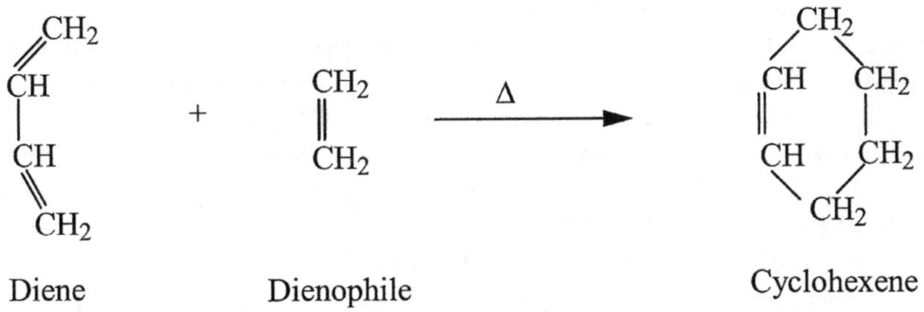

Diene Dienophile Cyclohexene

To prepare 3,4-epoxycyclohexylmethyl-3,4-epoxycyclohexanecarboxylate, 1,3-butadiene is reacted with acrolein to form an unsaturated cyclic aldehyde. Two molecules of the aldehyde are coupled by means of a Tischenko reaction to form a cyclic diene that can be oxidized with a peroxyacid, such as peracetic acid, to form the desired diepoxide.

[2] A Subsidiary of The Dow Chemical Company.

[3] Currently Vantico, Inc., which was created out of the Performance Polymers Division of Ciba Specialty Chemicals.

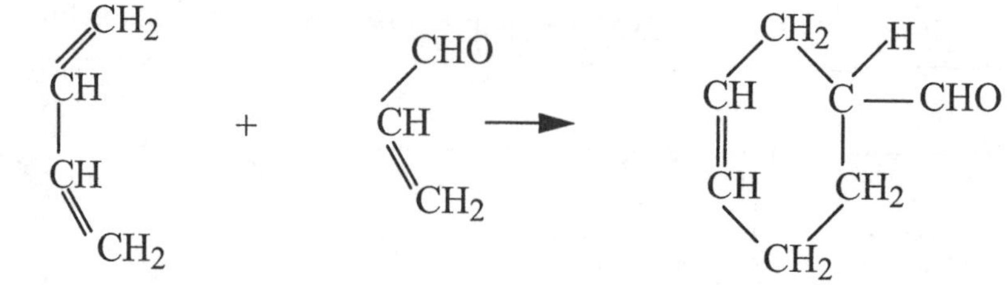

1,3-Butadiene + Acrolein → Unsaturated Cyclic Aldehyde

2 Unsaturated Cyclic Aldehyde 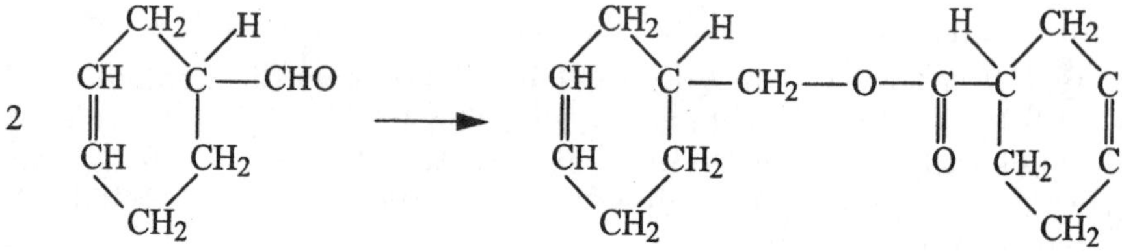 → Cyclic diene

This particular chemistry has been detailed with various α, β-unsaturated aliphatic aldehydes that form a number of diepoxycycloaliphatic esters [4]. The cyclic aldehydes have also been oxidized to the corresponding carboxylic acids, reacted with glycols, and then epoxidized [5]. The cyclic aldehydes have also been reduced to form the corresponding alcohols, reacted with dicarboxylic acid, and epoxidized [6]. Curable compositions from these epoxides were described and patented [7,8]. Recent technology describes the reaction of dienes with acrylate and methacrylate dienophiles and a broad variety of epoxides that can be prepared [9].

Certain aliphatic and cycloaliphatic epoxides are listed in various suppliers' literature as reactive diluents for cationic-cure systems, and these include 1,2-epoxyhexadecane (viscosity, 15 mPa·s at 25° C), mixtures of cycloaliphatic epoxides (viscosity 85–115 mPa·s at 25° C), limonene dioxide, the diglycidyl ethers of 1,4-butanediol and 1,6-hexanediol, as well as other low viscosity reactive compounds. Nonreactive solvents are also used to dilute systems to an application viscosity, but use of such solvents introduces volatile-organic-content and other difficulties encountered with solvents. Comparisons have been made of reactive diluents for radiation-curable formulations [10].

A number of specialty di- or triglycidyl ethers were marketed as diluent/flexibilizers for cationic ultraviolet cure formulations [11]. Some of these

TABLE 6.3. Multifunctional Glycidyl Ether Diluent/Flexibilizers [11].

GLYCIDYL ETHER	VISCOSITY, mPa·s, 25° C	EPOXIDE EQUIVALENT WEIGHT
Diglycidyl ether of 1,6-hexanediol	26	147
Diglycidyl ether of bisphenol-A ethoxylate	2200	331
Diglycidyl ether of neopentyl glycol propoxylate	44	220
Triglycidyl ether of glycerol propoxylate	155	186
Triglycidyl ether of trimethylolpropane ethoxylate	—	175

compounds, such as 1,6-hexanediol diglycidyl ether, had a low viscosity (26 mPa·s, presumably at room temperature), while others that were alkoxylates of diols and triols had viscosities that ranged from low to high as indicated in Table 6.3. A generic formulation suggested that the compounds should be used at a concentration of 10–30% with the remainder of the formulation a cycloaliphatic epoxide or one of two different diglycidyl ethers of Bisphenol A, a photoinitiator (3%) and a surfactant (0.5%). The lower viscosity diluent/flexibilizers significantly decreased application viscosity, and the cured systems had good mechanical properties.

Oxetanes or trimethylene oxide derivatives are another series of compounds that are beginning to be discussed in the literature as reactive diluents for cationic systems.

Trimethylene oxide

Oxetane itself is a low viscosity, low boiling point (50° C) clear liquid. However, derivatives of oxetane are higher boiling (lower vapor pressure) compounds that maintain a low viscosity and that have been shown to undergo cationic polymerization in general [12,13] and to polymerize when irradiated in the presence of cationic photoinitiators [14–16]. Oxetane was found to have a relatively long induction period for initiation of polymerization in comparison to that of the cycloaliphatic epoxides. However, if electron-donating groups are placed in the second position of the oxetane ring, its cationic reactivity is greatly enhanced. One of these compounds, 3-ethyl-3-hydroxylmethyl-oxetane, which has a viscosity of 22 mPa·s at 25° C, has been commercialized [2].[4]

[4] Union Carbide Corporation, CYRACURE™ UVR-6000.

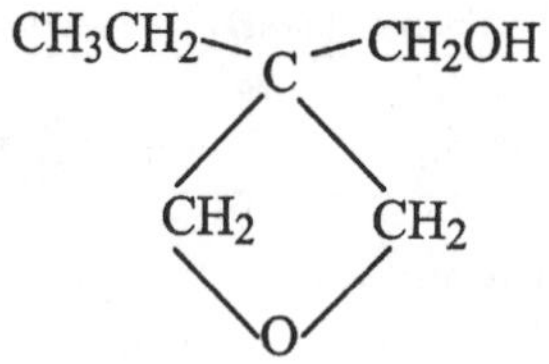

$$CH_3CH_2-C(-CH_2OH)(CH_2-O-CH_2)$$

3-Ethyl-3-hydroxymethyl-oxetane

The viscosity reducing effect on 3,4-epoxycyclohexylmethyl-3,4-epoxycyclohexane-carboxylate is given in Table 6.4. At low levels of less than about 10–15%, this oxetane did not deleteriously affect curing rate. However, at higher concentrations cure rate was significantly decreased. The decrease was thought to be related to the hydroxyl group on the compound that has the ability to chain transfer, and at high hydroxyl concentrations, the molecular chain length of the polymerized system could be decreased. Alternatively, it was felt that if the hydroxyl groups did not react, they could function as a tackifier. The difunctional oxetane shown below, as well as others, was also studied [15]. It had a low

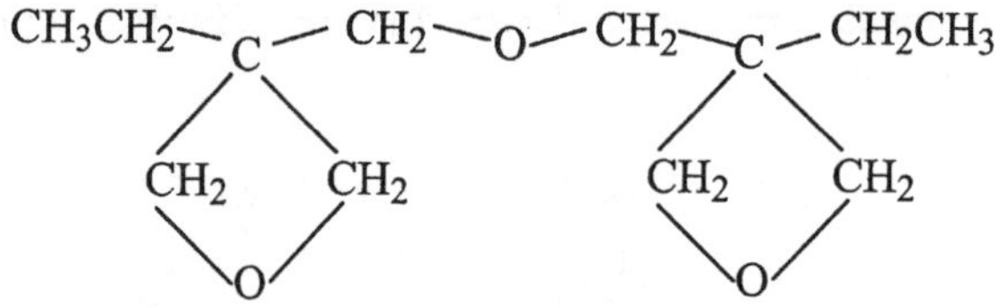

$$CH_3CH_2-C(CH_2-O-CH_2)-CH_2-O-CH_2-C(CH_2-O-CH_2)-CH_2CH_3$$

Di-(3ethyl-3methyl-oxirane)ether

viscosity of 13 mPa·s at 25° C, and when added to the diepoxide of Table 6.4 at a 20% and 40% level, resulted in a blend viscosity of 154 mPa·s and 109 mPa·s at 25° C without decreasing cure rate. However, this di-oxetane is not commercially available at this time.

TABLE 6.4. Viscosity Reducing Power of 3-ethyl-3-hydroxylmethyl-oxetane When Blended with 3,4-epoxycyclohexylmethyl-3,4-epoxycyclohexylcarboxylate. The Blends Had Three Parts per Hundred of an Arylsulfoniumhexafluorophosphate Solution Added as a Photoinitiator [Adapted from Reference 15].

WEIGHT PERCENTAGE OF CYCLOALIPHATIC DIEPOXIDE	WEIGHT PERCENTAGE OF OXETANE DERIVATIVE	VISCOSITY, mPa·s AT 25° C
100	0	380
90	10	214
85	15	170
80	20	138
75	25	112
70	30	93

Another study of 3-ethyl-3-hydroxymethyl-oxetane [17] found that 20% of it decreased the viscosity of 3,4-epoxycyclohexanemethyl-3,4-epoxycyclohexane carboxylate by about 50%; however, the neat viscosity of the cycloaliphatic epoxide was only about 210 mPa·s versus 380 mPa·s in the other study (Table 6.4) [15]. This study involved the curing of clear coatings as well as yellow- and blue-pigmented inks. The results indicated that the reactive diluent is effective in enhancing the curing performance of the coatings when it is used in the proper concentration range, which is a factor that may vary from formulation to formulation.

Oils

Epoxidized oils can be used in cationic-cure formulations. These oils may be typified by the following structure, which is closely related to that of epoxidized soybean oil. In addition to the radiation-cure coating area, such oils are used as acid

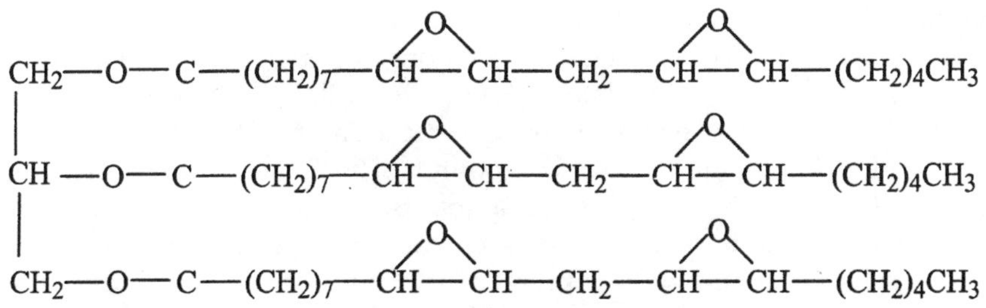

$$
\begin{array}{l}
CH_2-O-C-(CH_2)_7-\overset{O}{CH-CH}-CH_2-\overset{O}{CH-CH}-(CH_2)_4CH_3 \\
| \\
CH-O-C-(CH_2)_7-\overset{O}{CH-CH}-CH_2-\overset{O}{CH-CH}-(CH_2)_4CH_3 \\
| \\
CH_2-O-C-(CH_2)_7-\overset{O}{CH-CH}-CH_2-\overset{O}{CH-CH}-(CH_2)_4CH_3
\end{array}
$$

A representation of an epoxidized soybean oil epoxide (Actual oil on an average contains 51% linoleic acid, 25% oleic acid and 9% linolenic acid residues in an epoxidized form)

scavengers in vinyl inks and elsewhere, as oligomeric plasticizers for vinyl chloride films, and related uses. Although coatings that result from the use of only epoxidized oil and photoinitiator in a formulation are much too soft for any conventional usage, small amounts of these modified oils can be used as modifiers and flexibilizers for cycloaliphatic diepoxide based formulations [18–20]. Epoxidized oils from other sources, such as linseed oil and palm oil, are also being investigated in cationic formulations. Vernonia oil, a naturally occurring epoxidized oil, has been used in air-dry formulations [21] and in ultraviolet radiation-cured systems [22]. Vernonia is considered a weed and is in the same family as the dandelion and thistle and actually resembles a thistle without thorns. It is an unsaturated, epoxidized vegetable oil with a homogeneous molecular structure that has identical triglyceride molecules with three equal vernolic acid residues [23]:

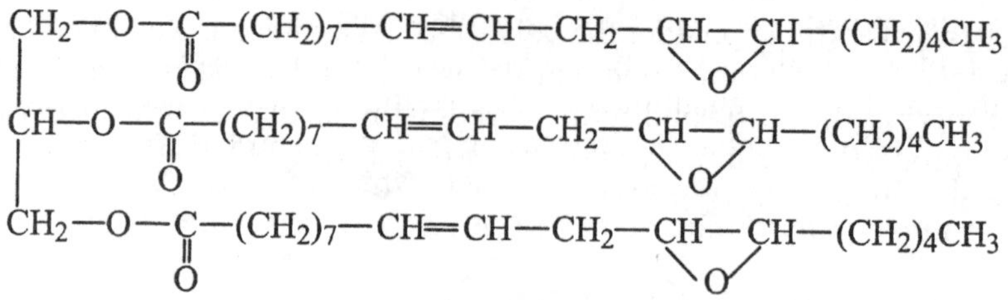

$$
\begin{array}{l}
CH_2-O-\underset{\underset{O}{\|}}{C}-(CH_2)_7-CH{=}CH-CH_2-\overbrace{CH-CH}^{}-(CH_2)_4CH_3 \quad (\text{epoxide } O \text{ bridging } CH-CH) \\
| \\
CH-O-\underset{\underset{O}{\|}}{C}-(CH_2)_7-CH{=}CH-CH_2-CH-CH-(CH_2)_4CH_3 \quad (\text{epoxide } O \text{ bridging } CH-CH) \\
| \\
CH_2-O-\underset{\underset{O}{\|}}{C}-(CH_2)_7-CH{=}CH-CH_2-CH-CH-(CH_2)_4CH_3 \quad (\text{epoxide } O \text{ bridging } CH-CH)
\end{array}
$$

Vernonia Oil

A group backed by the United Soybean Board has developed novel gallium-based photoinitiators [24]. These new photoinitiators produce thin, colorless clear-coats from epoxidized soybean oil. The initiator is considered a breakthrough and allows solvent-free coatings to be obtained at line speeds of more than 1000 ft/min. (305 m/min). Epoxidized soybean oil has also been used in cationic, thermally cured coatings [25].

The compound basic to most formulations, 3,4-epoxycyclohexylmethyl-3,4-epoxycyclohexanecarboxylate, is usually formulated with other ingredients that include other epoxides, oxetanes, acrylates, vinyl ethers, polyols [26], glycols, alcohols, and similar compounds. The "other" epoxides usually are aliphatic in nature, though small amounts of aromatic epoxides can be used, as well as blends of cycloaliphatic epoxides [27]. Aromatic epoxides are radiation absorbers and, thus, compete with the photoinitiator for available radiation energy. This makes it more difficult for the radiation to penetrate through the liquid coating and find the photoinitiator, with the result being a slowdown in polymerization or cure rate. The free radicals formed during photolysis of cationic photoinitiators will independently polymerize suitable unsaturated compounds, if present, and thus, form an interpenetrating network with the cycloaliphatic epoxide network. Vinyl ethers will copolymerize with the cycloaliphatic epoxides in both clear and pigmented formulations [28]. Polyols and glycols will chain transfer with the polymerizing epoxide molecules and can be used to increase cure rate, decrease system cost, and provide flexibility. Alcohols will chain transfer as the polyols and glycols do and can be used to control molecular weight by acting as chain stopping or termination agents. Each of these items will be discussed in detail.

Studies of the tensile and toughness properties of radiation-cured cycloaliphatic epoxide/polyol blends indicate that a broad range of mechanical properties can be achieved from these systems. Both polyether [29], in the form of trifunctional poly(propylene oxide)s, and polyester [30], in the form of di- and trifunctional polycaprolactones, polyols have been investigated.

Tensile strength, elongation, and secant modulus are a function of polyol content with modulus and strength decreasing and elongation increasing as the polyol content increases. Polyol molecular weight was also studied as a variable, from 702 to 4008 for the polyethers and from 300 to 900 trifunctional and 530 to 3000 difunctional for the polyesters. At a constant weight percentage of polyol in the formulation, tensile properties decreased as molecular weight increased, but flexibility increased as molecular weight increased. Overall, properties are improved over those that were obtained with a wide variety of urethane acrylates [31].

Mechanism of Polymerization

Cation formation was detailed in Chapter 4 and can be summarized as follows:

$$[C_6H_5]_N-(S_X)^+MF_6^- + R-H \xrightarrow[\text{RADIATION}]{\text{ULTRAVIOLET}} H^+MF_6^- + \text{Other Products}$$

Arylsulfoniumhexafluoro-metallic salt — Active hydrogen Source — Bronsted Acid (A protonic acid)

Thus, the cationic photoinitiator, an onium salt, interacts with an active hydrogen source under the influence of ultraviolet radiation to form a protonic acid that is the actual initiating species and other products of photolysis. Included in the other products are free radicals that are capable of initiating the polymerization of ethylenically unsaturated compounds such as the acrylates.

The protonic acid is capable of initiating the polymerization of epoxides, particularly cycloaliphatic epoxides (for simplicity, the initiation and propagation reaction are described with a monoepoxide).

Initiation

$$\text{3,4-Epoxy-1-methyl cyclohexane} + H^+MF_6^- \longrightarrow \text{Initiated Epoxide (HO, } MF_6\text{, } CH_3\text{)}$$

3,4-Epoxy-1-methyl cyclohexane — Initiated Epoxide

The hydroxyl group formed is very reactive, and the initiated epoxide then reacts with other epoxide molecules to form the polymerized product.

Propagation

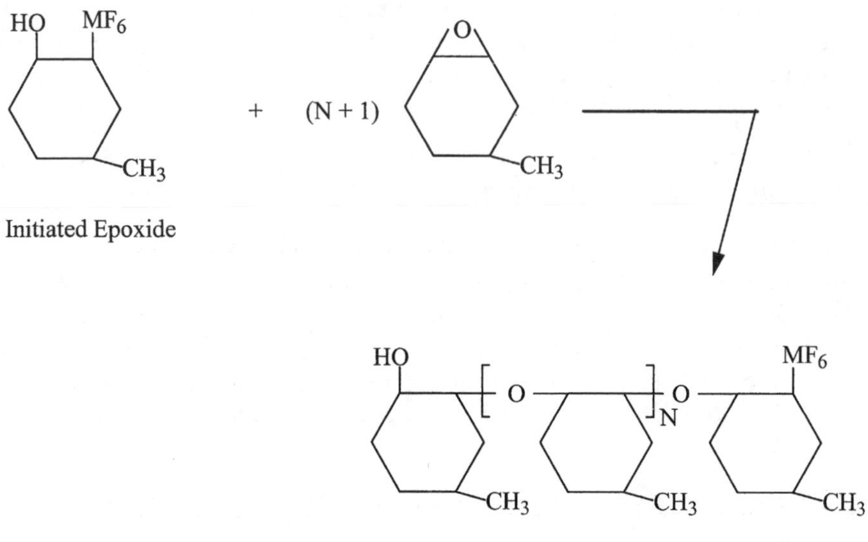

Initiated Epoxide + (N + 1) epoxide → Polymerized epoxide

Termination takes place when the monomer is depleted to a very low level or exhausted, the system is neutralized in some manner, or chain transfer occurs. The polymerized epoxide behaves somewhat as a living polymer, in that it remains active after the epoxide supply is exhausted [32]. However, since the epoxide polymer is acidic in nature, it can be neutralized if basic compounds are present. Interactive chain transfer is a means of terminating the reaction and amounts to a copolymerization of a polyol or other active hydrogen-containing compound with the epoxides. When this process takes place with oligomeric hydroxyl-containing compounds, it is a means of introducing flexibility into the polymerized mass.

Polyols

In the chain transfer with hydroxyl groups, the hydroxyl group is deprotonated, and the freed proton interacts with the anionic species associated with the growing molecular chain. This results in regeneration of the protonic acid, and it can initiate the polymerization of another molecular chain as described below [33].

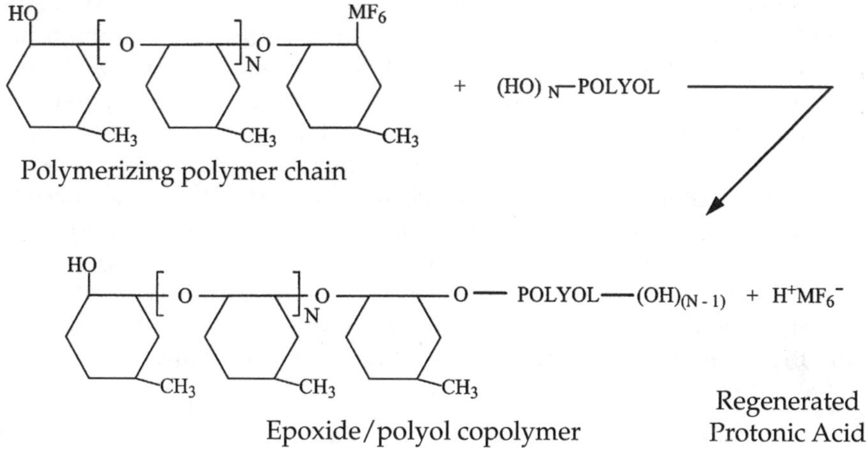

HO MF$_6$ O O N CH$_3$ CH$_3$ CH$_3$ + (HO)$_N$—POLYOL

Polymerizing polymer chain

HO O O N O— POLYOL—(OH)$_{(N-1)}$ + $H^+MF_6^-$ CH$_3$ CH$_3$ CH$_3$

Epoxide/polyol copolymer

Regenerated Protonic Acid

The value of *N* in this reaction sequence can range from one, when alcohols or other monohydroxyl-containing compounds are used, to six or eight, or even more. After the epoxide/polyol copolymer has formed, it can react with other growing epoxide chains or with epoxide monomers that are in the reacting mass. Thus, when the polyol has sufficient molecular weight, it can act as the soft segment or flexibilizing segment acts in a polyurethane and, thus, may increase the coating elongation, flexibility, and toughness, while decreasing tensile modulus and shear modulus [34,35]. A propagating chain of a diepoxide might appear as follows:

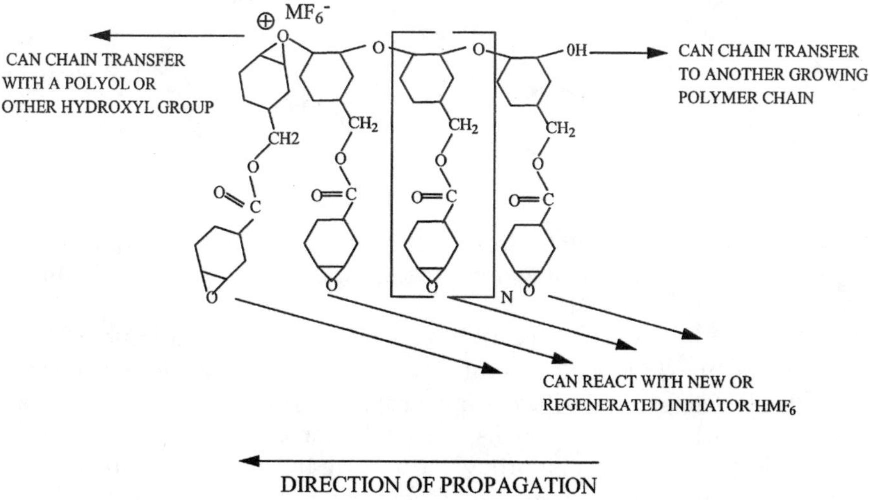

CAN CHAIN TRANSFER WITH A POLYOL OR OTHER HYDROXYL GROUP

⊕ MF_6^-

CAN CHAIN TRANSFER TO ANOTHER GROWING POLYMER CHAIN

CAN REACT WITH NEW OR REGENERATED INITIATOR HMF_6

DIRECTION OF PROPAGATION

The remaining epoxide groups on this polymer can react with new or regenerated initiator molecules to increase molecular weight and form branches and new hydroxyl groups. It is also possible for the epoxide groups to react with available hydroxyl groups. It is readily apparent that the polymeric mass can become a highly cross-linked species. If polymer molecules are present in the system, the hydroxyl groups on the polyol will undergo interactive chain transfer with the growing species. As a result, the copolymerized polyol will add either flexibility or flexibility and toughness, depending on the nature of the miscibility between the polyol and the epoxide polymer. Of course, the hydroxyl groups formed on the growing epoxide chain may also undergo interactive chain transfer with a growing epoxide species. Shortly after the above species formed, the growing species might appear as:

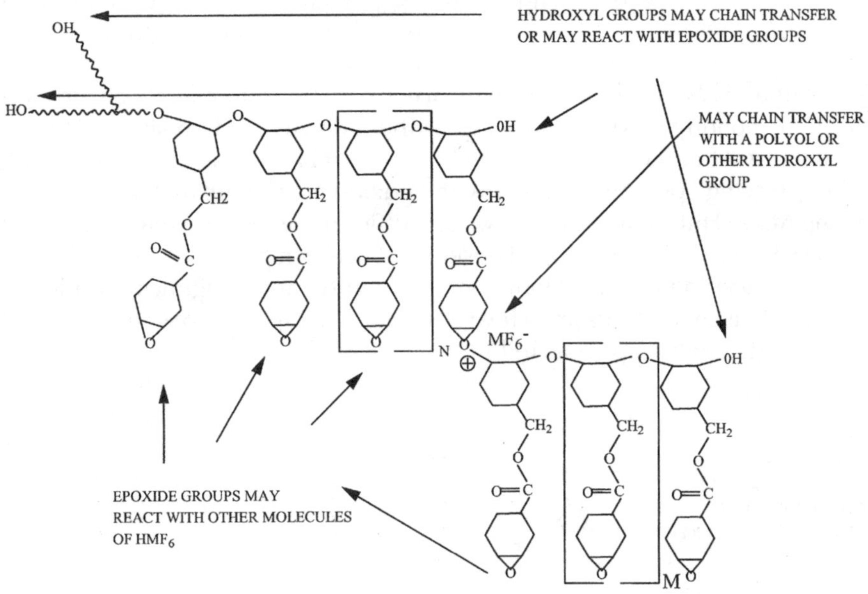

HYDROXYL GROUPS MAY CHAIN TRANSFER OR MAY REACT WITH EPOXIDE GROUPS

MAY CHAIN TRANSFER WITH A POLYOL OR OTHER HYDROXYL GROUP

EPOXIDE GROUPS MAY REACT WITH OTHER MOLECULES OF HMF_6

The polymeric mass resulting from such a polymerization is complicated in nature early in the polymerization process, and the complication continues until the reactants are exhausted.

It is readily apparent that the use of polyols in radiation-cured, cycloaliphatic epoxide-based formulations greatly increases the formulating latitude of coating manufacturers and, in almost every instance, will decrease cost. A noninclusive list of polyols useful in cationic radiation-cured coatings is given in Table 6.5. The polypropylene oxide, ethylene oxide-capped propylene oxide, and the poly(tetramethylene oxide) polyols are commercially available as diols, triols, and in certain instances higher hydroxyl-functional

TABLE 6.5. Partial List of Polyols that Can Be Used in Formulating Cationic, Radiation-Cured Coatings.

POLYOL	REFERENCE
Polyester adipates	[3]
Polycaprolactones	[34]
Poly(propylene oxide)s	[35]
Ethylene oxide capped poly(propylene oxide)s	[36, 37]
Poly(tetramethylene oxide)s	[38]

compounds [39–41]. In addition, they are available in a broad range of molecular weight, from less than a few hundred to several thousand. In a general sense, these are the polyols used to form the myriad of polyurethane products that are used as foams, coatings, elastomeric materials, and so on. Although not specifically mentioned, random or block propylene oxide/ethylene oxide copolymers are useful. Also useful in the radiation-cured systems are oligomeric and polymeric hydroxyl-containing compounds, such as the styrene/allyl alcohol copolymer, ethylene/vinyl acetate/vinyl alcohol copolymers, butyl acrylate/2-hydroxyethyl acrylate copolymers, as well as similar compounds.

When polyols, or other compounds for that matter, are formulated with cycloaliphatic epoxides, their acidity should be considered. Cycloaliphatic epoxides will react with acidic compounds, and they are used as acid scavengers in, for example, vinyl chloride based inks. Thus, if the acid number of a polyol is too high (a rather indeterminate term, but no amount has been established), the acidity may lead to storage instability or, at the very least, an increase in viscosity of the formulated product. This caution should be particularly considered in the case of polyester polyols, such as the diethylene glycol adipates, hexanediol adipates, and so on, that are usually prepared from a dicarboxylic acid or acid anhydride and a diol. Even if an excess of diol is used, a common practice to reduce acid number, in preparation of the polyester polyol, it is difficult to remove all carboxylic acid end groups. One method to reduce acidity in the polyesters might be to react the excess acid with a compound such as commercially available 1-vinyl-3,4-epoxycyclohexane:

~~~COOH + CH_2═C– [epoxycyclohexane ring, O] ⟶ ~~~C(═O)–O– [cyclohexane ring, HO] –C═CH_2

Acid group on a polyester — A monoepoxide — End-capped polyester with hydroxyl functionality

As would be expected, the various glycols, such as ethylene glycol, 1,4-butanediol, 1,6-hexanediol, and so on, can be used. The use of such compounds will increase reaction rate, but it should be kept in mind that use of
~~~

too much of these short molecules will lead to embrittlement by decreasing kinetic chain length. Also, the lower the molecular weight of the additive, the less it takes to react with all of the oxirane units, making it easier to lead to a loss of mechanical properties at even relatively low concentrations. It has been shown that as little as 4.5% ethylene glycol will increase reaction rate threefold [33].

Water, Humidity, and Temperature Effects

Water will react into the system. It copolymerizes into cationic-cure systems, and it can act as a chain stopper [42,43]. Cure rate can be decreased significantly, and the time to achieve a tack-free state is markedly increased if the humidity is high during cure [44]. When the relative humidity is greater than 80%, there is a drastic decrease in cure rate. At a relative humidity of 65% or less, there appeared to be little or no effect on the time needed for the coating to become tack free. If the substrate was warmed to about 45° C, the effect of high humidity was negated and cure rate returned to normal. Another investigator reported that water would copolymerize in cationic systems, and that at a relative humidity greater than about 70%, there is a marked decrease in the rate at which formulations cure [43].

The moisture content of paper substrates has been shown to have an effect on cure rate [45]. Dried paperboard was stored under various humidity conditions, and it was found that paperboard stored at 45% relative humidity cured in 1 s, whereas when it was stored at 95% relative humidity, it took more than 60 s to effect cure. This dried paperboard technique might be considered as a tool to test particular formulations if high relative humidity is expected during manufacturing.

In certain end uses, the presence of water can be very detrimental. In the field of linerless labels in the pressure sensitive adhesive area, the release composition must be essentially free of water [46]. That is, the composition should have less than 1.5% by weight of free water, which excludes bound water, such as water of hydration. If water were in excess of about 1.5%, the adhesive coating and release coating would "lock-up" when the linerless labels were rolled. In this case, the most preferred condition was to have less than 0.5% water in the formulation of epoxy–functional silicone, reactive diluent, photoinitiator, and wax-treated silica powder.

Apparently, water in the bulk of an epoxide-based formulation will copolymerize [26,27]. Small amounts of water, up to about 1%, might be added and will dissolve in 3,4-epoxycyclohexylmethyl-3,4-epoxycyclohexyl carboxylate containing an onium salt photoinitiator. (Note that this is in more or less agreement with the previous paragraph dealing with linerless labels.) When cured with ultraviolet radiation, the water doesn't appear to affect properties seriously. However, at a 3–4% addition, water became immiscible and an emulsion formed. When cured, there was a complete loss of mechanical properties and even some separation of liquid water from the cured film. Results such as these probably will be formulation dependent. Water is not a recommended additive, and there are many other choices of organic compounds that could be used if a very low molecular weight diluent is needed. Water certainly should not be used in a commercial operation, unless a careful study is made before proceeding.

Since small amounts of water copolymerize in cationic formulations but high relative humidity markedly affects cure, it would appear that adsorbed, or more particularly a chemisorbed, layer of water on a substrate acts as if it is alkaline in nature and, thereby, affects both cure rate and adhesion. This seemed to be particularly true on steel, and in industry, metal substrates are often given a very short flame treatment or infrared treatment to remove water, oils, etc.

Effect of Added Thermal Energy

Warming the substrate or the formulation will enhance cure rate. The diglycidyl ethers of bisphenol A are particularly difficult to cure with cationic, onium salt-initiated systems. It has been found [44] that a formulation based on this epoxide that required more than a min. to become tack free at 23° C was tack free in about 5 s at 50° C and in about 1 s at 60° C. A 65/35 diglycidyl ether of bisphenol A/cycloaliphatic epoxide formulation, initiated with an onium salt, was tack free in 4 s at 23° C, and in less than 1 s at 40° C. Other investigators have also studied the thermal acceleration of cationic photocure systems [44,47–51]. One of these studies [47] pointed out that a cycloaliphatic epoxide could be homopolymerized in the presence of an onium salt photoinitiator without preheating the substrate, i.e., at a temperature of 24° C (75° F) and a conveyor speed of greater than 50 ft/min (15.2 m/min). In contrast, when a diglycidyl ether of bisphenol A was cured, the tack-free time was 25 s, even when the conveyor speed was decreased to 25 ft/min (7.6 m/min). However, when the substrate metal panel was preheated to 49° C (120° F), the aromatic epoxide could be cured to a tack-free state in less than 1 s. This again points out the temperature sensitivity of the initiation/propagation reaction that takes place with cationic photocuring. Another study involved a study of curing efficiency by means of photoinitiator and wavelength efficiency [52].

Electron Beams and Epoxide Cure

Ionizing radiation from electron beams and other sources has been used to polymerize cycloaliphatic epoxides, diglycidyl ethers of bisphenol A, and the triglycidyl ether of tris(hydroxyphenyl) methane. This work has been conducted in the area of composites [53–56], negative resists in optical lithography [57], and coatings [58]. The studies have been carried out with and without cationic photoinitiators, such as the diaryliodonium salts and the arylsulfonium salts. The electron-beam radiation initiates reaction in the epoxide monomers and, as with ultraviolet radiation, this leads to polymerization of the molecules. The final product is a cross-linked, three-dimensional network with superior chemical, mechanical, and thermal properties [53].

Pigmentation

Cationic systems have a special concern regarding pigmentation, other than the difficulties arising through opacity, that are germane to all ultraviolet radiation cured systems. As described earlier, the active species after photolysis is an acid, a Bronsted or protonic acid, which initiates cure. If a pigment or other additive is basic in nature relative to the acid generated, it can neutralize the active initiator and negate or markedly slow down rate of cure. Pigment dispersions suitable for cationic cure systems are available [59]. Some pigments that are known to be chemically compatible with epoxy vehicles are listed in Table 6.6 [60]. In certain instances, certain pigments such as titanium dioxide are treated with amines or other bases to improve dispersibility. Such treated pigments should be avoided. Successfully cured pigmented or colored formulations have been published [61,64].

In a general sense, formulators of cationic systems should avoid compounds with functional groups or other moieties that will combine with cations or react with acids, i.e., in particular basic compounds. Examples of these include amines, amides, urethanes, and residues from some cleaning formulations, as well as others. Ionic materials should also be avoided.

TABLE 6.6. Pigments Chemically Compatible with Cationic Systems [adapted from [60].

PIGMENT	COMPANY LISTED*
Barium Litho Red DC-1151	Ridgeway Color & Chemical
Indofast Brilliant Scarlet R-6500	Allied Color Industies
Magenta RT-816	DuPont
Perylene Red RT-6812	Chemetron
Deep Cadmium Yellow	Hercules
Diarylide Yellow YB-5769	Sun Chemical Company
Indanthrene Golden Yellow	GAF
Microlith Yellow 3R-T	Ciba-Geigy
Microlith Blue 4G-T	Ciba Geigy
Milori Blue 4050	Harshaw Chemical
Monastral Violet RT-887	DuPont
Carbon Black Raven 22	Cities Service
Carbon Black Raven 1020	Cities Service
TiO2 White Horsehead A-430	New Jersey Zinc
TiO2 White Titanox 1070	N. L. Industries
TiO2 White Unitane 0-220	American Cyanamid

* Some company names may have changed from original listing.
Also see [64] for other pigments.

TABLE 6.7. Generalized Formulation for Clear Cationic Coating.

COMPONENT	PARTS PER 100 PARTS
Cycloaliphatic Epoxide	60–97
Polyol(s)	2–35
Onium Salt Photoinitiator	0.5 to 4
Surfactant	0 to 1

Formulation

Formulation of clear cationic-cure systems based on 3,4-epoxycyclohexanemethyl-3,4-epoxycyclohexane carboxylate is done in a relatively simple manner. Basically, one blends epoxide, polyol, and photoinitiator in a ratio that will produce the desired physical characteristics. A great deal of guidance can be found in the literature [29,30,64–68], including information about the effect of urethane acrylates on final properties [69]. A generalized formulation that specifies ranges that might be used in a formulation for clear coatings is given in Table 6.7. Addition of more or less of any component can be done to achieve particular effects. Usually, only simple stirring of the components under safe radiation conditions is required for blending.

References

[1] Koleske, J. V., *Cationic Radiation Curing*, Federation of Societies for Coating Technology, Blue Bell, PA, 1991.

[2] Union Carbide Corp., Brochure UC-958B, "CYRACURE™ Cycloaliphatic Epoxides, Cationic UV Cure," 1999.

[3] UCB Chemicals–Industrial Coatings, Internet site *www.chemicals.ucb-group.com* (03/20/2000).

[4] Patent: Frostick, F. C., Union Carbide and Carbon Corp., Diepoxides of Cycloaliphatic Esters. US 2,716,123, (1955).

[5] Patent: Phillips, B. and Starcher, P. S., Union Carbide and Carbon Corp., Diepoxides. US 2,745,847 (1956).

[6] Patent: Phillips, B. and Starcher, P. S., Union Carbide and Carbon Corp., Diepoxides. US 2,750,395 (1956).

[7] Patent: Phillips, B., Frostick, F. C., McGary, C. W., and Patrick, C. T., Union Carbide Corp., Compositions of Epoxides and Polycarboxylic Acid Compounds. US 2,890,194. (1959).

[8] Patent: Phillips, B., Starcher, P. S., McGary, C. W., and Patrick, C. T., Union Carbide Corp., Curable Composition Comprising a Diepoxide, a Polycarboxylic Anhydride and a Polyhydric Compound. US 2,890,196. (1959).

[9] Patent: Koleske, J. V., Argyropoulos, J. N., and Smith, O. W., Union Carbide Corp., Production of Unsaturated Cycloaliphatic Esters and Derivatives Thereof. US 5,268,489. (1993).

[10] Carroy, A., "Comparison of Reactive Diluents in Cationic UV-Curable Formulations," *Proceedings of RadTech Europe '93*, Fribourg, Switzerland, 2–6 May 1993, p. 489.

[11] Sinka, J. V. and Mazzoni, D., "Specialty Glycidyl Ethers for Cationic Ultraviolet Cure Systems," *Technical Conference Proceedings, RadTech '88*, New Orleans, LA, 24–28 Apr. 1988, p. 378.

[12] Rose, J. B., *Journal of the Chemical Society*, 1956, p. 542.

[13] Bucquoye, M. C. and Goethals, E. J., "Co-oligomers in the Copolymerization of Oxetane with 3,3′-Dimethyloxetane," *Makromolecular Chemie*, Vol. 182, 1981, p. 3379.

[14] Sasaki, H., Rudzinski, J. M., and Kakuchi, T. J., *Polymer Science: Part A*, Vol. 33, No. 11, 1995 p. 1807.

[15] Saski, H., "Oxetanes: Curing Properties in Photo-Cationic Polymerization," *Technical Conference Proceedings, RadTech 2000*, Baltimore, MD, 9–12 Apr. 2000, p. 61.

[16] Carter, W. and Lamb, K., "New Oxetane Derivative Reactive Diluent for Cationic UV Cure," *Proceedings of RadTech 2000*, Baltimore, MD, 9–12 Apr. 2000, p. 641.

[17] Carroy, A., Chomienne, F., and Nebout, J. F., "Performance Enhancement of Cationic Radiation Curing Systems," *Technical Conference Proceedings, RadTech Europe '99*, Berlin, Germany, 8–10 Nov. 1999, p. 499.

[18] Ragjavacjar, R., Sarnecki, G., Baghdachi, J. A., and Massingill, J. L., "Cationic UV Cured Coatings Using Epoxidized Soybean Oil," *RadTech Report*, Vol. 12, No. 5, 1998, p. 36.

[19] Desai, D., Zhong, B., Bai, X., Rahim, M., and Massingill, J. L., "Soy Oil Phosphate Ester Polyols in UV-Cured Coatings," *Proceedings of the International Coatings Exposition*, Oct. 1999.

[20] Desai, D., Rahim, M., and Massingill, J. L., Jr., "Cationic UV Cured Coatings Using Epoxidized Soybean Oil," *Technical Conference Proceedings, RadTech 2000*, Baltimore, MD, 9–12 Apr. 2000, p. 22.

[21] Muturi, P., Wang, D., and Dirlikov, S., *Progress in Organic Coatings*, Vol. 25, 1994, p. 85.

[22] Crivello, J. V., *Macromolecular Reports*, A33 (Supplements 5 & 6), 1996, p. 251.

[23] Dirlikov, S., Islam, M. S., Frischinger, I., Lepkowski, T., and Muturi, P., "Increasing High-Solids With a 'Weed'," *Industrial Finishing*, Vol. 68, No. 2, Feb. 1992, p. 17

[24] Greissel, M., "Soy-Based UV-Curable Coatings on the Horizon," *Industrial Paint and Powders*, Vol. 77, No. 3, March 2001, p. 36.

[25] Raghavachar, R., Sarnecki, G., Baghdachi, J., and Massingill, J., *Journal of Coatings Technology*, Vol. 72, No. 909, 2000, p. 125.

[26] Patent: Koleske, J. V. and Kwiatkowski, G. T., Union Carbide Corp., Photocopolymerizable Compositions Based on Hydroxyl-containing Organic Materials and Substituted Cycloaliphatic Monoepoxide Reactive Diluents. US 4,812,488. (1989).

[27] Patent: Koleske, J. V. and McCarthy, N. J., Jr., "Blends of Epoxides and Monoepoxides," Union Carbide Corp., US 4,622,349. (1986).

[28] Carroy, A., Chomienne, F., Schrof, W., Binder, H., Schwalm, R., and Weiguny, S., "Cationic UV Curing of Clear and Pigmented Systems Based on Cycloaliphatic Epoxide/Vinyl Ether Blending," *Proceedings of RadTech 2000*, Baltimore, MD, 9–12 Apr. 2000, p. 10.

[29] Koleske, J. V., "Mechanical Properties of Cationic Ultraviolet Light-Cured Cycloaliphatic Epoxide Systems," *Proceedings of RadCure Europe '87*, Munich, Germany, 4–7 May, 1987, p. 9–43.

[30] Koleske, J. V., "Copolymerization and Properties of Cationic, Ultraviolet Light-Cured Cycloaliphatic Epoxide Systems," *Proceedings of RadTech '88—North America*, New Orleans, LA, 24–28 Apr. 1988, p. 353.

[31] Koshiba, M., Kirk, K. Huang, S., Foley, S. K., Yarusso, D. J., and Cooper, S. L., "Properties of Ultra-violet Curable Polyurethane Acrylates," *Journal of Materials Science*, Vol. 17, No. 5, 1982, 1447.

[32] Goethals, E. J., "Factors Influencing the Living Character of Cationic Ring-Opening Polymerizations," *Journal of Polymer Science: Symposium No. 56*, 1976, p. 271.

[33] Crivello, J. V., Conlon, D. A., Olson, D. R., and Webb, K. K., "The Effects of Polyols as Chain Transfer Agents and Flexibilizers in Photoinitiated Cationic Polymerization," *Journal of Radiation Curing*, Vol. 13, No. 4, 1986, p. 3.

[34] Koleske, J. V., "Mechanical Properties of Cationic Ultraviolet Radiation-Cured Cycloaliphatic Epoxide Systems", *Proceedings of RadCure Europe '87*, Munich, Germany, 4–7 May, 1987, p. 9.

[35] Koleske, J. V., "Copolymerization and Properties of Cationic, Ultraviolet Radiation-Cured Cycloaliphatic Epoxide Systems," *Proceedings of RadTech '88—North America*, New Orleans, 24–28 Apr. 1988, p. 353.

[36] Union Carbide Corp., "Formulating Ultraviolet Radiation-Cured CYRACURE Cycloaliphatic Epoxide Coatings," Bulletin F-60590, September 1987.

[37] Patent: Koleske, J. V., Union Carbide Corp., Photocopolymerizable Compositions Based on Epoxy and Hydroxyl-Containing Organic Materials Having Primary Hydroxyl Content, US 4,818,776. (1989).

[38] Patent: Koleske, J. V., Union Carbide Corp., Conformal Coatings Cured with Actinic Radiation, US 5,043,221. (1991).

[39] Patent: Gerkin, R. M. and Comstock, L., Union Carbide Corp., Polyepoxide-caprolactone Polyols and Coatings Based Thereon, US 3,896,303. (1975).

[40] Patent: Taller, R. A. and Elder, D. K., Union Carbide Corp., Low Viscosity Poly(epoxide-caprolactone) Polyols, US 4,045,474. (1977).

[41] Patent: Koleske, J. V., Union Carbide Corp., Low Viscosity Adducts of a Poly(active hydrogen) Organic Compound a Polyepoxide, US 4,707,535. (1987).

[42] Rose, J. B., "Cyclic Oxygen Compounds," *The Chemistry of Cationic Polymerization*, P. H. Plesch, Ed., The Macmillan Co., New York, NY, 1963, p. 433.

[43] Brann, B. L., "The Effects of Moisture on UV Curable Cationic Epoxide Systems," *Conference Papers, RadTech '89*, Florence, Italy, 9–11 Oct. 1989; *RadTech Report* Vol. 4, No. 1, January/February 1990, p. 11.

[44] Watt, W. R., "Photosensitized Epoxides as a Basis for Light-Curable Coatings," *Epoxy Resin Chemistry*, R. S. Bauer, Ed., ACS Symposium Series, No. 114, American Chemical Society, Washington, D.C., 1979, p. 17.

[45] Watt, W. R., *Journal of Radiation Curing*, Vol. 13, No. 4, 1986, p. 7.

[46] Patent: Kline, J. R., Monarch Marketing Systems, Inc., Silicone Release Coating Composition, US 6,231,922. (2001).

[47] Alm, R. R. and Carlson, R. C., "Heat Accelerated Cationic Photocuring of Epoxy Coatings," *Conference Proceedings, Radiation Curing VI*, Association for Finishing Processes of SME, 14-2, Chicago, IL, 20–23 Sept., 1982.

[48] Koleske, J. V., "Cationic Ultraviolet Light-Cure Chemistry," Paper FC85-798, *Finishing '85*, Detroit, MI, 16–19 Sept., 1985.

[49] Crivello, J. V., Lam, J. H. W., Moore, J. E., and Schroeter, S. H., *Journal of Radiation Curing*, Vol. 5, No. 1, 1978, p. 2.

[50] Patent: Smith, G. H., Minnesota Mining and Manufacturing Co., Complex Salt Photoinitiator, US 4,231,951. (1980).

[51] Patent: Smith, G. H., Minnesota Mining and Manufacturing Co., Process of Using Photocopolymerizable Compositions Based on Epoxy and Hydroxyl-Containing Organic Materials, US 4,318,766. (1982).

[52] Carroy, A., "Cationic UV-Curing Efficiency of Cycloaliphatic-Epoxide-Based Systems through Photoinitiator and UV Wavelength Selection," *Proceedings of Aspects of Photoinitiation*, Paint Research Association, Egham, England, 19–20 Oct., 1993, p. 65.

[53] Cabage, B., "E-Beam Curing: With Electron Beams, the Heat's Off For Advance Composite Resin Processing, Internet, *www.ornl.gov/publications/labnotes/oct96/e-beam.htm* (1996).

[54] Goodman, D. L. and Byrne, C. A., "Composite Curing with High Energy Electron Beams: Novel Materials and Processes," *Technical Paper, 28th International SAMPE Technical Conference*, Seattle, Washington, 4–7 Nov., 1996.

[55] Patent: Crivello, J. V., Polyset Corp., Electron Beam Curable Epoxy Compositions, US 5,260,349. (1993).

[56] Patent: Janke, C. J., Lopata, V. J., Havens, S. J., Dorsey, G. F., Moulton, R. J., Lockheed Martin Energy Systems, High Energy Electron Beam Curing of Epoxy Resin Systems Incorporating Cationic Photoinitiators, US 5,877,229. (1999).

[57] Patent: Shu, J. S., Covington, J. B., Lee, W., Venable, L. G., Varnell, G. L., Texas Instruments Inc., Plasma Developable Negative Resist Compositions for Electron Beam, X-ray and Optical Lithography, US 4,657,844. (1987).

[58] Carroy, A. C., "EB Induced Polymerization of Cycloaliphatic Epoxides-Based Systems," *Conference Proceedings of 19th Technology Days on Radiation Curing Process and Systems*, Le Mans, France, 27–28 May, 1998.

[59] Penn Color Co., Technical Information Bulletin, "Pennco® Radiation Curable Epoxy Dispersions, Oxirane Functional Color Concentrates for UV-Curable Inks, and Coatings," 1985.

[60] Commercial Chemical Division, 3M Company, "3M Brand UV Activated Epoxy Curatives," Brochure UVGC(68.06)R1, (undated, ~1983).

[61] Manus, P. J. M., "Coating Performance and Formulation Parameters of Cationic Systems," *Proceedings of Radiation Europe '89 Conference*, Florence, Italy, 9–11 Oct. 1989, p. 535.

[62] Carroy, A., "Efficient Photo-Induced Curing of Opaque Cationic White Base Coatings," *Proceedings of RadTech '94—North America*, Orlando, FL, 1–5 May, 1994, p. 462.

[63] Carroy, A. C., "Cationic Pigmented Systems and UV Light Sources," *Technical Paper, Proceedings of Recent Developments in UV-Curing Technology for the Printing and Coating Industry*, I.O.M/Fusion UV Systems Conference, Leipzig, Germany, 25–26 Nov. 1998.

[64] Union Carbide Corp., "CYRACURE® Cycloaliphatic Epoxides: Cationic UV Cure," Brochure UC-958B, August 1999.

[65] Koleske, J. V., "Cationic Ultraviolet Light-Cure Chemistry," *Technical Paper FC85-798, Finishing '85*, AFP/SME Meeting, Detroit, MI, 16–19 Sept. 1985.

[66] Carroy, A., "Cationic Radiation Cured Coatings: Physico-Chemical Properties Related to Formulation and Irradiation," *Proceedings of RadTech Europe '91*, Edinburgh, Scotland, 29 Sept.–2 Oct. 1991, p. 265.

[67] Carroy, A., "New Developments in the Formulation of Cationic UV Curing Systems," *Proceedings of RadTech Europe '95*, Maastricht, The Netherlands, 25–27 Sept. 1995, p. 523.

[68] Carroy, A. C., Chomienne, F., and Nebout, J. F., "Recent Progress in the Photoinitiated Cationic Polymerization of Cycloaliphatic Epoxy Systems," *Proceedings of RadTech Europe '97*, Lyon, France, 16–18 June 1997, p. 303.

[69] Manus, P. J. M., "Cationic UV-Curable Coatings: Contribution of Acrylate Oligomers to Final Film Properties," *Proceedings of Radiation Curing of Polymers II*, Manchester, England, 12–14 Sept. 1990, p. 284.

7 Vinyl Ethers

Introduction

VINYL ETHERS HAVE BEEN known since the early 1900s, but a commercial type synthesis for them and their polymers was not described until the late 1920s [1]. However, it was not until the early 1980s that the vinyl ethers were first suggested as new monomers for cationic radiation curing [2]. In the same decade after that first paper by Crivello was published, other studies were undertaken to show the viability of vinyl ethers in ultraviolet radiation curing [3–5] and electron beam curing [6,7].

The first practical synthesis of vinyl ethers was carried out by Reppe, who reacted monomeric hydroxyl-containing compounds with acetylene [8]:

$$X{-}R{-}OH + HC{\equiv}CH \underset{KOH}{\overset{\Delta}{\rightleftharpoons}} CH_2{=}CH{-}O{-}R{-}X$$

wherein *R* is an aliphatic, hydroaromatic, aromatic radical, or aralkyl radical and *X* is hydrogen, carboxyl, hydroxyl (and so on) group. It is readily apparent that when *X* is a hydroxyl group, divinyl ethers will be formed. Even though this method was useful, it was limited in yield, due to the fact that it is an equilibrium process and in certain instances cyclic acetals can form rather than the desired product.

Crivello [2] has described a laboratory process that can be used to synthesize vinyl ethers. In this process, a hydroxyl-containing compound, such as an alkane diol, can be reacted at moderate temperatures with 2-chloroethyl vinyl ether in the presence of sodium or potassium hydroxide and a phase transfer catalyst such as a tetra-alkyl ammonium salt.

$$CH_2{=}CH{-}O{-}(CH_2)_2{-}Cl \quad + \quad X{-}R{-}OH \quad + \quad KOH \xrightarrow[60\text{-}70^\circ C]{\text{PHASE TRANSFER CATALYST}}$$

$$CH_2{=}CH{-}O{-}(CH_2)_2{-}O{-}R{-}X \quad + \ NaCl \quad + H_2O$$

With this procedure, it was possible to synthesize a wide variety of mono-, di-, tri-, and higher functional vinyl ethers.

Vinyl ethers are known to homopolymerize rapidly in the presence of cationic compounds [9], and care must be taken to moderate the polymerization and decrease the opportunity for an explosive reaction [1]. The chemistry involved in this high reactivity is due to the fact that vinyl ether unsaturation is very electron rich, and because of electron character, vinyl ethers will very readily undergo acid-initiated polymerization. Because of the high reactivity, homopolymerizations are carried out in solution where the solvent can dissipate the heat of polymerization. In the radiation-cure polymerization of thin-film coatings, homopolymerizations probably would be successful, since the substrate would absorb the heat of polymerization.

In contrast to this behavior, vinyl ethers have only sluggish reactivity with free radicals, and usually, only low molecular weight products are formed. However, vinyl ethers will readily copolymerize with electron acceptor monomers in a free radical environment. Illustrative examples of such copolymers include methyl vinyl ether/maleic anhydride copolymers [10], methyl vinyl ether/ethylene copolymers [11], and isobutyl vinyl ether/vinyl acetate [12].

The main radiation-polymerization or co-polymerization mechanisms for vinyl ethers are free radical, cationic, hybrid, and charge-transfer reactions. A number of commercial and developmental vinyl ethers are listed in Table 7.1.

Free Radical Polymerization

Free radical radiation polymerization of vinyl ethers is initiated through a free radical, FR•, that is formed from a photoinitiator in the presence of suitable radiation. The initiated vinyl ether then propagates through its ethylenic unsaturation as it does with acrylates. The same fragmentation and hydrogen-abstraction photoinitiators as are used with acrylates are used in vinyl ether systems.

$$n\ CH_2{=}CH(O{-}R{-}X) \xrightarrow[\text{Free Radical Photoinitiator}]{h\nu} FR\cdot + n\ CH_2{=}CH(O{-}R{-}X) \xrightarrow{\text{Initiation}} FR{-}CH_2{-}\overset{H}{C}\cdot(O{-}R{-}X) + (n-1)\ CH_2{=}CH(O{-}R{-}X)$$

$$\xrightarrow{\text{Propagation}} FR{-}CH_2{-}\overset{H}{C}(O{-}R{-}X){-}\left[CH_2{-}\overset{H}{C}(O{-}R{-}X)\right]_{(n-2)}{-}CH_2{-}\overset{H}{C}\cdot(O{-}R{-}X) \longrightarrow \text{TERMINATION BY COMBINATION OF VARIOUS SPECIES}$$

TABLE 7.1. A Tabulation of Some Commercial and Developmental TSCA Listed Vinyl Ethers [13].

Vinyl Ether	Structure	Viscosity, cP @ 25°C	Acute Oral Toxicity, mg/kg	Dermal Toxicity, mg/kg	Skin Irritation
Butanediol divinylether	$CH_2{=}CH{-}O{-}CH_2{-}CH_2{-}CH_2{-}CH_2{-}OH$	No data	No data	No data	No data
Cyclohexanedimethanol divinylether	$CH_2{=}CH{-}O{-}CH_2{-}(\text{cyclohexane-1,4-diyl}){-}CH_2{-}O{-}CH{=}CH_2$	5.0	>5,000	>2,000	Moderate
Diethyleneglycol divinylether	$CH_2{=}CH{-}O{-}CH_2{-}CH_2{-}O{-}CH_2{-}CH_2{-}O{-}CH{=}CH_2$	No data	No data	No data	No data
Dodecylvinylether	$CH_2{=}CH{-}O{-}CH_2{-}(CH_2)_{10}{-}CH_3$	2.9	7,500	No data	No data
Hydroxybutylvinylether	$CH_2{=}CH{-}O{-}CH_2{-}CH_2{-}CH_2{-}CH_2{-}OH$	5.4	2,050	No data	Mild
Propenylether of propylene carbonate	$CH_3{-}CH{=}CH{-}O{-}CH_2{-}(\text{cyclic carbonate ring: }O{-}C({=}O){-}O)$	5.0	5,000	No data	Nonirritating
Triethyleneglycol divinylether	$CH_2{=}CH{-}O{-}CH_2{-}CH_2{-}O{-}CH_2{-}CH_2{-}O{-}CH_2{-}CH_2{-}O{-}CH{=}CH_2$	2.67	>5,000	>2,000	Minimal

It can be easily seen that if *X* is a vinyl or divinyl group, a highly cross-linked system will readily develop. As mentioned earlier, this is a sluggish reaction and the curing or homopolymerization of vinyl ethers by use of free-radical radiation chemistry is not usually commercially done.

Cationic Polymerization

The cationic or Bronsted acid polymerization of vinyl ethers takes place through the usual steps of initiation, propagation, and termination [13]. An onium salt photoinitiator and radiation are used to generate the protonic acid, HMF_6, as was described for the cationic polymerization of cycloaliphatic epoxides. This acid then interacts with the double bond on a vinyl ether molecule to form the initiated species, which then reacts with other vinyl ether molecules in the propagation step. Polymerization, as was the case with the cationic polymerization of epoxides, takes place until all monomer has been consumed.

$$\left[\begin{array}{l}\text{ARYLSULFONIUM OR IODONIUM HEXA-}\\ \text{FLUOROMETALLIC SALT PHOTOINITIATOR}\end{array}\right] + h\nu \longrightarrow HMF_6 + FR\cdot$$

$$\underset{\substack{|\\O\\|\\R\\|\\X}}{CH_2{=}CH} \xrightarrow[\text{Free Radical Photoinitiator}]{h\nu} n\ \underset{\substack{|\\O\\|\\R\\|\\X}}{CH_2{=}CH} \xrightarrow[\text{tion}]{\text{Initia-}} FR{-}CH_2{-}\underset{\substack{|\\O\\|\\R\\|\\X}}{\overset{\substack{H\\|}}{C}}\cdot + (n-1)\ \underset{\substack{|\\O\\|\\R\\|\\X}}{CH_2{=}CH}$$

Propagation

$$\begin{array}{l}\text{TERMINATION BY}\\ \text{COMBINATION OF}\\ \text{VARIOUS SPECIES}\end{array} \longleftarrow FR{-}CH_2{-}\underset{\substack{|\\O\\|\\R\\|\\X}}{\overset{\substack{H\\|}}{C}}\left[CH_2{-}\underset{\substack{|\\O\\|\\R\\|\\X}}{\overset{\substack{H\\|}}{C}}\right]_{(n-2)}CH_2{-}\underset{\substack{|\\O\\|\\R\\|\\X}}{\overset{\substack{H\\|}}{C}}\cdot$$

When monomer is depleted, the molecule may stay active or it may be neutralized if it comes in contact with an alkaline material.

The early study of cationic photocuring [2] demonstrated that various arylsulfoniumhexafluoroarsenates would cure diethylene glycol divinyl ether at rates of 150–300 ft/min (~46–92 m/min). Arylsulfonium hexafluorophosphate salts and arylsulfonium perchlorate salts cured the monomer at

500 ft/min (~152 m/min) when a concentration of 0.062 weight percent photoinitiator was used. As the concentration of photoinitiator was decreased, the phosphate salt yielded a more rapid cure rate than the perchlorate salt. At a concentration of 0.015%, the cure rate with the phosphate salt was 200 ft/min (~61 m/min), which was twice as fast as the 100 ft/min (~30.5 m/min) with the perchlorate. When three divinyl ethers, diethylene glycol divinyl ether, butanediol divinyl ether, and hexanediol divinyl ether, were individually blended with a triacrylate, trimethylolpropane triacrylate, the cure time was markedly decreased from >240 s for the acrylate alone to 5 s or less when 20% of the vinyl ether was added. The photoinitiator, diphenyl-4-thiophenoxyphenylsulfonium hexafluorophosphate, was cationic in nature, but also generated free radicals. In the case of epoxide/vinyl ether blends and using the same photoinitiator, the same general trend in cure time was observed.

Schrof and coworkers [14] found that triethylene glycol divinyl ether was an excellent reactive diluent for 3,4-epoxycyclohexylmethyl-3,4-epoxycyclohexane carboxylate. It was far more effective in decreasing viscosity than 1,6-hexanediol diglycidyl ether. At a 10% level, the divinyl ether decreased the viscosity by about 50%, whereas the aliphatic epoxide decreased it by about 25%. Above about 20% divinyl ether there were only small changes in viscosity. Blends of the epoxide containing up to 20% of the divinyl ether were formulated with an arylsulfonium hexafluorophosphate cationic photoinitiator and cured with ultraviolet radiation. They found that, in addition to being an excellent reactive diluent, the vinyl ether improved reactivity, which resulted in improved methylethyl ketone and ethanol resistance. The optimum divinyl ether concentration appeared to be in the 10–20% concentration range.

A listing of available mono and polyfunctional vinyl ethers, epoxides, and glycidyl ethers has been detailed [15]. A number of these compounds are only available in small quantities, but the listing is useful for those interested in evaluating the effect of molecular structure—alkyl, cyclic, aromatic, aliphatic; type of functionality—cycloaliphatic, glycidyl, vinyl ether, etc.; the number of functional groups present; and molecular weight. All compounds listed were thought to be polymerizable by cationic means.

Cyclic vinyl ethers of the 3,5-dihydropyran-2-methanol family have also been used in ultraviolet radiation-cured formulations using diazonium salts as the photoinitiator [16]. The simplest member of the family is shown below.

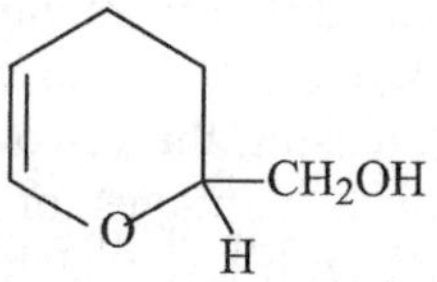

CH_2OH
O H

3,4-Dihydropyran-2-methanol

When a 3,4-dihydropyran-2-methyl-3,4-dihydropyran-2-carboxylate was cured with ultraviolet radiation in the presence of a diazonium salt, a hard film (3H pencil hardness) with excellent solvent resistance ($>$100 acetone double rubs) was obtained. Blends based on 2-methoxytetrahydropyran and cycloaliphatic epoxides [17] and similar cyclic vinyl ethers with urethane acrylates [18] are known to give films with improved impact resistance and flexibility. These compounds are all based on acrolein chemistry that forms dimers and then can form tetramers in a Tischenko reaction.

2 [dihydropyran–CHO] —TISCHENKO REACTION→ [dihydropyran–C(=O)–O–CH_2–dihydropyran]

Acrolein Dimer — Acrolein Tetramer

Cyclic vinyl ethers such as these can be combined with urethane acrylates and cured with ultraviolet radiation [19].

Hybrid Polymerizations

It has been found that vinyl ether/acrylate free radical polymerizations can be markedly increased, if a combination of free radical and cationic photoinitiators are used [20–22]. The reason given for the enhanced reactivity is that free radicals are generated from the cationic photoinitiator and from the free radical photoinitiator that will initiate polymerization of acrylate functionality, as well as some vinyl ether functionality. In addition, the cations initiate a rapid polymerization of the vinyl ether molecules. Most of the work in these studies centered around two vinyl ether molecules—triethylene glycol divinyl ether and 1,4-cyclohexane dimethanol divinyl ether (Table 7-1)—and an acrylated epoxide. The investigations pointed out that low molecular weight acrylate diluent monomers could be eliminated from acrylate formulations. That is, the divinyl ethers could be substituted for the acrylate reactive diluents in formulations that contained high viscosity acrylated epoxides and urethane acrylates [23]. In addition, the systems cured rapidly and did not require nitrogen inerting. Both ultraviolet and electron beam radiation-cured systems resulted in films with excellent physical properties. A somewhat similar study [24] indicated that triethylene glycol divinyl ether could be used as a replacement reactive diluent for compounds such as tripropylene glycol diacrylate for decreasing the viscosity of epoxy acrylates or unsaturated polyesters. The main advantages for the vinyl ether are good miscibility and diluency, coupled with low toxicity. Other advantages of vinyl ether coatings are listed in Table 7.2 [25]. In the acrylate formulations, up to

TABLE 7.2. Advantages of Vinyl Ether-Based Coatings [25].

Good miscibility with other compounds
High conversion of vinyl functionality
High cure speeds
High vinyl group conversion
Low odor (many compounds, but not all)
Low toxicity
Low viscosity, which provides good diluency
Non-acrylate technology
Ultraviolet irradiation is not oxygen inhibited

10% of the divinyl ether was completely converted. In the polyester formulations, up to 50 mole percent could be used.

In a continuing effort to develop monomers for photoinitiated, cationic polymerization, a series of "hybrid" monomers have been synthesized [26]. The goal in investigating these hybrid monomers was to prepare new more reactive monomers that contain both a cycloaliphatic epoxide group and a vinyl ether group. When these dual-function monomers and monomers containing only one of the functional groups were cured using a diaryliodonium hexafluoroantimonate and ultraviolet radiation, the hybrid compounds were shown to be markedly more reactive. Especially reactive was a novel monomer that contained both cycloaliphatic epoxide functionality and 1-propenyl ether functionality. The acceleration of the polymerization was attributed to two different effects. First, there is a rapid conversion of growing vinyl ether chains into epoxide chain ends. This results in a preferential polymerization of the epoxide groups. Second, the vinyl ether functionality interacts with the photoinitiator and, through a redox process, generates a large number of propagating cationic species. Hybrid monomers that contain the 1-propenyl vinyl ether group and a glycidyl ether group have also been investigated [27,28]. Since vinyl ethers have infrared absorption bands due to the C=C group at 1616 and 1635 cm^{-1}, they can be used to follow disappearance and reaction of the vinyl ether [6,25]. The 1616 cm^{-1} group lies in a clear or unobstructed portion of the spectrum and is easier to follow than the 1635 cm^{-1} band, which overlaps the response of other groups.

Donor/Acceptor or Charge-Transfer Polymerizations

The probability of free radical reactions taking place between two different molecules is increased when there is a large difference in charge between the different molecules. Another way of saying this is that an electron rich molecule will donate an electron to an electron deficient molecule making a reaction take place. A mechanism of this type is known as a charge-transfer or donor (electron rich)/acceptor (electron deficient) reaction. Since vinyl ethers (donor) are very electron rich, they will readily react with electron deficient compounds, such as maleic anhydride (acceptor), to form 1:1 alternating vinyl ether/maleate copolymers [29] when irradiated in the presence of a free radical-generating photoinitiator that yields the free radical FR•.

$$n\ \text{Maleate} + n\ CH_2{=}CH{-}O{-}R{-}X \xrightarrow[\text{FREE RADICAL-GENERATING PHOTOINITIATOR TO YIELD FR}\bullet]{h\nu}$$

Maleate   Vinyl Ether

Vinyl ether/maleate alternating copolymer

The propagating free radical species then terminates by means of a combination reaction, as described earlier. If *X* is a vinyl or divinyl group, cross-linked compositions can be obtained. These non-acrylate formulations offer "lower toxicity," lower irritation systems [30].

N-substituted maleimide and vinyl ethers are an interesting donor/acceptor pair in that the maleimide acts not only as a comonomer, but also as a photoinitiator. Such combinations do not have photoinitiator residues in the final film, which is a factor that may have importance in certain applications. The reaction between typical components takes place as described below:

$$\text{Hydroxypentyl maleimide} + CH_2{=}CH{-}CH_2{-}(CH_2)_2{-}CH_2OH \longrightarrow \text{Alternating copolymer}$$

Hydroxypentyl maleimide   Hydroxybutyl vinyl ether   Alternating copolymer

Previous studies indicated that transferable hydrogen atoms are very important to the initiation process leading to polymerization, but a recent study undertook investigation of the process in the absence of readily available transferable hydrogen atoms [31]. Various vinyl ethers and maleimides were used, and the radiation source was a high-pressure mercury-xenon lamp. Results of the photolysis indicated that quenching of the maleimide by the vinyl ether takes place by means of an electron transfer process that is independent of the presence of transferable hydrogen atoms. However, real-time FTIR [32] results indicated labile or transferable hydrogen atoms are required for rapid polymerization [31,33]. The latter factor indicated that the anionic radical of the maleimide produces the free radicals capable of initiating rapid polymerization by means of proton transfer. Laser flash transient spectroscopy has been used to carry out a mechanistic investigation dealing with formation and lifetimes of the N-alkylmaleimide triplet states [34].

Maleimide/acrylate systems containing sensitizer mixtures have been studied [35]. A typical sensitizer system was a benzophenone/N-ethylmaleimide/diethanolamine combination that was used to enhance the polymerization rate of 1,6-hexanediol diacrylate. Both monofunctional and multifunctional maleimides have been studied [36]. Other sensitized systems involving multifunctional acrylates and N-substituted maleimides indicate that isopropylthioxanthone, 4-benzoylbiphenyl, and benzophenone are effective sensitizers for acrylate polymerization [37]. Amine synergists were employed in the studies. The synthesis of novel bismaleimides and their cure characteristics with styrene and vinyl ethers are available [38], as well as studies with charge-transfer complexes [39]. Other investigations have looked at the structure of the amines used to initiate polymerization [40].

Unsaturated polyesters are usually reacted with styrene to form useful materials, as was described in Chapter 5. The double bonds contained in unsaturated polyesters are electron deficient and will act as acceptors when reacted with vinyl ethers in a charge-transfer reaction [41]. The best results are obtained when a one-to-one ratio of vinyl ether to unsaturated polyester double bonds are used. In addition, it has been found that the Norrish Type I free radical photoinitiators that function by alpha-cleavage of alkylaryl ketones are preferred over the Norrish Type II photoinitiators that require a bimolecular reaction to take place. This technology provides a basis for nonstyrene containing unsaturated polyester systems.

It is preferred that the vinyl ether content be less than 50 mole percent, since no homopolymerization of the vinyl ether was found in a study by Schwalm and coworkers [22]. These investigators used a hydroxylalkyl phenyl ketone and an onium salt as the photoinitiators. They blended various vinyl ethers with an ethylene glycol/fumaric acid unsaturated polyester and cured the systems with ultraviolet radiation. The analysis suggested that vinyl ethers tended to undergo an alternating copolymerization with the polyester. Cured films had the same properties as one would obtain with acrylates, but the vinyl ethers provided better diluting properties and scratch resistance coupled with higher reactivity. The use of vinyl ethers with unsaturated polyesters to prepare radiation-curable compositions was also disclosed much earlier [42]. These systems have various benefits, such as formulations that do not contain styrene, and thus, offer greater formulation latitude, and that can be completely acrylate free while offering versatile chemistry for innovation.

Oligomeric Vinyl Ethers

Oligomeric vinyl ethers provide increased formulating latitude for vinyl ethers and for cationic radiation curing in general. Such oligomers will increase cure speed, since fewer reactions are required to achieve a cross-linked or fully cured coating, and they have a low odor character. Oligomers based on urethane technology or polyester technology will impart the broad range of desirable properties usually associated with the technologies, both of which allow a wide variety of products to be made.

Urethane Vinyl Ethers

Vinyl ether terminated urethane compounds have been prepared by reacting hydroxyl terminated vinyl ethers with multifunctional isocyanates, such as toluene diisocyanate, 1,4-dicyclohexylmethyl diisocyanate, and the like [43–46][1]. When

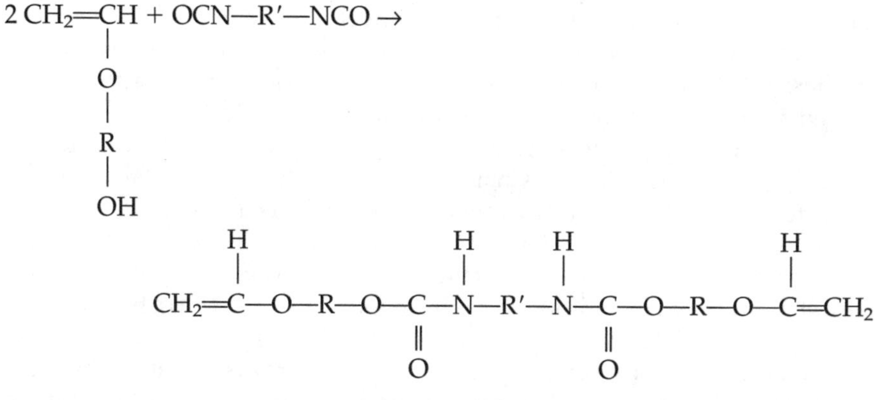

$$2\,CH_2{=}CH(\text{—O—R—OH}) + OCN\text{—}R'\text{—}NCO \rightarrow$$

$$CH_2{=}CH\text{—O—R—O—}C({=}O)\text{—}NH\text{—}R'\text{—}NH\text{—}C({=}O)\text{—O—R—O—}CH{=}CH_2$$

the oligomer is formulated with onium salt photoinitiators and used in ultraviolet and electron beam curable formulations, these compounds cure to clear, tack-free coating compositions. In addition to the simplistic urethane vinyl ether described above, other urethane vinyl ethers have been described [7]. In the above expression, *R′* can take on a wide variety of characteristics that would include the isocyanate-terminated prepolymers of polylactone, polyester,

[1] Vinyl oligomers were introduced by Allied Signal Corporation at the RadTech '90-North America conference that was held in Chicago during 25–29 March 1990 [36]. These oligomers were marketed as the VEctomer™ product line. Recently (June 2000), Morflex, Inc. of Greensboro, NC took over and is commercializing this product line [47].

polyether, acrylate, vinyl, and other polyols, as well as such prepolymers from low molecular weight diols, triols, tetraols, and higher functionality hydroxyl-containing compounds. In effect, the myriad of compounds possible for polyurethane or urethane acrylate synthesis can be made with vinyl ether termination and used in cationic and free radical radiation formulations.

A urethane vinyl ether based on triethylene glycol divinyl ether reacted with 1,4-diphenylmethane diisocyanate was formulated with an arylsulfonium hexafluoroantimonate photoinitiator (2%) and then cured with ultraviolet radiation [7]. The resultant coatings were hard (2H Pencil), solvent resistant (>100 methylethyl ketone double rubs), and impact resistant (>160 in.-lbs, >0.202 N·m, forward and >50 in.-lbs, >0.063 N·m, reverse). As might be expected with the aromatic diisocyanate used, there was a yellowish discoloration after irradiation. When the same urethane vinyl ether was exposed to electron beam radiation without a photoinitiator present, no cure was obtained. However, when the photoinitiator was added at the same 2% level and exposed to the electron beam radiation, cure was obtained and hard (3H), solvent resistant, impact resistant coatings resulted.

Jönsson and coworkers [48] have investigated the initiation of vinyl ethers with a cationic species derived from the redox reaction between onium salt photoinitiators and excited state photosensitizers or free radicals generated from photolysis. They [13,49] have also generated a table of acceptors and donors that would be useful for those interested in studying charge-transfer reactions.

Ester Vinyl Ethers

Series of ester vinyl ethers were prepared by reacting 4-hydroxybutyl vinyl ether or 1,4-cyclohexane dimethanol monovinyl ether with various multifunctional carboxylic acid derivatives [50]. For example, bis(4-vinyloxybutyl)adipate was prepared by transesterifying 4-hydroxybutyl vinyl ether with dimethyl adipate in the presence of titanium tetraisopropoxide and removing methanol.

$$2\ CH_2{=}CH{-}O{-}(CH_2)_4{-}OH \quad + \quad CH_3{-}O{-}\underset{\displaystyle O}{\underset{\|}{C}}{-}(CH_2)_4{-}\underset{\displaystyle O}{\underset{\|}{C}}{-}O{-}CH_3 \xrightarrow[\text{HEAT}]{\text{CATALYST}}$$

4-Hydroxybutyl vinyl ether — Dimethyladipate

$$\left[CH_2{=}\underset{H}{\underset{|}{C}}{-}O{-}(CH_2)_4{-}O{-}\underset{\displaystyle O}{\underset{\|}{C}}\right]{-}(CH_2)_4 \quad + \quad 2\quad CH_3OH$$

Bis(4-vinyl oxybutyl)adipate — Methanol

The above adipate had a room temperature viscosity of less than ten centipoises, and other synthesized compounds ranged from similar viscosities or low viscosities to low melting solids. The compounds were good diluents for a vinyl ether-terminated urethane oligomer, and when such blends were cured with ultraviolet or electron beam radiation, strong, flexible, solvent resistant coatings resulted.

Vinyl Ether-Silicone Blends

In the release coatings area, it has been well known for a number of years that many commercial radiation curable release coatings are based on epoxide-functional polydimethylsiloxane polymers, in which some of the methyl groups have been replaced with cycloaliphatic epoxy, acrylate, or other radiation reactive moieties [51–55]. Radiation curing offers this huge volume business a cost effective way to cure the thin films that are used for release purposes. Eckberg and coworkers [56] investigated blends containing two levels of various vinyl ethers in epoxy-modified polydimethylsiloxane polymers. The purpose was to determine the viscosity reducing power of the vinyl ethers, their cure response, the application characteristics of the blends, and the release characteristics of the cured films. The vinyl ethers investigated were the monovinyl ether of 2-ethyl-1-hexanol, the monovinyl ether of 1-dodecanol, and the divinyl ether of 1,4-cyclohexanedimethanol. Other investigations [57] dealing with formulations have demonstrated enhanced shelf stability of catalyzed formulations when they contain vinyl ethers, and that these additives have utility in areas other than release coatings as, for example, in an end use where smooth, low-friction topcoats are needed. It was also found that vinyl ethers improve miscibility of photoinitiators, photosensitizers, and other additives in silicone-based coating systems.

In general, the cationic photoinitiators have poor solubility/miscibility in the type siloxane polymers used for release coatings. For example, a mixture of bis(dodecylphenyl)iodonium hexafluoroantimonate/isopropylthioxanthone mixture in alkylglycidyl ether diluent will form a turbid, hazy mixture that phase separates within 24 h after blending [57]. The monofunctional vinyl ethers improved photoinitiator solubility in the epoxy-modified siloxane polymers. When used at a 20% level, clear and stable (>six months) blends were obtained. These monomers had similar dilution effects with 10% of the vinyl ether decreasing the viscosity by a factor of about two. The release properties of cured films indicated there were small but significant increases in release when the vinyl ethers were used and that the cured monomers do not cause release instability when aged. Another study [57] showed that a variety of photosensitizers, such as anthracene, di-iodofluorescein, and di-iodobutoxy fluorone, had improved solubility and increased cure speed in the presence of vinyl ethers.

Divinyl ethers will increase the release properties of modified dimethylsiloxane polymers [56,58]. The divinyl ether was miscible with the siloxane polymers, but it decreased the miscibility of the system with the cationic initiator. Thus, formulators should consider the use of both monovinyl and divinyl ether in products. Atomic Force Microscopy indicated that cross-linked domains of the divinyl ether were apparent in the cured films, and analysis of the results point out that the incompatible domains are important to film-release characteristics. Overall, the results indicate that cost-effective performance advantages can be obtained when vinyl ethers are blended and cured with the epoxy silicone release polymers. These include extended pot-/shelf-life stability, viscosity reduction and, thus, improved application characteristics, and release-property modification including coating smoothness and low friction characteristics.

Vinyl Ether-Epoxide Blends

Decker and coworkers [59] have developed an excellent review of the recent advances in radiation curing of vinyl ethers. It deals with the topics of the cationic polymerization of vinyl ethers and vinyl ether/epoxide blends, the free radical polymerization of vinyl ether/maleate blends, and the polymerization of vinyl ether/maleimide blends. It points out that the structure of the vinyl ethers has an influence on the reactivity, with the reactivity following this scheme:

Aliphatic Ether > Aromatic Ester > Aliphatic urethane >
Aromatic ester > Aromatic ether

when the compounds are irradiated with ultraviolet rays in the presence of an arylsulfonium hexafluorophosphate photoinitiator. An aryl iodonium hexafluoroantimonate was found to be the best photoinitiator for initiation of vinyl ethers.

Other facets of the study included the ultraviolet radiation curing of vinyl ether/epoxide (both aromatic and cycloaliphatic epoxides) blends, as well as numerous other factors. An interesting and surprising aspect of the investigation of vinyl ether/aromatic epoxide blends is that vinyl ether monomer actually inhibited the polymerization, rather than enhancing it as is the case with cycloaliphatic epoxides. This was particularly true in the presence of air or oxygen, even though the vinyl ether polymerization is not inhibited by air. Three potential explanations for the behavior were given:

- The vinyl ether's rapid polymerization resulted in a three-dimensional mass, and motion of the aromatic epoxide was restricted, resulting in very slow if any polymerization of the epoxide.
- The protonic acid species used for initiation was not available for use by the aromatic epoxide, even though they caused very rapid polymerization of the vinyl ether.
- The vinyl ether carbocation was inactive with respect to the aromatic epoxide, and even if an oxonium ion was formed, it would prefer to interact with the vinyl ether.

A real-time infrared analysis, a technique developed by Decker and Moussa [32], indicated that the vinyl ether rapidly polymerized, but even after 1 min. of irradiation, the aromatic epoxide content was essentially unchanged.

In related technology and in a search for non-acrylate radiation curable materials, the photoinitiated polymerization of 4-methylene-1,3-dioxolanes was investigated [60]. The results indicate that the methylenedioxolanes undergo rapid cationic polymerization and can be direct replacements for or complementary to vinyl ethers in photocure systems. A synthesis for the dioxolanes is given.

References

[1] Field, N. D. and Lorenz, D. H., "Vinyl Ethers," Ch. 7, *Vinyl and Diene Monomers*, Part 1, E. C. Leonard, Ed., Wiley-Interscience, New York, 1970, p. 365.

[2] Crivello, J. V., Lee, J. L., and Conion, D. A., "New Monomers for Cationic UV-Curing," *Proceedings of Radiation Curing VI*, Association for Finishing Processes of SME, Chicago, IL, 20–23 Sept. 1982, pp. 4–28.

[3] Dougherty, J. A., Vara, F. J., and Anderson, L. R., "Vinyl Ethers for Cationic UV Curing," *Proceedings of RadCure '86*, Baltimore, MD, 8–11 Sept. 1986, pp. 15-1.

[4] Dougherty, J. A. and Vara, F. J., "Triethylene Glycol Divinyl Ether as a Reactive Diluent for Cationic Curing," *Proceedings of RadCure Europe '87*, Munich, Germany, 4–7 May, 1987, p. 5–1.

[5] Dougherty, J. A. and Vara, F. J., "1,4-Cyclohexanedimethanol Divinyl Ether as a Reactive Diluent for Cationic Curing," *Proceedings of RadTech '88—North America*, New Orleans, LA, 24–28 Apr. 1988, p. 372.

[6] Lapin, S. C., "Electron Beam Activated Cationic Curing of Vinyl Ethers," *Proceedings of RadCure '86*, Baltimore, MD, 8–11 Sept. 1986, p. 15-15.

[7] Lapin, S. C., "Vinyl Ether Functionalized Urethane Oligomers: An Alternative to Acrylate-Based Systems," *Proceedings of RadTech '88—North America*, New Orleans, LA, 24–28 Apr. 1988, p. 395.

[8] Patent: Reppe, W., I. G. Farbenindustrie A.G., "Production of Vinyl Ethers," US 1,959,927. (1934).

[9] Vandenberg, E. J., *Journal of Polymer Science*, A-1, Vol. 4, 1966, p. 1609.

[10] Patent: Verberg, R. M., General Aniline & Film Corp., US 2,782,182. (1959).

[11] Patent: Strauss, H. W., E. I. du Pont de Nemours and Co., US 3,033,840. (1962).

[12] Patent: Merz, P. L., Witzel, F., and Burhans, A. S., Beech-Nut Packing Co., US 2,662,016. (1953).

[13] International Specialty Products, "RAPI-CURE® Vinyl Ethers: Reactive Agents for Radiation Curing Systems," Brochure RAP01.0995, 1995; Brochure 2302-296, (1996).

[14] Schrof, W., Binder, H., Schwalm, R., Weiguny, S., Chomienne, F., and Carroy, A., "Cationic UV Curing: Benefits of Formulating Epoxides and Vinyl Ethers Together," *Proceedings of RadTech Europe '99*, Berlin, Germany, 8–10 Nov. 1999, p. 95.

[15] Bloch, D. R., "Vinyl Ether and Epoxide Monomers for Radiation Curing," *Modern Paint and Coatings*, Vol. 84, No. 9, August 1994, p. 44

[16] Patent Application: Wang, A. E., Knopf, R. J., and Osborn, C. L., Union Carbide Corp., United Kingdom, GB 2,073,760 (1980).

[17] Patent: Koleske, J. V. and Kwiatkowski, G. T., Union Carbide Corp., US 4,977,199. (1990).

[18] Patent: Koleske, J. V. and Osborn, C. L., Union Carbide Corp., US 4, 920,156. (1990).

[19] Patent: Koleske, J. V. and Osborn, C. L., Union Carbide Corp., Blends of Cyclic Vinyl Ether Containing Compounds and Urethane Acrylates, US 4,920,156. (1990).

[20] Decker, C. and Decker D., *Proceedings of RadTech '94—North America*, Orlando FL, 1994, p. 602.

[21] Dougherty, J. A. and Vara, F., J., "Versatility Using Vinyl Ethers in Concurrent Cationic/Free Radical Formulations," *Proceedings of RadTech '90—North America*, Chicago, IL, Vol. 1, 25–29 Mar., 1990, p. 402.

[22] Schwalm, R., Binder, H., Funhoff, D., Lokai, M., Schrof, W., and Weiguny, S., "Vinyl Ethers in UV Curing: Copolymers with Acrylates and Unsaturated Polyesters," *Proceedings of RadTech Europe '99*, Berlin, Germany, 8–10 Nov. 1999, p. 103.

[23] Dougherty, J. A. and Vara, F., J., "Vinyl Ethers Show Versatility in Hybrid Radiation-Cure Coatings," *Modern Paint and Coatings*, Vol. 81, No. 6, June 1991, p. 42.

[24] Seufert, M., Binder, H., Funhoff, D. J. H., Lokai, M., Reich, W., Schrof, W., Schwalm, R., and Weiguny, S., "Vinyl Ethers in Radical Initiated UV Curing," *Proceedings of RadTech 2000*, Baltimore, MD, 9–12 Apr. 2000, p. 631.

[25] Lapin, S. C., *RADTECH Report*, Vol. 3, No. 4, July/August 1989, p. 17.

[26] Mowers, W. A., Rajaraman, S. K., Liu, S., and Crivello, J. V., "The Design and Photopolymerization of Monomers Bearing Epoxy and Vinyl Ether Groups," *Proceedings of RadTech 2000*, Baltimore, MD, 9–12 Apr. 2000, p. 45.

[27] Crivello, J. V. and Bi, D., *Journal of Polymer Science: Polymer Chemistry Edition*, Vol. 13, No. 12, 1993, p. 3109.

[28] Crivello, J. V. and Kim, W. G., Part A, Polymer Chemistry *Journal of Polymer Science*, Vol. 32, No. 9, 1994, p. 1639.

[29] Noren, G. K., Tortorello, A. J., and Vandenberg, J. T., "New Non- Radical Free Radical Radiation Cure Technology, New Maleate End-Capped Oligomers," *Proceedings of RadTech '90—North America*, Chicago, IL, Vol. 2, 25–29 Mar., 1990, p. 201.

[30] Schrantz, J., *Industrial Finishing*, Vol. 69, No. 5, May 1993, p. 20.

[31] Yang, D., Viswanathan, K., and Hoyle, C. E., "Studies of Initiation Mechanism of Maleimide/Vinyl Ether Copolymerizations," *Proceedings of RadTech 2000*, Baltimore, MD, 9–12 Apr. 2000, p. 221.

[32] Decker, C. and Moussa, K., "Real-Time Monitoring of Ultrafast Curing by UV-Radiation and Laser Beams," *Conference Papers, RadTech Europe '89*, Florence, Italy, 9–11 Oct., 1989, p. 1.

[33] Decker, C., Bianchi, C., Morel, F., Jönsson, S., and Hoyle, C., "Light Induced Polymerization of Photoinitiator-Free Donor/Acceptor Monomer Systems," *Proceedings of RadTech Europe '99*, Berlin, Germany, 8–10 Nov. 1999, p. 447.

[34] Jönsson, S., Viswanathan, K., Hoyle, C. E., Clark, S. C., Miller, C., Morel, F., and Decker, C., "Acceptor Monomers as Efficient Hydrogen Abstracting Photoinitiators," *Proceedings of RadTech Europe '99*, Berlin, Germany, 8–10 Nov. 1999, p. 461.

[35] Hoyle, C. E., Nguyen, C., Johnson, A., Clark, S. C., Viswanathan, K., Miller, C., Jönsson, S., Hill, D., Zhao, W., and Shao, L., "Sensitized Photopolymerization of Maleimide/Acrylate Systems," *Proceedings of RadTech Europe '99*, Berlin, Germany, 8–10 Nov. 1999, p. 455.

[36] Dias, A. A., Jansen, J. F. G. A., and van Dijck, M., "Mono and Multifunctional Maleimides as Photoinitiators in UV Curing Formulations," *Proceedings of RadTech Europe '99*, Berlin, Germany, 8–10 Nov. 1999, p. 473.

[37] Nguyen, C. K., "Sensitized Photopolymerization of Multifunctional Acrylates Using Substituted Maleimides," *Proceedings of RadTech 2000*, Baltimore, MD, 9–12 Apr. 2000, p. 196.

[38] Ericsson, J., Nilsson, M., Lundmark, S., Svensson, L., Jönsson, S., and Lindgren, K., "Synthesis and Photoinduced Copolymerization of Novel Maleimides combined with Vinyl- and/or Styrene-Ethers," *Proceedings of RadTech 2000*, Baltimore, MD, 9–12 Apr. 2000, p. 173.

[39] Cole, M. C., Bachemin, M., Jönsson, S., and Hall, H. K., "Photoinitiatorless Photopolymerizations That Form Charge Transfer Complexes," *Proceedings of RadTech 2000*, Baltimore, MD, 9–12 Apr. 2000, p. 211.

[40] Viswanathan, K., Hoyle, C. E., and Jönsson, S., and Hasselgren, C., "Structural Effects on the Efficiency of Amines in Initiating Acrylate Polymerization," *Proceedings of RadTech 2000*, Baltimore, MD, 9–12 Apr. 2000, p. 231.

[41] Noren, G. K., "Non-Acrylate Curing Mechanisms: Investigations of the Reactivity of Unsaturated Polyesters in the Amine-ene Reaction," *Proceedings of RadTech '90—North America*, Chicago, IL, Vol. 1, 25–29 Mar. 1990, p. 191.

[42] Patent: Prucnal, P. J. and Friedlander, C. B., PPG Industries, Inc., Process for Applying Polyester-acrylate Containing Ionizing Irradiation Curable Coatings, US 3,874,906. (1975).

[43] Patent: Lapin, S. C. and House, D. H., Allied-Signal, Inc., Vinyl Ether Terminated Urethane Resins, US 4,751,273. (1988).

[44] Lapin, S. C., "Radiation Induced Cationic Curing of Vinyl Ether Functionalized Urethane Oligomers," *Proceedings of the ACS Division of Polymeric Material: Science and Engineering*, Vol. 60, No. 233, Dallas, TX, Spring 1989.

[45] Brautigam, R. J., Lapin, S. C., and Snyder, J. R., "New Vinyl Ether Oligomers and Diluent Monomers for Cationic Curing," *Proceedings of RadTech '90—North America*, Chicago, IL, Vol. 1, 25–29 Mar. 1990, p. 99.

[46] Burlant, W. J., Plotkin, J., and Vara, F., "Vinyl Ethers: Versatile Monomers for Coating Applications," *Journal of Radiation Curing/Radiation Curing*, Vol. 18, No. 14, Fall 1991.

[47] News Bulletin of Morflex, Inc., "VEctomer® News–Vinyl Ether Monomers and Oligomers–Latest Developments," 110 High Point Road, Greensboro, NC, 30 June 2000.

[48] Jönsson, S., Sundell, P. E., and Hult, A., "Photoredox Induced Cationic Polymerization of Divinyl Ethers," *Proceedings of RadTech '90–North America*, Chicago, IL, Vol. 1, 25–29 Mar. 1990, p. 417.

[49] Jönsson, S., Sundell, P. E., and Hult, A., *Proceedings of RadTech '94—North America*, Orlando, FL, 1994, p. 194.

[50] Lapin, S. C. and Snyder, J. R., "Vinyl Ether Terminated Ester Monomers: New Reactive Diluents for Cationic Curing," *Proceedings of RadTech '90—-North America*, Chicago, IL, Vol. 1, 25–29 Mar. 1990, p. 410.

[51] Patent: Cavezzan, J. and Priou, C., "Cationically Crosslinkable Polyorganosiloxanes and Anti-adhesive Coatings Produced Therefrom," Rhone-Poulenc Chemie, US 5,340,898. (1994).

[52] Patent: Curatolo, B. S. and Fox, T. J., Avery Dennison Corp., Radiation-Curable Release Coating Compositions," US 5,888,649. (1999).

[53] Patent: Eckberg, R. P., General Electric Co., Ultraviolet Radiation-Curable Silicone Release Compositions with Epoxy and/or Acrylic Functionality," US 4,640,967. (1987).

[54] Patent: Eckberg, R. P., General Electric Co., Ultraviolet Radiation-Curable Silicone Release Compositions with Epoxy and/or Acrylic Functionality," US 4,576,999. (1986).

[55] Patent: Eckberg, R. P. and LaRochelle, R. W., General Electric Co., Ultraviolet Curable Epoxy Silicone Coating Compositions," US 4,279, 717. (1981).

[56] Eckberg, R. P., Krenceski, M., Mejiritski, A., and Neckers, D., "Radiation Cured Vinyl Ether-Silicone Blends: A Detailed Study," *Proceedings of RadTech Europe '99*, Berlin, Germany, 8–10 Nov. 1999, p. 507.

[57] Eckberg, R., Rubinsztajn, S., Krenceski, M., Hatheway, J., and Griswold, R., "Novel Radiation Curable Vinyl Ether-Epoxysilicone Compositions and Coatings," *Proceedings of RadTech 2000*, Baltimore, MD, 9–12 Apr. 2000, p. 69.

[58] Patent: Eckberg, R. P. and O'Brien, M. J., General Electric Co., "UV Curable Epoxysilicone Blend Compositions," US 5,650,453. (1997).

[59] Decker, C., Bianchi, C., Decker, D., and Morel, F., "Recent Advances in UV-Curing of Vinyl Ethers," *Proceedings of RadTech 2000*, Baltimore, MD, 9–12 Apr. 2000, p. 741.

[60] Davidson, S. R., Howgate, G. J., Lester, Fiona H. A., and Mead, C. J., "The Photoinitiated Polymerization of 4-Methyl-1, 3-Dioxolanes," *Proceedings of RadTech Europe '99, Berlin, Germany*, 8–10 Nov. 1999, p. 483.

8

Ultraviolet Radiation-Curable Powder Coatings

Introduction

POWDER COATING ROSE to prominence in the early 1970s when there was both an energy shortage accompanied with high costs and a move by the U. S. Environmental Protection Agency to reduce markedly the amount of certain organic solvents, now known as VOC or volatile organic content, in coating formulations. Powder coatings are solvent-free coatings that can be thermoplastic or thermoset in nature. Thermoplastic powders can be applied to substrates by fluidized bed or electrostatic spray techniques. These powders only require heat to liquefy the particles and make them flow into a smooth coating. The coated object then cools to ambient conditions to form the coating.

Thermoset powders have the same requirements as thermoplastic powders, plus a need for a cross-linking mechanism that would allow liquefaction and flow before cross-linking was effected to any large degree. The latter requirement is much easier to say than to accomplish. If cross-linking occurred before flow took place, the cured coating would have a sandpaper-like finish. The cross-linking systems were usually quite reactive in nature, and the powder manufacturer had to guard against reaction taking place during storage and transport. Even small amounts of reaction during storage could result in particles that would not flow. A variety of thermoset powder coatings exist, such as acrylic/anhydrides, blocked-isocyanate urethanes, catalyzed epoxides, vinyl chloride copolymers, and others, all of which require relatively long bake times of 30 min. at about 120–160° C. The molten, flowed, and cured powder coatings provided end users with high quality coatings, and this technology has taken a firm foothold in the coating industry during the past three decades. In 1995, powder coatings comprised about 6% of the North American coating market [1] and about 11% of the international coating market [2].

In a general sense, powder coatings offer several advantages. These include the facts that they are solvent free, and thus, there is very low or near zero nil VOC, application of one coat yields a high quality coating, and powder over spray can be reclaimed and recycled. However, these coatings also have several disadvantages. These include difficulty in forming thin coatings, high temperature and long times needed for flow and for cure, and an inability to coat heat sensitive substrates. If powders could be made to flow at low temperatures and cross-link rapidly at low temperatures, there is a potential for a significant market share increase. The increase would be in the finishing of heat sensitive substrates, particularly in coating substrates such as plastics, electronic components, medium-density fiberboard, wood, and others. It is thought that coupling radiation curing with powder coating may help solve this difficulty and decrease energy requirements. Powder coatings would offer radiation curing the opportunity to have a system that has a lower skin and eye irritation potential, because of lower vapor pressure of the ingredients, as well as less shrinkage during cure, since fewer reactions would be taking place. *Coatings World* [3] published an article that indicated a new product line of radiation-curable powder coatings was available. These powder coatings would cure by holding them 2 min. at 100° C in the melt-and-flow step of the process, and then following this by a 5-s exposure to ultraviolet radiation to effect the cross-linking portion of cure. As indicated above, a cure schedule such as this would be very useful for assembled components, plastics, wood, and other heat sensitive substrates.

In the last five years or so of the Twentieth Century, some people in the coating industry began to talk about and develop radiation-curable powder coatings [4]. The main idea was to develop a thermoset powder system that would have a cure mechanism that was basically independent of the long bake times required for thermoset thermal cures. The powder coating would act as a thermoplastic powder coating. That is, when the powder was exposed to thermal energy, it would liquefy (melt if crystalline) and flow into a smooth film without cross-linking [5]. However, to be a thermoset powder, the system would need to have functional groups and these groups would:

- Be essentially independent of temperature,
- React and lead to cross-linking at a temperature at or near the liquefaction temperature,
- React while the powder was still in a liquid state to form a cross-linked coating.

The cross-linking would occur when the coating was subjected to a non-thermal and different energy source—namely radiation. It is important that essentially no cross-linking takes place until the liquid mass is irradiated.

The requirement of cross-linking in a liquid state ensures that the radiation-reactive portion of the powder has sufficient mobility to find other reactive or reacting groups and, thus, effect cross-linking. It has been reported that, above 80° C, there is sufficient mobility to ensure acceptable curing [6], but in another study it was found necessary to have the film in a molten form for effective curing [7]. Further, with very little or no curing taking place during the liquefaction process, the radiation-curable powder coating can have

TABLE 8.1. Advantages of Radiation-Cured Powder Coatings.

ADVANTAGE	RESULT OR REASON FOR ADVANTAGE
Very low VOC or nil solvent	Environmentally compliant system
Reduced health hazards potential from certain low molecular weight liquids	Higher molecular weight materials used
Good film build	Achieved with a one-coat application
Good flow at relatively low temperatures	Ability to coat heat sensitive substrates—wood, plastics, alloys, composite metal/plastic objects
Inability to react until subjected to radiation	Good stability during manufacture—melt mixing, extrusion, and powder formation
Good storage stability—thermally stable	Long shelf life and less concern about storage temperature
New technology potential	Vinyl ether/maleate systems [4]
Relatively thick films easily attained	Five mil (125 μm) clear coats [8]

the optimum time to flow and form a smooth coating without orange peel. Radiation also offers the potential for good storage stability, since photoinitiator photolysis and subsequent polymerization will not take place if the powder is shielded from radiation, and particularly from ultraviolet radiation. Storing the powder at reasonable temperatures is also important to storage stability and is a factor that depends on the nature of any particular formulated powder. Tables 8.1 and 8.2 are listings of certain advantages of and benefits obtained from radiation-cured powder coatings.

The result of these developments is an ultraviolet radiation-curable powder coating market segment, a segment that has taken the fancy of many in the industry over the past few years. For example, at the RadTech 2000 Exhibition, there was a complete powder coating/ultraviolet radiation-curing line in operation. At this RadTech Conference and at RadTech Europe held in

TABLE 8.2. Technological Benefits of Ultraviolet Radiation-Cured Powder Coatings.

BENEFIT	ACHIEVEMENT
Application	Electrostatic spray allows both flat and three dimensional objects to be coated
High utilization yields material savings and various economic benefits	Powder over-spray is easily collected and recycled to attain >95% utilization. Also energy and labor savings can be attained.
Outstanding protection qualities	Powder coatings meet many national and international standard specifications
Preparation and handling	Easier for powder than for liquid
Solvent free—ideal to meet environmental requirements	Almost no VOC (volatile organic content) emitted during cure cycle
Various finishes easily attained in one-step	From high to low gloss coatings with a broad variety of colors. Can also easily obtain hammer tone and textured finishes
Good sandability	Important to wood substrates—can be done immediately after radiation curing
Good general properties	Edge coverage, chemical resistance, and adhesion
Controllable final appearance	Controlled by varying heating profile

Berlin, Germany, in 1999, separate presentation sections devoted to this technology were made available to attendees.

Polymer Systems

One way to achieve powder coatings with good storage/transportation characteristics, a relatively low liquefaction temperature, and good flow characteristics is to use crystalline materials of low or moderate (oligomeric) molecular weight [9,10]. Crystalline polyesters with methacrylate end groups suitable for use as an ultraviolet radiation cured powder coating have been made by both a two-step and a three-step process [10]. In the two-step process, a carboxyl-terminated, crystalline polyester is first made in a bulk condensation polymerization process. This process involves reacting 1,4-cyclohexanedimethanol or other suitable diol with an excess of a suitable difunctional carboxylic acid, such as adipic acid, in the presence of a catalyst.

n $HO-CH_2-C_6H_{10}-CH_2-OH$ + (n + 1) $HOOC-(CH_2)_4-COOH$ →

1,4-Cyclohexanedimethanol — Adipic acid

$HOOC-(CH_2)_4-C(=O)-[O-CH_2-C_6H_{10}-CH_2-O-C(=O)-(CH_2)_4-C(=O)]_n-OH + H_2O$

Cyclohexanedimethanol adipate (crystalline polyester)

In the second step, the molten polyester is cooled to 140° C, at which point glycidyl methacrylate in an amount sufficient to react with all of the carboxyl groups, plus a small excess, is added, along with a suitable catalyst for the acid-glycidyl reaction.

$CH_2=C(CH_3)-C(=O)-O-CH_2-CH-CH_2$ (epoxide, O)

Glycidyl methacrylate

$CH_2=C(CH_3)-C(=O)-O-CH_2-CH(OH)-CH_2-O-C(=O)-(CH_2)_4-C(=O)-[O-CH_2-C_6H_{10}-CH_2-O-C(=O)-(CH_2)_4-C(=O)]_n-O-CH_2-CH(OH)-CH_2-O-C(=O)-C(CH_3)=CH_2$

Methacrylyl end-capped cyclohexanedimethanol adipate (crystalline polyester)

The final oligomeric product (number-average molecular weight of 5600 in a patent example [10] and "melting point" of 130° C) is a methacrylate-capped, crystalline polyester that functions well as an ultraviolet radiation curable powder coating.

The three-step process for producing similar crystalline polyesters might involve, first, making the same polyester from 1,4-cyclohexanedimethanol and adipic acid, but in this case, using an excess of the diol rather than the carboxylic acid. In this way, a hydroxyl-terminated polyester is formed. This is followed by reaction with an anhydride or a dicarboxylic acid, such as isophthalic acid to form a carboxyl-terminated polyester. The third step involves reaction with glycidyl methacrylate to form again the methacrylate-capped, crystalline polyester.

Crystalline vinyl ether-terminated oligomers have been prepared by reacting a glycol with a diisocyanate, preferably isophorone diisocyanate, to form isocyanate-terminated adducts, as described in the following reactions [11].

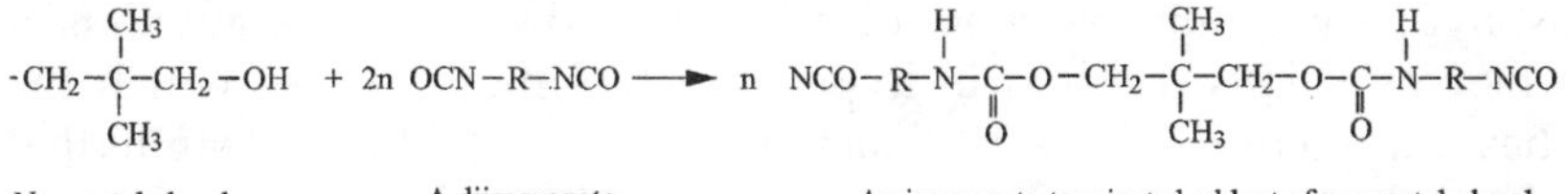

$$\text{-}CH_2-C(CH_3)_2-CH_2-OH + 2n\ OCN-R-NCO \longrightarrow n\ NCO-R-NH-C(=O)-O-CH_2-C(CH_3)_2-CH_2-O-C(=O)-NH-R-NCO$$

Neopentyl glycol — A diisocyanate — An isocyanate-terminated adduct of neopentyl glycol

This adduct is then reacted with a hydroxy vinyl ether such as 4-hydroxybutyl vinyl ether to form the crystalline vinyl ether-terminated adduct that is useful in ultraviolet radiation-cured powder coatings.

$$+\ CH_2=CH-O-(CH_2)_4-OH$$

4-Hydroxybutyl vinyl ether

$$n\ CH_2=CH-O-(CH_2)_4-O-C(=O)-NH-R-NH-C(=O)-O-CH_2-C(CH_3)_2-CH_2-O-C(=O)-NH-R-NH-C(=O)-O-(CH_2)_4-CH=CH_2$$

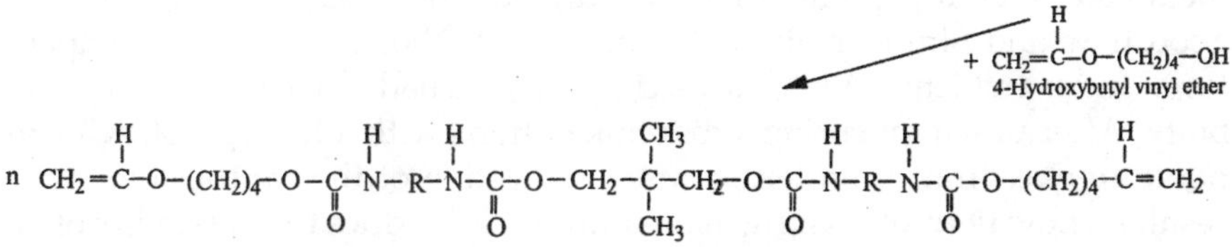

Vinyl ether end-capped crystalline urethane adduct of neopentyl glycol

This vinyl ether adduct was then melt blended with a commercial polyester based on fumaric acid, terephthalic acid, and 1,6-hexanediol, photoinitiators, flow control agent, and pigment to form an exudate. The exudate was chipped and ground into a powder that would pass through a 140-mesh screen. A cured coating from the powder on fiberboard was adherent and resistant to methylethyl ketone, had a pencil hardness of HB, and had 20°/60° gloss of 38/86.

Wood Substrates

Wood is a particularly temperature sensitive substrate to coat, and ultraviolet radiation-cured powder coatings may be particularly useful if formulations that cure at sufficiently low temperatures can be developed. In the wood industry, it is understood that heating wood above 93° C will be detrimental for many end uses. For example, both wood and wood products contain volatile components, and if these materials are heated above 93° C, the volatiles will try to escape and cause undesirable imperfections such as pits and bubbles into the coating. Clear coatings based on acrylated urethanes, unsaturated polyesters, free radical-type photoinitiators, and other additives [12,13] and on epoxides, vinyl ethers, polyols, cationic-type photoinitiators, and other additives [14,15] are described in the recent patent literature. These powder coatings will flow at temperatures of less than 100° C and can be used on temperature sensitive substrates such as wood.

An approach involving alteration of molecular architecture into a dendritic structure to obtain improved flow while achieving storage stability by use of crystalline areas in the polymer has been investigated [16][1]. Two partially crystalline, hydroxyl-functional, aliphatic polyesters with an average of 16 or 64 hydroxyl groups per molecule were reacted with various amounts of ϵ-caprolactone. This resulted in hyperbranched molecules in which the branches were comprised of ϵ-caprolactone units and of different lengths that were dependent on the amount of monomer used and the kinetics of the reaction. The poly-ϵ-caprolactone branches were of sufficient length (from an average of about 10 to 40 units) that they would be crystalline in nature. These dendritic structures were then reacted with methacrylic anhydride in a base-catalyzed transesterification reaction to produce methacrylate moieties that will respond to free-radical radiation curing. The resulting polymers had "melting" points that ranged from 36–50° C and exhibited very good flow and film formation at about 10–20° above the melting temperature. Under ambient temperatures, the polymers had satisfactory storage stability. After radiation curing, cross-linked films with a low ($\sim -50°$ C) temperature glass transition temperature indicative of the caprolactone units resulted. Low flow and curing temperatures also reduce the possibility of defects such as orange peel [8].

Metal Substrates

In a general sense, thermoplastic and thermoset powder coatings are well adapted to coating metal. However, the curing temperature is quite high and curing time is quite long. One study [17] pointed out that thermoset powders are cured at temperatures of about 140–250° C for times of 30 min. to about 1 min. when a convection oven, infrared plus convection oven, or infrared oven are used, with the longest times corresponding to the lowest temperatures. Induction heating also can be used to cure the coatings in less than a minute at temperatures of 240–300° C. The investigation used a commercial

[1] The introduction in this reference [16] contains an interesting historical section about polymer chemistry that may stimulate research in people with such a bent.

ultraviolet light curable powder coating of undisclosed composition and found that it needed a heating cycle of 1–2 min. in a convection oven at 90–140° C and an irradiation time that was measured in s. This powder was utilizable on metal, wood, plastics, and other heat sensitive substrates.

An unsaturated maleate polyester was combined with a crystalline vinyl ether and further formulated with a flow control agent, titanium dioxide, and a combination phosphine oxide/α-hydroxy ketone photoinitiator to form a radiation curable powder coating [8]. This powder plus another formed from an acrylated unsaturated polyester, flow control agent, and the same pigment/photoinitiator system were evaluated for adhesion and smoothness. Adhesion depended on photoinitiator concentration with 3% yielding good adhesion. With the first described formulation, acceptable smoothness could be obtained in 3 min. at 121° C, in 10 min. at 100° C, and in 15 min. at 93° C. At 85° C, it took more than 20 min. to obtain acceptable smoothness. The second formulation required a higher flow temperature and longer times to achieve similar degrees of smoothness, for example, 10 min. at 121° C.

Fiberboard Substrates

Results with a primer designed for medium-density fiberboard (MDF) and a topcoat designed for veneer-covered MDF, veneer-covered plywood, and Beech wood panels have been described [18]. These radiation-curable powder coatings are cationic in nature and based on formulations containing diglycidyl ethers of bisphenol-A. In addition to the temperature sensitive substrates, soft aluminum, hard aluminum, Bonderized steel, and tin-plated steel panels were coated and cured with good results.

Summary

With new technologies such as this, there is only limited information available in the literature regarding formulation, and even this information is in a state of flux. In addition to the specific technical details described above, interested readers may find useful details about polymeric species, photoinitiators, pigments, stabilizers, and additives in the literature [19–21]. Different ways of liquefying the powder include conventional ovens, dielectric chambers, infrared radiation, and infrared radiation coupled with conventional ovens [17]. Also under active investigation are the parameters of ultraviolet radiation lamps, the actual radiation dosage, and peak irradiance on the liquefied powder in the step independent of thermal heating [22]. Other studies are investigating how to optimize the curing of these powders by a differential photocalorimetry technique [23]. Preliminary studies indicate the manner in which titanium dioxide absorbs ultraviolet radiation and decreases the degree of cure. Also found was the reduced rate of cure at elevated temperatures was not due to photoinitiator loss but was related to a thermally dependent, reversible reaction equilibrium that was favored at high temperatures. For the system studied, the optimum rate of cure was found to be 110° C.

In addition to the use of radiation-curable powder coatings on wood and metals as discussed above, new areas are being investigated. These include overprint varnishes on cardboard and full color digital prints, and wallpaper and in the printing area where partial coating of paper substrates takes place [24]. The latter end use has the capability of opening up the copier toner market to this technology.

References

[1] Biller, K. M. and MacFadden, B., "The Ultimate Low Temperature Cure in Powder Coatings," *Modern Paint and Coatings*, Vol. 86, No. 9, September 1996, p. 34.

[2] Skinner, D., "UV Curable Powder Coatings–A Partnership for the Future," *Proceedings of RadTech Europe '99*, Berlin, Germany, 8–10 Nov. 1999, p. 599.

[3] Esposito, C. C., "The Powder Coatings Market," *Coatings World*, Vol. 6, No. 9, September 2001, p. 28.

[4] Witte, F. M., de Jong, E. S., and Misev, T. A., "UV Curable Powder Coating Systems: A Promising Future," *RADTECH Report*, Vol. 10, No. 5, September/October 1996, p. 13.

[5] Buysens, K., "UV Curable Powder Coatings, A New Coating Technology for MDF," *Proceedings of RadTech Europe '99*, Berlin, Germany, 8–10 Nov. 1999, p. 622.

[6] Udding-Louwrier, S., de Jong, E. S, and Baijards, R. A., *Proceedings of RadTech '98—North America*, Chicago, IL, 19–22 April, 1998, p. 106.

[7] Udding-Louwrier, S., Baijards, R. A., *Journal of Coatings Technology*, Vol. 72, No. 904, 2000, p. 71.

[8] Griese, C. and Carlson, B., "Developments in UV-Curable Powder Coatings," *Proceedings of RadTech 2000*, Baltimore, MD, 9–12 Apr. 2000, p. 658.

[9] Johansson, M., Falkén, H., Irestedt, A., and Hull, A., "On the Synthesis and Characterization of New Low Temperature Curing Powder Coatings Cured with Radiation," *Journal of Coatings Technology*, Vol. 70, No. 878, 1998, p. 57.

[10] Patent: Moens, L., Loutz, J. M., Maetens, D., Loosen, P., and Van Kerckhove, M., UCB S.A., Powder Composition of Crystalline Polyesters Containing End Methacrylyl Groups, US 5,639,560. (1997).

[11] Patent: Shah, N. B. and Daly, A. T., Solid Vinyl Ether Terminated Urethane Curing Agent, US 6,028,212. (2000).

[12] Patent: Biller, K. M. and MacFadden, B. A., Radiation Curing of Powder Coatings on Wood, *Herbert's Powder Coatings, Inc.*, US 5,824,373. (1998).

[13] Patent: Biller, K. M. and MacFadden, B. A., Radiation Curable Powder Coatings for Heat Sensitive Substrates, *Herbert's Powder Coatings, Inc.*, US 5,877,231. (1999).

[14] Patent: Biller, K. M. and MacFadden, B. A., Herbert's Powder Coatings, Inc., Radiation Curing of Powder Coatings on Heat Sensitive Substrates: Chemical Compositions and Processes for Obtaining Coated Workpieces, US 5,789,039. (1998).

[15] Patent: Biller, K. M. and MacFadden, B. A., Herbert's Powder Coatings, Inc., Radiation Curing of Powder Coatings on Heat Sensitive Substrates: Chemical Compositions and Processes for Obtaining Coated Workpieces, US 5,935,661. (1999).

[16] Hult, A., Johansson, M., Jansson, A., and Malmström, E., "UV Curable Powder Coatings Based on Hyperbranched Polymers," *Proceedings of RadTech Europe '99*, Berlin, Germany, 8–10 Nov. 1999, p. 634.

[17] Buysens, K., "UV Curable Powder Coatings: Benefits and Performance," *Proceedings of RadTech 2000*, Baltimore, MD, 9–12 Apr. 2000, p. 669.
[18] Reisinger, M., Dr., "UV-Curable Coatings for Temperature Sensitive Substrates," *Proceedings of RadTech Europe '99*, Berlin, Germany, 8–10 Nov. 1999, p. 628.
[19] Misev, L., Schmid, O., Udding-Louwrier, S., de Jong, E. S., and Baijards, R., *Journal of Coatings Technology*, Vol. 71, No. 891, 1999, p. 37.
[20] Bender, J., Laver, H., Lehmann, K., Margraf, R., and Schmid, O., "The Formulation of UV-Curable Powder Coatings," *Proceedings of RadTech Europe '99*, Berlin, Germany, 8–10 Nov. 1999, p. 615.
[21] Javanovic, Z., Lahaye, J., Laver, H., Megert, S., and te Walvaart, C., "UV-Curable Powder Coatings: A Guide to Successful Formulation," *Proceedings of RadTech 2000*, Baltimore, MD, Apr. 9–12, 2000, p. 687.
[22] Udding-Louwrier, S., Baijards, R. A., and Feima, N. W., "New Developments in Radiation Curable Powder Coating II," *Proceedings of RadTech Europe '99*, Berlin, Germany, 8–10 Nov. 1999, p. 607.
[23] Padaki, S. and Buehner, R. W., "Optimizing Cure of UV Powder Coatings Using Differential Photocalorimetry," *Proceedings of RadTech 2000*, Baltimore, MD, 9–12 Apr. 2000, p. 698.
[24] Udding-Louwrier, S., Baijards, R. A., de Jong, E. Sjoerd, and Binda, P. H. G., "Application of UV-Curable Powder Coatings on Paper-Like Substrates," *Proceedings of RadTech 2000*, Baltimore, MD, 9–12 Apr. 2000, p. 650.

9

Dual-Cure Mechanisms

THERE ARE TIMES when radiation curing alone may be limited in its ability to effect cure. The technique depends on "line-of-sight" contact of radiation with molecules that can be activated to initiate polymerization. If an object has shadowed areas, as for example, under the chips or other components on a printed circuit assembly, no cure will take place in the shadowed area unless a means of directing radiation under the component is devised. Similar difficulties can be encountered on other three-dimensional objects. The main reason for seeking a system that can be cured with radiation, followed by another means of curing or vice-versa, is to overcome or alleviate such difficulties, as well as to improve properties. The use of two or more mechanisms to effect cure is known as dual cure or hybrid cure. The usual goal of such cure systems is to take advantage of the extremely rapid cure rate of radiation induced curing and obtain a tack-free system, and ultimate properties are obtained in the second step, typically, a thermally induced cure.

In the previous chapter, one type of dual mechanism, radiation-cured powder coatings, was discussed. This system was devised to allow powder coatings, which usually cure at relatively high temperatures, to be used on temperature sensitive substrates, and thus, allow powder-coating technology to expand its use applications.

Free Radical/Cationic Systems

Cationic photoinitiators generate both cations and free radicals when they are irradiated. However, both of these mechanisms are activated by the ultraviolet radiation, and cationic and free radical curing take place simultaneously. Such a dual mechanism does not give the formulator control over the polymerization, nor does it do away with the line-of-sight requirement for curing. It broadens the scope of photocuring and expands the range of materials that can be used in cationic systems by making available all of the various acrylates or other ethylenically unsaturated molecules capable of rapid polymerization [1].

Investigators have found that mixtures of cycloaliphatic epoxides and aromatic epoxides can be combined with, for example, glycidyl acrylate, 2-ethylhexyl acrylate, diethylene glycol diacrylate, and trimethylolpropane triacrylate and an onium salt photoinitiator to form useful coating systems [2]. However, even though these mixtures would cure when exposed to ultraviolet radiation, it was found that if a free radical photoinitiator such as 2,2-dimethoxy-2-phenyl-acetophenone were added to the mixture, improved properties resulted. The coatings could be cured in air and hardness improved. In addition, thicker tack-free films could be made. In one instance, the thickness that could be obtained in a tack-free state increased from 0.19 cm to 0.25 cm when the free radical photoinitiator was present (exposure time was 105 s under a 5000 W arc lamp). Results similar to these were found in an independent investigation when carbonyl-type photoinitiators, such as benzophenone, were combined with onium salt photoinitiators [1].

Synergistic effects in film properties were found when a cycloaliphatic epoxide, polyol, and onium salt photoinitiator were combined with various multifunctional acrylates [3]. The improved properties depended on the particular acrylate used. In accordance with the above described studies, improved properties were obtained when a free radical photoinitiator, benzophenone, was incorporated into the formulations. Related and similar results have been found by other investigators [4].

Another example of this type of coating is given by a hybrid urethane acrylate-based pressure sensitive adhesive [5]. A polycarbonate urethane diacrylate oligomer was first blended with isobornyl acrylate and vinyl ether diluents. To this mixture, a cycloaliphatic epoxide, free radical and cationic photoinitiators, tackifiers, and an adhesion promoter were added. The ingredients were warmed to form a homogeneous mixture and cured to form a tacky coating that was useful as a pressure sensitive adhesive for digital optical discs.

Vinyl ethers have been combined with urethane acrylates and acrylated epoxides to form dual-cure coatings [6]. The vinyl ethers were good reactive diluents for the high viscosity acrylates, and the formulated coatings, which contained both a free radical and a cationic photoinitiator, cured rapidly. The systems did not require inerting or thermal post cure to achieve improved physical properties over those of the oligomeric acrylates alone.

Radiation/Thermal Cures

One of the early dual-cure systems involved ultraviolet radiation/thermally curable plastisols [7]. Plastisols are dispersions of solid polymer particles, often porous poly(vinyl chloride) particles made by a bulk polymerization process, in high-boiling or essentially non-volatile plasticizers. The plastisols were fluxed by heating at elevated temperatures. Plasticizers such as adipates, polyesters, and phthalates were used. Usually, the plastisols were highly filled and pigmented. They were made dual curable by replacing a portion of the plasticizers with a polyfunctional acrylate, such as trimethylolpropane triacrylate and adding both a free radical photoinitiator and a thermally activated initiator for ethylenically unsaturated double bonds. Thermal initiators were compounds such as benzoyl peroxide, azobisisobutyronitrile and the like, and these were combined with commercial-free radical photoinitiators. The final product contained about 20% polyfunctional acrylate. The modified plastisols were first flowed onto the substrate and then exposed to ultraviolet radiation to form a tack-free skin or surface cure on the coating to hold the system in place. They were then heated in a forced-air oven to the fusion temperature of the polymer/plasticizer and the decomposition temperature of the thermal initiator. The products were held at temperature for about 20–30 min. to allow fusion and thermal cross-linking. The cured products expanded the use range of plastisols to include automotive sealants and electronic encapsulants, such as conformal coatings and electrical component insulators. Research and product development in this area is ongoing.

Recently, a similar radiation/thermal dual-cure system was used in the production of a coated abrasive fabric [8]. Irradiating the sample began the free radical polymerization. Then, as the temperature increased sufficiently due to the heat of reaction, the thermal initiator was activated and curing was completed. Another free-radical system involved first cure of acrylates with radiation and then thermal cure of isocyanate/hydroxyl moieties in the system [9]. Cationic systems were also first partially cured with ultraviolet radiation and then cured by exposure to elevated temperatures [10]. Hydroxy-functional acrylates were combined with cycloaliphatic epoxides and an onium salt photoinitiator to form systems that would undergo both free radical and cationic photopolymerization. During the process, the hydroxyl groups also entered into the cross-linking scheme by chain transfer [11]. Concurrent cationic/free radical cure systems have also been discussed in a review paper [12].

Combinations of ultraviolet and infrared radiation curing systems were also suggested as a way to supply thermal energy to radiation cured systems [13]. Such combinations would be useful in the above described plastisols formulations, coated abrasive fabric, and free radical/cationic dual-cure systems.

Dual-cure systems can also be based on urethane acrylates that contain both acrylate and isocyanate functionality.

$$OCN{-}R{-}N(H){-}C({=}O){-}O\sim\sim\sim\sim\sim O{-}C({=}O){-}N(H){-}R{-}N(H){-}C({=}O){-}O{-}(CH_2){-}O{-}C({=}O){-}CH{=}CH_2$$

Free isocyanate end group and diisocyanate residue | Polyol residue | Diisocyanate residue | 2-Hydroxyethyl acrylate residue

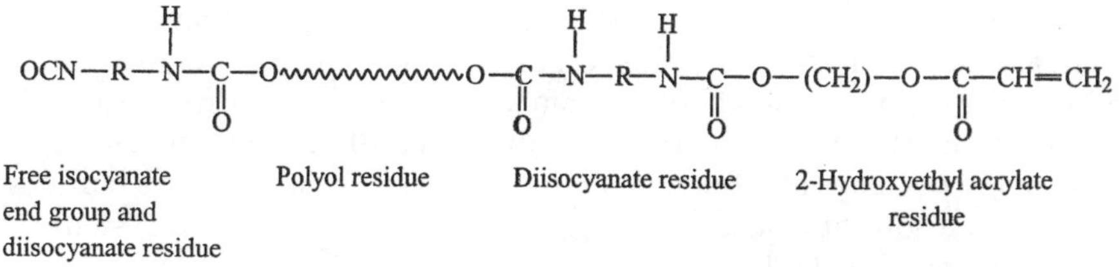

A urethane acrylate with free isocyanate functionality

These molecules have been combined with polyols to form coating systems that will first cure with ultraviolet radiation, and then they achieve final properties by heating to promote the isocyanate/hydroxyl reaction [14,15] and achieve a highly cross-linked state. As might be expected, such products have a limited pot life and are two-package systems.

Formulations that were solventborne and waterborne in nature were investigated, and after cure, the coatings were tested for scratch resistance in a laboratory car-wash system. Cure was effected by keeping the coatings at room temperature for 5 min, then heating for 10 min at 80° C. After the thermal treatment and while the coatings were warm, they were exposed to ultraviolet irradiation at 1.1 m/min under two 80-watt/cm lamps. Both type coatings had a car-wash resistance that was as good as a conventional two-package polyurethane system. In addition, the coatings had a high degree of solvent and chemical resistance and 20° gloss of 89 and 88, respectively. Curing time for the coatings was markedly shorter than that required for conventional polyurethane coatings.

This dual-type molecule with both acrylate and isocyanate functionality has also been blended with a polyol such as 1,4-butanediol or trimethylolpropane triol and exposed to radiation in the first step [16]. Then, in the second step, the system is heated at 150° C for 30 min to facilitate formation of urethane linkages through the isocyanate/hydroxyl reaction. The final product is a cross-linked acrylate/urethane system. When a triol was used as the cross-linking agent, the final film had a "Young's Modulus" of 60 kg/cm^2, a tensile strength of 70 kg/cm^2, and an elongation of 140%. This film was more flexible and deformable than the control, which only contained acrylate functionality and was only radiation cured, that had a Young's modulus of 110 kg/cm^2, a tensile strength of 70 kg/cm^2, and an elongation of 100%.

Another dual-cure system of this general type involves radiation-curable acrylates and hydroxyl groups that can be thermally cross-linked with methoxymelamines, benzoguanamines, and other amino cross-linking agents [17]. The hydroxyl-containing compounds may contain both acrylate and hydroxyl functionality, as is the case with the ϵ-caprolactone derivatives of 2-hydroxyethyl acrylate as well as multifunctional compounds such as trimethylolpropane or diethylene glycol. These materials are formulated with a free radical generating photoinitiator and compounds capable of generating an acid that will catalyze the cure of hydroxyl and amino functionalities at elevated temperatures. Hard (up to about 8H pencil hardness), solvent resistant (> 100 acetone rubs) coatings were achieved after exposure to ultraviolet radiation and about 3–5 min at 100–150° C. After irradiation, the coatings had a hardness of < HB and only passed 10 acetone rubs. However, after aging two days at room temperature, the same coatings had a hardness of 4H and a solvent resistance of > 100 acetone rubs. A related system has also been described [17].

Interpenetrating polymer network technology has been examined as a means of achieving dual-cure systems [18]. Interpenetrating polymer networks arise when two polymerizable compositions are mixed and are independently reacted, with the resultant polymeric mass containing two distinct, intertwining, continuous polymeric domains. For example, one of the domains might be a radiation-curable acrylate formulation and the other might be a thermally curable, two-package urethane formulation. The latter portion may be catalyzed, if desired, with stannous octanoate, dibutyltindilaurate, or similar compounds used for the isocyanate/hydroxyl reaction. The investigators demonstrated the technology with a variety of polyols and isocyanates for the polyurethane portion of the network and multifunctional, low molecular weight acrylates. They found that such systems combined in a synergistic way to yield cured films with better overall properties than either component. The radiation-cured acrylates gave the system a fast cure mechanism, and the urethane portion gave toughness to the film. Overall, the films had optical clarity, hardness was increased along with impact strength, and the network had improved solvent resistance. The weathering character was that of the individual components. Other examples of dual-cure interpenetrating compositions can be found in the literature [19].

Acrylated melamines offer another potential for a combination radiation and thermal cure system [20]. Acrylated melamines may be synthesized by reacting acrylamide with etherified melamine compounds [21]. Any of the six reactive sites on the triazine ring of melamine can be acrylated. As an example, a methoxymelamine compound with three of the sites reacted is shown below:

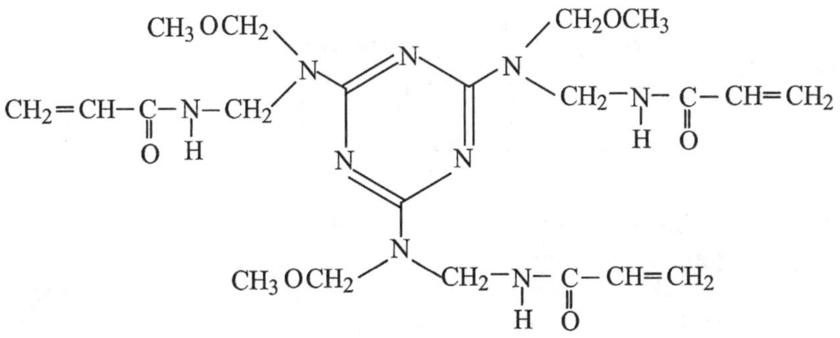

CH_3OCH_2 CH_2OCH_3
N N N
$CH_2{=}CH{-}C({=}O){-}N(H){-}CH_2$ $CH_2{-}N(H){-}C({=}O){-}CH{=}CH_2$
N N
N
CH_3OCH_2 $CH_2{-}N(H){-}C({=}O){-}CH{=}CH_2$

An acrylated melamine

The methoxymelamine groups will react with hydroxyl functionality on polyols, etc., in the usual manner (thermally), and the acrylate groups will react with themselves or with added acrylates of various kinds. Thus, molecules of this family may be first radiation cured, and this partially cured coating may be followed by a thermal cure to achieve maximum properties. Or, the molecules may be first thermally cured and then radiation cured.

A particular melamine acrylate was formulated with tripropyleneglycol diacrylate, a free radical photoinitiator, and a p-toluene sulfonic acid catalyst [20]. When only cured with ultraviolet radiation (two 200 W/in., ~80 W/cm, lamps at 100 ft per min., ~30 m per min.), the resulting film had excellent adhesion and passed 196 methylethyl ketone double rubs. When a thermal post cure of 5 min at 150° C was used after the radiation exposure, the adhesion remained excellent and the coating was not affected by more than 300 methylethyl ketone double rubs.

Another system involved the use of hydrogen-abstraction photoinitiator systems in combination with a thermal initiator such as peroxides or azo compounds [22]. When the formulations were exposed to ultraviolet radiation, polymerization was initiated in the surface region of the coating. The polymerization generates heat, and this energy activated the thermal initiator and caused polymerization to proceed through the entire coating.

Radiation/Moisture-Cure Urethane Cures

The dual-cure system, that involves a urethane acrylate in which one end of the urethane acrylate molecule is replaced with an isocyanate group as described in the previous section and shown below,

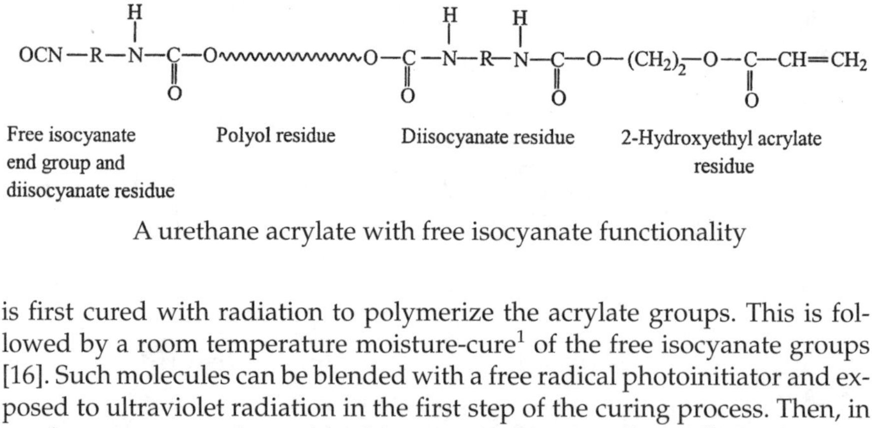

$$OCN-R-\overset{H}{\overset{|}{N}}-\underset{\underset{O}{\|}}{C}-O\sim\sim\sim\sim O-\underset{\underset{O}{\|}}{C}-\overset{H}{\overset{|}{N}}-R-\overset{H}{\overset{|}{N}}-\underset{\underset{O}{\|}}{C}-O-(CH_2)_2-O-\underset{\underset{O}{\|}}{C}-CH=CH_2$$

Free isocyanate end group and diisocyanate residue | Polyol residue | Diisocyanate residue | 2-Hydroxyethyl acrylate residue

A urethane acrylate with free isocyanate functionality

is first cured with radiation to polymerize the acrylate groups. This is followed by a room temperature moisture-cure[1] of the free isocyanate groups [16]. Such molecules can be blended with a free radical photoinitiator and exposed to ultraviolet radiation in the first step of the curing process. Then, in a subsequent step, the residual isocyanate functionality will slowly react with adventitious moisture in air or a controlled environment at room temperature to form mainly urea groups between the isocyanate molecules, with a cross-linked acrylate/urea system resulting. After undergoing radiation curing and being held at 100% relative humidity for 24 h, the resulting film

[1] Moisture-cure polyurethanes are also known as ASTM Type II polyurethanes. See [23] for the ASTM descriptions of the six "Types" polyurethanes.

had a "Young's Modulus" of 90 kg/cm^2 compared to 100 kg/cm^2 for the control that only contained acrylate functionality and was only radiation cured, a tensile strength of 98 kg/cm^2 compared to 70 kg/cm^2 for the control, and an elongation of 130% compared to 100% for the control. The moisture-cured film was stronger, more extensible, and easier to deform than the control. Before the moisture-curing step, the modified product was very flexible, Young's modulus 27 kg/cm^2, weak, tensile strength of 12 kg/cm^2, but quite extensible, elongation 82%. From these data, the effect of moisture-cure and urea formation is readily apparent. Pigmented ultraviolet radiation/isocyanate dual-cure systems have been studied [24]. Good curing was obtained in layers as thick as 50 microns.

Although it is not a coating application, polyolefinic foams with improved creep, elastic recovery, fatigue resistance, and tensile properties have been prepared by a radiation-activated first stage cure that coincides with or follows the foaming operation [25]. This is followed by a moisture-cure stage that completes the curing operation. The radiation-cure stage is accomplished with an electron beam or similar radiation source.

Radiation/Epoxide Cure

The radiation cure of half acrylated diglycidyl ether of bisphenol A[2] has been

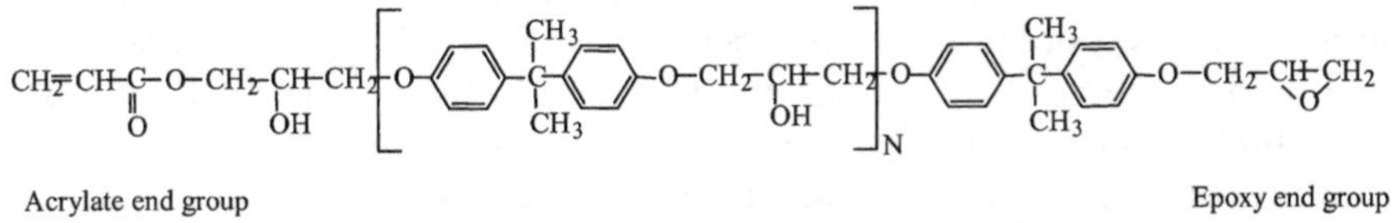

$$CH_2{=}CH{-}C({=}O){-}O{-}CH_2{-}CH(OH){-}CH_2{-}\left[O{-}C_6H_4{-}C(CH_3)_2{-}C_6H_4{-}O{-}CH_2{-}CH(OH){-}CH_2\right]_N{-}O{-}C_6H_4{-}C(CH_3)_2{-}C_6H_4{-}O{-}CH_2{-}\underset{O}{CH{-}CH_2}$$

Acrylate end group | Epoxy end group

Half acrylated adduct of the diglycidyl ether of bisphenol A

studied alone and in combination with other acrylates in the presence of a variety of cationic and free radical photoinitiators [26]. The molecule described above is designed to react with cationic photoinitiators through the epoxy end group, with chain transfer taking place by means of the hydroxyl groups on the molecule or with added hydroxyl-functional compounds, while the acrylate group is able to undergo free radical polymerization. A large number of formulations were investigated, and it was concluded that the novel epoxy/acrylate adduct caused a decrease in cure speed in comparison with the fully acrylated version, but it had better adhesion to nonporous, rigid substrates. Good adhesion and solvent resistance were obtained on metal substrates. As might be suspected, formulations containing the adduct were sensitive to the ultraviolet radiation dosage, the type of photoinitiators used, and the effect of added acrylates. The slowness of cure with the adduct was

[2] Ebecryl™ 3605, UCB Radcure Group.

thought to be related to its acting mainly as a monofunctional compound with the different type photoinitiator systems.

Radiation/Radiation Cure

Because of the absence of solvent in ultraviolet radiation-curable coatings, it is almost always difficult to obtain flatting or to control the degree of flatting. A novel radiation dual-cure process, in which radiation of different intensities was used to develop different degrees of gloss (flatting) in coatings, is useful in various applications, and particularly, in the wood coating area [27,28][3]. A commercial 35% unsaturated polyester/65% styrene mixture was formulated with coated, micronized silica gel flatting agents and 2,2-dialkoxy-2-phenylacetopheneone photoinitiator. Various amounts of flatting agent were used. The radiation/radiation dual-cure process was carried out as follows. Films of 50–200 μm were applied to black glass plates with a drawdown bar. They were first exposed to ultraviolet radiation from two 80W/cm medium pressure mercury vapor lamps at 10 m per min. After this pregelling step, the films were then passed under two 80-W/cm medium pressure mercury vapor lamps at 3.5 m per min. In effect, the second step was about five to six times slower with markedly greater radiation exposure than in the pre-gelling step. For comparison purposes, a direct-cure or one-step process, in which the coatings were exposed to two 80W/cm medium pressure mercury vapor lamps at 2.5 m per min, was used. The total radiation exposure in the direct-cure process is equivalent to that of the dual-cure process. In all instances, the dual-cure process resulted in lower gloss values than the direct-cure process. The flatting efficiency of the process increased with increasing flatting agent concentration, which varied from about 3% to about 10%. Flatting efficiency increased as film thickness increased.[4] The potential for similar gloss control, involving a dual-cure process with medium pressure mercury vapor lamps that emit radiation of different wavelengths, was described. As might be anticipated by those familiar with the complexity of the flattening process, the optimum concentration and type of flattening agent and the curing conditions must be determined by experimentation.

Radiation/Air-Drying Cures

These dual-cure systems are based on acrylates and unsaturated compounds that will react with the oxygen present in air, known as air-drying com-

[3] References [26] and [27] contain a good description of the flatting process and of why and how it differs in radiation-cure coating formations in comparison to that from solvent-based coating formulations. Also see [28] for further information about flatting UV systems.

[4] Other investigators have found the opposite to be true, i.e., flatting efficiency decreased as film thickness increased.

pounds [16]. The air-drying compounds used in the art are alkyds or unsaturated polyesters [30] and drying oils [31]. The oxidative drying reaction that takes place under ambient conditions is usually catalyzed with metallic salts such as salts of cobalt, which promotes surface drying, salts of manganese, which effects both surface and internal or through drying, and salts of other metals [32]. The curing or drying takes place by oxygen absorption followed by peroxide formation. The peroxide then decomposes into free radicals, and these cause cross-linking and cure over a period of time, usually hours or days.

The unsaturated compounds investigated [16] were unsaturated polyesters and mono-, di-, and tri-"oxygen sensitive" compounds. None of the compounds were structurally described. These compounds were formulated with one or more multifunctional acrylates, a mixed functionality urethane compound, free radical generating photoinitiators, and a cobalt salt. The formulated systems were first exposed to ultraviolet radiation and then allowed to "age" or further cure under room conditions. Solvent resistance, in general, increased with aging time, and greater than fifty acetone rubs were achieved with all systems over an aging time of 0 h to 1 week. The investigators indicated that control coatings without a cobalt salt did not show any significant change in properties. However, it should be pointed out that the cobalt salts usually promote surface cure, and a solvent-rub test might not be sensitive to any change the salt caused. Hardness and gel content also increased as a function of time at room conditions after radiation exposure.

References

[1] Perkins, W. C., "New Developments in Photo-Induced Cationic Polymerizations," *Journal of Radiation Curing*, Vol. 8, No. 1, January 1981, p. 16; *Proceedings of Radiation Curing V*, AFP/SME meeting, Boston, MA, 23–25 Sept. 1980, p. 145.

[2] Patent: Tsao, J. H. and Ketley, A. D., W. R. Grace & Co., Photocurable Epoxy-Acrylate Compositions, US 4,156,035. (1979).

[3] Manus, P. J. M., "Coating Performances and Formulation Parameters of Cationic Systems," Conference Papers, *RadTech Europe '89*, Florence, Italy, 9–11 Oct., 1989.

[4] DeVoe, R. J., Brown-Wensley, K. A., Holmes, G. L., Mathis M. D., McCormick, F. B., Palazzotto, M. C., and Spurgeon, K. M., "Dual Cure Photocatalyst Systems," Vol. 63, *Proceedings of the ACS Division of Polymeric Material: Science and Engineering*, Washington, D. C., Fall 1990, p. 941.

[5] Patent: Ha, C. T. M. and Sullivan, M. G., Cationic and Hybrid Radiation Curable Pressure Sensitive Adhesives for Bonding of Optical Discs, DSM n.v., US 6,180,200. (2001).

[6] Vara, F. J. and Dougherty, J. A., "Concurrent Cationic/Free Radical Polymerization of Vinyl Ethers with Acrylate Functional Oligomers," Conference Papers, *RadTech Europe '89*, Florence, Italy, 9–11 Oct., 1989.

[7] Morgan, C. R., "Dual UV/Thermally Curable Plastisols," Technical Paper FC83-249, *RadCure '83 Conference*, Lausanne, Switzerland, May 9–11, 1983.

[8] Patent: Gaeta, A. C. and Swei, G. S., Norton Co., Enhanced Radiation Cure, US 6,187,070. (2001).

[9] Hall, R. H., Van Loon, B. H. C. M., and Van der Sanden, J. B. J., "Dual Cure—The Best of Both Worlds," Technical Paper FC87-278, *Radcure Europe '87*, Munich, Germany, May 4–6, 1987.

[10] Patent: Green, G. E. and Irving, E., Ciba-Geigy Corp., Photopolymerizable and Thermally Polymerizable Compositions, US 4,299,938. (1981).

[11] European Patent Application: Nagy, F.A., Mobil Oil Corp., Radiation Curable Epoxy/Acrylate-Hydroxyl Coating Compositions, EP 82,603. (1983); Chemical Abstracts 99:124202f, 1983.

[12] Pappas, S. P. "Photoinitiation of Cationic and Concurrent Radical-Cationic Polymerization, Part V," *Progress in Organic Coatings*, Vol. 13, No. 1, 1985, p.1.

[13] Barisonek, E. M. and Froehlig, G., "Radiation Curing Hybrid Systems," Technical Paper FC83-254, *RadCure '83 Conference*, Lausanne, Switzerland, May 9–11, 1983.

[14] Demarteau, W., Herze, Py., and Loutz, Jm., "Dual UV/Thermally Curable Formulations," *Proceedings of RadCure '84*, Atlanta, GA, 10–13 Sept. 1984, p. 1-1.

[15] Fischer, W., Meier-Westhues, U., and Hovestadt, W., "Dual-Cure, New Possibilities of Radiation Curing Coatings," *End User Conference, RadTech 2000*, Baltimore, MD, 9–12 Apr. 2000, p. 38.

[16] Peeters, S., Loutz, J. M., and Philips, M., "Overview of Dual Cure Possibilities in UV Coatings," *Proceedings of RadTech '88*, New Orleans, LA, 1988, p. 79.

[17] Boeckeler, R. H., "One-Component, Dual-Cure Coatings," *Proceedings of RadCure '86*, Baltimore, MD, 8–11 Sept. 1986, p. 16-1.

[18] Roesler, R. T., "An Interpenetrating Polymer Network," *Proceedings of RadCure '86*, Baltimore, MD, 8–11 Sept. 1986, p. 16–13.

[19] Patent: Skinner, E., Emeott, M., Jevne, A., Henkel Corp., Interpenetrating Dual Cure Resin Compositions, US 4,247,578. (1981).

[20] Gummeson, J. J., "Acrylated Melamines in UV Curable Coatings," *RadTech Report* Vol. 5, No. 2, March/April 1991, p. 17.

[21] Patent: Strazik, William F., LeBlanc, John R., and Santer, J. Owen, Monsanto Co., "Unsaturated Melamine Condensates," US 4,293,461 (1981).

[22] Patent: Dixon, George D., Westinghouse Electric Corp., "In Depth Curing of Resins Induced by UV Radiation," US 4,222,835. (1980).

[23] ASTM Standard D 16: Terminology Relating to Paint, Varnish, Lacquer, and Related Products, *ASTM Book of Standards* Vol. 06.01, ASTM International, West Conshohocken, PA.

[24] Noomen, A., "Pigmented UV Dual Cure Coatings," *Journal of Radiation Curing*, Vol. 9, No. 4, 1982, p.16.

[25] Patent: Walton, K. L. and Karande, S. V., The Dow Chemical Co., Crosslinked Polyolefinic Foams with Enhanced Physical Properties and a Dual Cure Process of Producing Such Foams, US 6,124,370. (2000).

[26] Ravijst, J. P., "Hybrid Cure of 'Half Acrylates' by Radiation," *Proceedings of RadTech '90—North America*, Chicago, IL, Vol. 1, 25–29 Mar. 1990, p 278.

[27] Garratt, P. G., "The Flatting of Radiation Curable Paints Based on Unsaturated Acrylic Binders, *Proceedings of RadTech '90—North America*, Chicago, IL, Vol. 1, 25–29 Mar. 1990, p. 268.

[28] Garratt, P. G., "Ultraviolet Radiation Dual-Cure Processes for the Production of Low-Gloss Films," *Proceedings of RadCure Europe '87*, Munich, Germany, 4–7 May 1987, p. 10–23.

[29] Bussell, L., Vega, N., and Christmas, B., "The Use of Flatting Agents in UV Curable Formulations," *Proceedings of RadTech '90—North America*, Chicago, IL, Vol. 1, 25–29 Mar. 1990, p. 257.

[30] Heitkamp, A. and Pellowe, D., "Alkyd and Polyesters," Ch. 7, *Paint and Coating Testing Manual*, J. V. Koleske, Ed., ASTM International, West Conshohocken, PA 1995.

[31] Koleske, J. V., "Drying Oils," Ch. 4, Paint and Coating Testing Manual, J. V. Koleske, Ed., ASTM International, West Conshohocken, PA, 1995.

[32] Schnall, M. J., "Driers and Metallic Soaps," Ch. 5, *Paint and Coating Testing Manual*, J. V. Koleske Ed., ASTM International, West Conshohocken, PA, 1995.

Adhesives

10

Introduction

THE USE OF RADIATION to cure adhesives has been known for over 40 years, and what appears to be the first review of the use of irradiation as a process to make pressure sensitive adhesives appeared in 1977 [1]. The early pressure sensitive adhesives were mainly based on monofunctional alkyl acrylates with four or more carbon atoms in the alkyl chain and vinyl esters [2]. Although this early interest in radiation-cured adhesives existed, it has only been since about 1990 that interest in the technology has grown and become very strong [3–5]. The main driving forces for using this technique to cure adhesives were manufacturing costs and environmental regulations that forced producers to investigate solvent free or very low solvent systems. This was coupled with the need to have final products that did not detract from required performance characteristics.

As would be expected, solvent-based adhesives were and still are very well established in the industry. Such formulations can be made for a wide variety of end uses. But, solvent and energy costs have increased, and regulation has added the expense of solvent recovery for many operations using conventional technology. In addition, solvent removal from adhesive coatings, particularly from thick films, results in relatively slow line and cure speeds. Radiation curing with its many advantages of rapid curing and cross-linking of polymers from the monomeric stage with nil volatile organic compound loss offers great advantages for the adhesive field [6–8]. In 1995, barriers to this technology were considered along with some of the ecological advantages [9]. With the success that has taken place in the area, one must surmise that many of the technical and economic barriers that were detailed have been surmounted. Yet, it is worth being aware of these if one is beginning an effort in the area.

In 1982, it was reported that the radiation-cured adhesive market was merely four million dollars in 1981 [10]. However, it was expected to reach $34 million by 1987. This represents a growth rate of about 7.7%, in comparison to a 5.3% growth rate predicted for the specialty adhesive market a few years later [11]. At about the same time, the use of ultraviolet radiation to cure structural aerobic acrylic adhesives was beginning to gain acceptance [12] as the industry began to surmount the difficulty in obtaining adhesion to smooth, nonporous substrates with low energy surfaces, such as glass and plastics. In 1999, the global radiation-curable market was estimated at $1.3 billion of which only a small portion, 1.7% or about $22 million, is in the adhesives area [13].

The annual growth rate in the adhesives area is expected to be 7.3%, a number that is in fair agreement with the number predicted about a decade earlier. While the growth rate might agree with the earlier number, the size of the market seems too small. An article [14] quoting a Frost & Sullivan report states that the United States radiation-curable market had sales of $97 million in 2000, and this was projected to grow to $150 million by 2007. The growth rate over this time period, if the projection is taken to mean the end of 2006 (i.e., by 2007), would be 7.5% in basic agreement with the other growth rates. The impetus for this growth is attributed to the U. S. Environmental Protection Agency regulations. The regulations are forcing the manufacturers to use more environmentally favorable products, such as radiation-cured adhesives.

The monomers, mono- and multifunctional, and oligomers described in earlier chapters can be used to formulate adhesives. Many of these have been described in terms of activity, hydrophilicity/hydrophobicity, and glass transition temperature and the importance of these factors to adhesives [15].

Pressure Sensitive Adhesives

Pressure sensitive adhesives (PSAs) are all around us today in the form of labels, tapes of various kinds, postage stamps, decorative and functional stickers, laminates, posting notes, and on and on. Such adhesives are characterized by three main properties: cohesion, adhesion, and tack. Cohesion is the internal strength of the adhesive, and it allows the adhesive layer to resist splitting when used in packaging operations or when the adhesive is removed from the substrate to which it is adhering. Adhesion is usually thought of as the property that describes the ability of the adhesive to remain attached to the surface to which it is attached and to do so without slippage at the interface between substrate and adhesive. Tack is the stickiness of the adhesive, and it is related to the aggressiveness of the adhesive for attachment to other surfaces. Variation of these properties leads to a wide variety of pressure sensitive products. A discussion of these three properties and their interrelationship with radiation curing can be found in the literature [16,17].

In a general sense, as the amount of ultraviolet radiation used in the cure of a PSA increases, adhesion decreases and cohesion increases for acrylate-based formulations [18]. Test descriptions are available along with some ranges that might be expected for commercial PSAs [19]. For example, shear values vary from 20 to about 35 000 min for a variety of adhesives. PSAs depend on a balance of elastic and viscous properties to achieve performance characteristics, and this balance defines the character of a PSA. Tackifiers are compounds that provide the viscous portion of this needed viscoelastic-effect balance [20].

Pressure sensitive adhesives are available in a variety of forms. The main ones are solventborne, waterborne, hot melt, and radiation cured. In 1986, water-based PSAs comprised about 60.2%, solvent-based and hot melt PSAs each about 19.8%, and radiation-cured PSAs about 0.2% of the dollar value of the PSA market [21]. Even though radiation-cured PSAs had only a small fraction of the market, energy costs, environmental concerns, and productivity made this technology one that sparked the interest of many investigators. In Europe, hot melts are a significant portion (~ 20%) of the PSA market, and it is felt that the radiation-curable hot melt technology will be an important growth factor in this area [22]. It is expected that the hot melt area will grow at an annual 7% rate versus 1% for solutions and 5% for waterborne. Today, it is an active area of product research and development.

Advantages and disadvantages of the four technologies have been delineated, and the properties of tack, peel, and shear for some commercial products and some ultraviolet radiation-cured products are given in Table 10.1. This market has grown, and in 2000, it was reported that there were about 16 suppliers with 30 products for the PSA area [23]. Ten of the suppliers had 15 products for PSAs that could be applied as liquids at room temperature, and six of the suppliers had 15 products that could be applied as warm or hot melts. These numbers do not take into account suppliers and products for the conventional hot melt PSAs.

The use of an electron beam for curing and cross-linking pressure sensitive adhesives has been described [24,25]. Costs, as related to both electron beam and ultraviolet radiation curing of PSAs and how these can lead to an integrated manufacturing operation, have been published [26, 27]. Coating of an ultraviolet radiation-curable silicone release and a waterborne acrylic PSA

TABLE 10.1. Properties of Commercial Pressure Sensitive Tapes and Some Ultraviolet Radiation-Cured Products of Undisclosed Composition [19].

TYPE OF TAPE	TACK, g/cm2	PEEL, lbs/linear in. (g/25 mm)	SHEAR, min
Commercial cellophane tape	460	1.6 (290)	600+
Ultraviolet radiation cured tape	300	1.4 (254)	600+
Commercial removable label	300	1.0 (181)	20
Ultraviolet radiation cured tape	200	2.0 (362)	40
Commercial high performance tape	930	4.0 (726)	600+
Ultraviolet radiation cured tape	1300	4.0 (726)	600+

suitable for use in the tape and label area can be done by a tandem coating process [28].

Early uses of radiation in this field often dealt with alteration of the cohesive strength factor. An electron beam was used to improve this characteristic in mixtures of natural rubber latex crepe (polyisoprene) and a pentaerythritol ester of wood rosin; natural rubber latex, styrene-butadiene copolymer, and a terpene oligomer; poly(vinyl ethyl ether); and other systems [29]. The goals were to cross-link the adhesive and improve internal strength properties. The cohesive property of hot (49° C or 120° F) shear resistance was improved by radiation, while the adhesive properties remained unchanged.

Adhesives based on butyl rubber have poor cohesive strength, in comparison to natural rubber based adhesives. Butyl rubber will degrade under electron beams and ultraviolet radiation. However, a novel compound known as conjugated diene butyl (CDB) will cross-link when exposed to an electron beam [30]. When CDB was compounded with commercial aliphatic and cyclic tackifiers and irradiated with either an electron beam or ultraviolet radiation, it exhibited high reactivity either alone or in a formulated product [31]. Good quality PSAs could be made. The cohesive strength and holding power were markedly improved by irradiation. Adhesiveness had only a minor increase. The aliphatic tackifier tended to provide the best adhesive character, and the cyclic tackifier tended to provide the best cohesive properties. Other basic PSA components include polyisoprene [32,33], epoxidized polyisoprene [34] and polybutadiene [32], and isobutylene copolymers with isoprene or p-methylstyrene and halogenated versions of these compounds [35].

Although styrenic block copolymers have been widely used as pressure sensitive adhesives for some time, PSAs made from these block copolymers are unsuitable for use in areas where they must withstand aromatic solvent or exposure to high temperatures. These difficulties arise from the fact that such block copolymeric adhesives are thermoplastic in nature and obtain their rubbery character due to physical cross-linking by means of microphase separation of styrene end blocks during the cooling process. To overcome this difficulty, the styrenic block copolymers were made with a multitude of dangling double bonds on the flexible block [36]. These polymers could be applied easily by warm or hot melt techniques and then covalently cross-linked by a free radical mechanism when exposed to electron beam or ultraviolet radiation. The same study discussed development of a liquid poly(ethylene/butylene) copolymer with a primary hydroxyl group on one end. The other end of the hydroxyl-terminal copolymer was an epoxidized polyisoprene block. This block copolymer would cure by cationic means in the presence of an onium salt photoinitiator. Details of cure studies were given and both new PSAs functioned well as adhesives that maintained their adhesion at elevated temperatures (~95° C). The dark cure that takes place with cationic systems can be used to improve ultraviolet radiation-cured adhesives [37].

Liquid butadiene/isoprene/styrene block copolymers with epoxide and hydroxyl functionality have been investigated as PSAs [38]. These copolymers do not exhibit microphase separation. They are star-shaped in structure and have aliphatic epoxy functionality that is localized on specific blocks and/or have primary hydroxyl functionality at one or both ends of the copolymers. The heat stable polymers can be rapidly cured if they are formulated with cationic, onium-salt photoinitiators and exposed to ultraviolet radiation. After cure, the PSAs had excellent high temperature adhesion, a balance of adhesive properties, aggressive tack, and shear adhesion of greater than 175° C.

An acrylic formulation of undisclosed composition involved an acrylic copolymer with a chemically bound photoreactive group that is applied in hot-melt form [39]. In addition to good adhesive properties, the adhesives have excellent resistance to aging and to deformation at high temperatures. This paper contains a generalized cost comparison between ultraviolet radiation-cured adhesives and conventionally applied and cured adhesives.

The use of electron beam radiation to cure acrylate monomers on flexible substrates into a PSA tape was also done in an early invention [40]. Low molecular weight polyacrylates based on 2-ethylhexyl acrylate and vinyl acetate with minor amounts of monomers such as dimethylaminoethyl methacrylate, acrylic acid, or an acryl amide have been applied to a flexible substrate as a hot melt and irradiated with high voltage electrons [41]. The resultant irradiated products had improved shear strength over similar conventional products.

Hot melt PSAs have disadvantages such as poor performance at elevated temperatures and to aging (light, oxygen, and heat resistance). A recent paper [42] described a combination of hot melt and radiation curing technology for the development of low performance, general purpose PSAs and of permanent PSAs. As would be expected, adhesion varied with both formulation and substrate, but the new products had good adhesion to both low and high-density polyethylene, polypropylene, poly(vinyl chloride), and glass. All three products were shown to cure at high speeds, and they had excellent cohesive strength with static shear resistance of >24 h, with no movement when the adhesive was applied to stainless steel and tested at 100° C. Dynamic mechanical analysis indicated that the glass transition temperature was unchanged but the use temperature or plateau region in the dynamic shear modulus was extended well above the normal flow region by crosslinking. Plasticizer migration resistance was excellent.

Solutions of a 50/50 low molecular weight 2-ethylhexyl acrylate-vinyl acetate copolymer, dissolved in 2-ethylhexyl acrylate with and without a multifunctional acrylate, were cured with ionizing radiation and then subjected to an artificial weathering environment for 300 h [43]. After this cure and treatment, the adhesives had good adhesive release characteristics from polished metal substrates. While the cohesive strength and release characteristics were acceptable with all systems, both properties improved when the multifunctional acrylates were present.

Another study [44] also sought to improve the cohesive strength, and thereby, the creep resistance of the adhesive. In this case, ultraviolet radiation of various intensities was used, with and without an added photoinitiator. Natural rubber was formulated with various ingredients, optionally including a photoinitiator such as benzophenone or similar compounds. In each case, after irradiation the creep resistance was markedly improved over that of a control.

The feasibility of grafting acrylates such as neopentyl glycol diacrylate, pentaerythritol triacrylate, and acrylated oils to poly(vinyl ethyl ether) and poly(vinyl methyl ether) in the presence of benzophenone has been studied [45]. The reactions were initiated with ultraviolet radiation in an air atmosphere. Cohesive strength improved as the concentration of multifunction acrylate increased, as indicated by higher degrees of cross-linking. The latter property was noted by increased gel content and low swelling indices. Thus, cohesive strength was improved. The cross-linked compositions contained some unreacted poly(vinyl ether), which functioned as a tackifier. The grafting of monofunctional acrylates onto the same polymers was also examined, and useful PSAs resulted [46,47].

A correlation between curing and formulation parameters indicated that the most important parameters that affected performance were acrylate and photoinitiator concentration and curing time [48]. Screen printable adhesives have also been investigated [49]. The reaction between an acrylate polymer and monomer that was irradiated with a low-energy electron beam indicated that introduction of acryloyl groups into the network improved peel and dead load strength [50]. If polar monomers, i.e., those containing hydroxyl or carboxyl groups, were added to the reaction mixture, the adhesive properties improved.

A highly flexible urethane acrylate (50%) has been compounded with four different monofunctional acrylates (40%), benzophenone (4%), methyldiethanolamine (3%), and a commercial photoinitiator of undisclosed chemical composition (4%) [51,52]. The latter photoinitiator was used to ensure good reactivity in the bulk of the adhesive, and the combination of benzophenone and amine synergist was used to ensure good surface cure of acrylates and, thus, ensure no uncured monomer was present in the surface of the adhesive. Absence of uncured monomer would mean the system should be odor-free, and also that surface tack or slipperiness would not be attributable to uncured monomer. Two of the monofunctional acrylates, ethoxyethoxyethyl acrylate and the four-mole ethoxylate of nonyl phenol acrylate, were polar in nature and two, isodecyl acrylate and a monohydroxyl caprolactone acrylate, were nonpolar in nature. Formulations of ethoxylated trifunctional acrylates yielded cured film properties that had no adhesion and were hard in nature. While these studies do not provide high quality adhesives, they can be used as a guidance tool that will aid formulators in devising radiation-curable PSA tapes.

A related formulation study investigated systems with aromatic and aliphatic epoxy acrylates with various acrylate monomers, free radical photoinitiators, and tackifiers [53]. The study proved that tackifier content and glass transition temperature were critical to PSA performance. If tackifier content was too high, the final film had poor adhesion, due to cross-linking causing the film to lose tack. Addition of certain monofunctional acrylates, as ethoxyethoxyethyl acrylate, to a formulation such as the one used in the study, increased the adhesive character and peel strength. By varying the tackifier and monomer content effectively, any peel strength between 0.1 and 6 lb/in. (~18 and 1090 g/25 mm) could be obtained.

An effort has been made to develop a variety of ultraviolet radiation curable PSAs with characteristics ranging from light duty to those that would permanently bond to a substrate [54]. Within this program, a method was devised for screening tackifiers, and the most promising candidates were used. The tackifier screening formulation consisted of 66% tackifier, 20% 2-ethylhexyl acrylate, 7% trimethylolpropane triacrylate, 3% hexanediol diacrylate, and 4% of a commercial-free radical photoinitiator. Statistical design analysis was employed, and a number of potential PSA formulations were evaluated. Of the PSAs evaluated, several ultraviolet radiation curable adhesives were found that had performance characteristics as good as or better than commercial products. A permanent PSA was not defined from the test results.

A recent paper [55] pointed out that in 1990 at the RadTech '90 radiation curing conference, there were no technical papers that dealt directly with electron beam or ultraviolet radiation cured PSAs. In contrast, at the RadTech 2000 conference ten years later, seven papers dealt directly with the topic and were presented at a session, focused on radiation-cured adhesives. The reporting investigators evaluated three types of commercial ultraviolet radiation curable PSAs—a liquid PSA, a warm-melt (~93° C), and a hot melt PSA (~ 149° C). These adhesives were studied on silicone release coatings of various types to best ascertain the potential processing advantages for the systems and to guide potential end users of the technology. The study found that the radiation-cured PSAs performed well on four different silicone release substrates—namely, radiation-cured epoxy-silicones and acrylate-silicones, and thermal-cured solventless and emulsion silicones. The data indicate that in-tandem coating of, first, the release coating, followed by the PSA, could offer manufacturing advantages. Very low voltage electron beam curing of acrylic PSAs has been considered [56].

In 1986 and later, polyesters with one terminal acrylic double bond were described as PSA candidates [57,58]. The other terminal group on the polymer was nonreactive when the system was irradiated. These low molecular weight (3000–6000) compounds had glass transition temperatures of about–10 to -50° C. The use of tackifiers was found to be optional, and addition of monomeric or other reactive diluents was undesirable. Blending polyesters of different molecular weight, glass transition temperaure, and associated properties were used to vary and control adhesive properties. The PSAs could be cured rapidly with electron beam or ultraviolet radiation. Good adhesive properties resulted, and end uses such as double-sided tapes, electrical tapes, medical tapes, decals and labels, and protective films were mentioned as end uses. Polyester technology offers an alternative to the rubber-based and acrylate-based products that are the main factors in the radiation-cured PSA market. The main advantages of the polyesters were low skin irritation, high cure speed, low photoinitiator concentrations, and wide formulation latitude. The disadvantages were high viscosity, which required handling and application at elevated temperature, limited polymer compatibility, and some coloration. Cost may also have been a disadvantage.

Release or Anti-Adhesive Coatings

It is difficult to discuss pressure sensitive adhesives without discussing release coatings, which were briefly mentioned in the previous section. Release coatings are the substrate to which pressure sensitive adhesives are applied and from which the final product—decal, envelope, label, personal hygiene item, protective liner in graphic arts, stamp, tag, tape, etc.—is removed. A simple label product might have the following construction:

Label Stock
Adhesive Coating
Silicone Release Coating
Release Layer Stock

Just as it is difficult to discuss pressure sensitive adhesives without discussing release coatings, it is difficult to discuss release coatings without discussing silicone coatings. Silicone coatings have a number of attractive features such as a low energy surface, excellent electrical properties, low moisture uptake, both thermal stability and low temperature performance, flexibility and elasticity, and easy handling characteristics [59–61]. In addition to their use as release coatings for PSAs, silicone coatings are important in the electronics area as ultraviolet radiation-curable conformal coatings, which are coatings that protect printed circuit assemblies from hostile environments.

Polyurethane casting is another important end-use area for release coatings [62]. Polyurethane castings are used in the coated fabric industry to produce leather substitutes used as furniture upholstery, shoes, automobile seat covers, women's handbags, suitcases, attaché cases, and similar items. Release coatings for such products can yield films with high gloss to high matte finishes, and in a variety of textures such as cowhide, pigskin, reptile hides, and so on. Other special uses for release coatings include three-piece diaper tapes, in which three distinct levels of release are needed, protection of hot melt coatings that must be stored, and transfer tapes [63].

Qualitative tests are widely used in testing release coatings to determine the completeness of cure. "Migration" is detected with a cellophane tape test. In this test, No. 610 Scotch tape is first firmly pressed onto the silicone coating. It is then removed and doubled back on itself. If the tape sticks to itself after this treatment, the coating is considered to be well cured and migration-free. "Rub-off" is a term that describes whether or not a silicone coating fails to adhere to the substrate and can be rubbed off in tiny balls of silicone on application of gentle finger pressure. "Smear" is a term to describe the effect that occurs when a finger is firmly pressed and drawn across the silicone film. If an obvious, permanent streak remains, smear is detected and the film is considered to be incompletely cured.

Traditionally, the silicones used as release coatings were applied from low solids, solvent- or aqueous emulsion-based liquids, or as 100% solids reactive systems. These thermal-activated compounds were cross-linked through either a condensation process using a tin catalyst or an addition process using a platinum or rhodium catalyst. Although these products were amply serving the market, in about 1984 there were market pressures such as increased productivity needs, energy costs, manufacturing space, and ecological concerns that led the way to new silicone systems cured by radiation [64]. Another important factor was the need to cure the coatings on temperature sensitive substrates that could not withstand oven temperatures. These features and other advantages and disadvantages of the various techniques including a comparison of free radical and cationic radiation-cured release coatings have been compiled [59]. When release coatings are designed, it is important that reliable and controlled release from the pressure sensitive adhesive is taken into account. The interfacial and viscoelastic contributions to such reliable and controlled release have been discussed in terms of the network set up when radiation-cured systems are used [65,66]. Epoxy silicones have been shown to have more stable release characteristics against acrylic adhesives [67].

Silicones with functional groups suitable for radiation curing have been known for some time, and the chemistry involved in deriving these compounds has been reviewed and a new acrylated epoxy silicone described [64]. The new silicone was based on the use of limonene oxide and a polymethylhydrogensiloxane. This adduct is then reacted with acrylic acid to form the following acrylated epoxy silicone that can be rapidly cured with ultraviolet radiation under inert conditions.

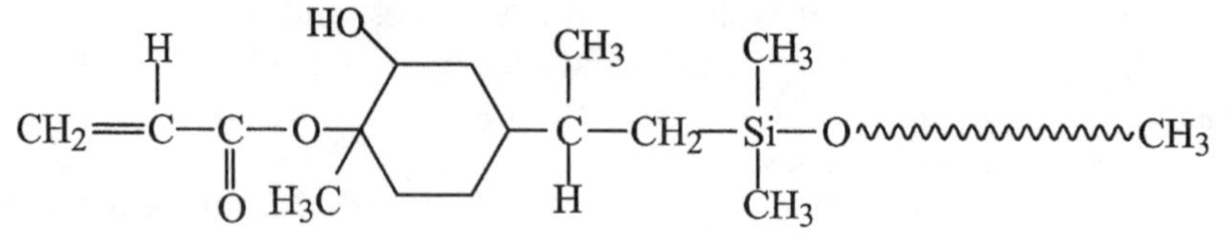

$$CH_2{=}C(H){-}C({=}O){-}O{-}[\text{cyclohexane ring with HO and } H_3C]{-}C(CH_3)(H){-}CH_2{-}Si(CH_3)_2{-}O\sim\sim\sim\sim CH_3$$

A number of different molecular weight epoxy silicones of this nature were prepared and evaluated with 2,2-diethoxyacetophenone and/or combinations of benzophenone and N-methyldiethanolamine photoinitiators. Curing was very rapid (> 400 ft/min, 122 m/min) under nitrogen inerted conditions. Formulation compatibility was achieved with a diacrylate-terminated alkane, a mixture of the diacrylate-terminated alkane and hexanediol diacrylate, N-vinylpyrrolidone, and 2-ethylhexyl acrylate. Hexanediol diacrylate, trimethylolpropane triacrylate, and pentaerythritol triacrylate alone or in combination gave either hazy translucent mixtures or opaque unstable mixtures that would separate on standing. When cured, the stable mixtures gave products that varied from soft and fragile to hard and/or tough films.

Cationic curable compounds were prepared from the precursor limonene epoxide-silicone used to make the above acrylated epoxide silicone. A number of epoxy silicones were prepared from glycidyl acrylate, vinyl cyclohexane monoxide, vinyl norbornene oxide, limonene oxide, and dicyclopentadiene monoepoxide. These compounds were formulated with an onium salt photoinitiator and cured with ultraviolet radiation. The results were variable, but excellent; no smear, no rub-off was obtained with the cyclohexyl epoxy functionalized silicones, which confirmed the excellent cure response of these epoxides in comparison to the glycidyl ethers or dicyclopentadiene epoxides. Useful information for those interested in this technology can be found in the literature [68,69], which contains, for example, a description of release coatings, definitions used, release coating characteristics, and performance requirements.

In the 1990s and later, a number of papers dealing with release coatings have appeared. The advantages of free radical-initiated radiation-cured silicones with details about performance properties important to this particular chemistry have been discussed [70]. The coatings described had good re-adhesion, did not require a post cure, and had release characteristics that could be controlled. Acrylate-functional silicones combined with reactive diluent vinyl ethers have also been investigated [71]. Acrylate-terminated siloxane polymers useful as release coatings have also been prepared by capping of silanol-terminated polydimethylsiloxanes with cycloxyalkylchlorodimethylsilane [72].

Conventional iodonium salts available at the beginning of the time period did not have good miscibility with the silicone epoxides that were being investigated and developed as release coatings. Modified iodonium salts that had alkyl end groups such as the following [60], which were miscible with the organofunctional siloxanes, were considered to be important factors in the development of current day technology:

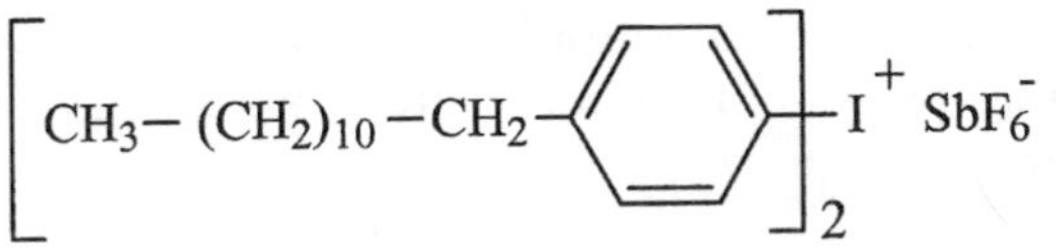

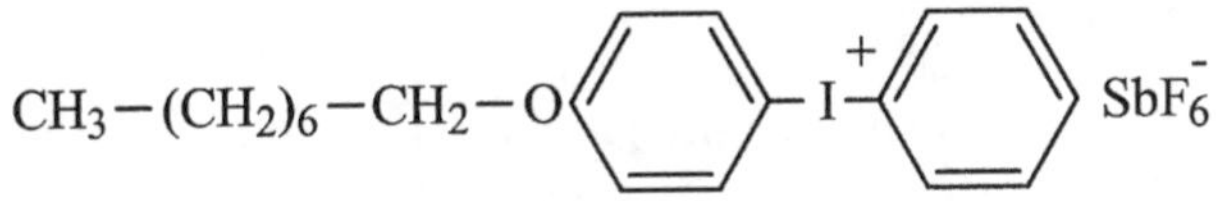

$$\left[CH_3-(CH_2)_{10}-CH_2-C_6H_4 \right]_2 I^+ \ SbF_6^-$$

$$CH_3-(CH_2)_6-CH_2-O-C_6H_4-I^+-C_6H_5 \quad SbF_6^-$$

These photoinitiators were used in combination with novel epoxy-functional organosilicones as well as organosiloxane monomers and polymers that were prepared by hydrosilylation techniques. Polymerization of these compounds was readily initiated with the alkylaryl iodonium salts. Film properties could be modified if co-curable alcohols were employed. Improved adhesion of the silicone coating to a substrate was obtained when a dicycloaliphatic epoxide with connecting methoxysilane groups of the following form was used:

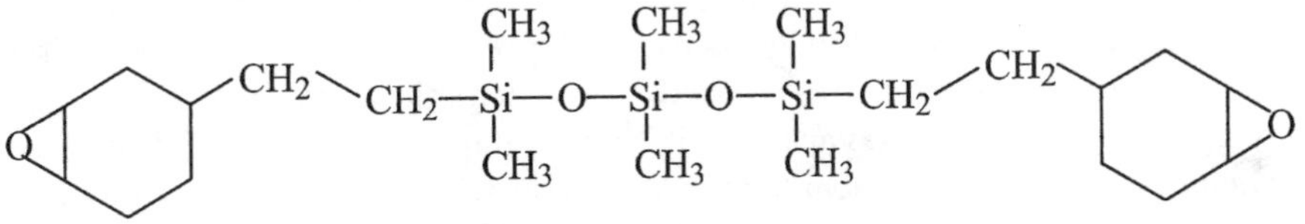

$$C_6H_9O-CH_2-CH_2-Si(CH_3)_2-O-Si(CH_3)_2-O-Si(CH_3)_2-CH_2-CH_2-C_6H_9O$$

bis(3,4-epoxycyclohexylethyl)trimethoxysilane

These epoxides and a number of other epoxy-functional monomers for cationic radiation curing, as well as their properties for coatings and for other applications, have been described in the literature [74–78]. Epoxy-functional silicone fluids that contain chloroalkyl groups attached to silicon by means of carbon-silicon bonds can be photocured with iodonium salts [79]. These compounds are useful as release coatings.

A special area in the field of release coatings is linerless labels. Such labels do not have the disposable, release-coated liner associated with them. Rather, linerless labels are composed of a release-coated top face and a PSA-coated bottom face, with each applied to a paper substrate. Such a configuration allows the labels to be wound in roll form and, subsequently, unwound without the adhesive sticking to the release coating. A release coating based on an epoxy-functional silicone dissolved in a reactive diluent, and formulated with an onium salt photoinitiator and a wax-treated silica powder, has been found useful as a low-gloss release coating for linerless labels [80,81]. Reactive diluents are compounds such as 1,4-cyclohexanedimethanol divinyl ether, 3,4-epoxycyclohexylmethyl-3,4-epoxycyclohexane carboxylate, the diglycidyl ether of hexahydrophthalic acid, and like compounds. The wax-treated silica powder is commercially available and has a particle size of about 2–10 microns.

Other Adhesives

The chemistry, reasons and benefits, and existing and potential applications of radiation-cured adhesives have been outlined [82,83]. The outline also points out that, in addition to cationic and free radical radiation curing, anionic curing also exists. Cyanoacrylates cure by an anionic mechanism. There are a number of other areas in which radiation-cured adhesives are used. These include laminating adhesives for transparent and translucent or opaque materials that use cationic curing technology [13,84–86]. However, more strictly speaking, laminating adhesives are those that are used to form a unitized structure from similar or dissimilar materials. Examples of such laminates include what have been termed "homo laminates," which include the bonding of polyester to polyester, polyethylene to polyethylene, etc., and "hetero laminates," which include paper to polyester, paper to metallized polyester, polyester to nylon, glass to vinyl, polyester, or polycarbonate, etc. [85]. These adhesives may be used to bond clear films, such as a thin wear layer of polypropylene to various foil, film, or paper substrates. The adhesive is usually applied to the clear film, combined in a nip roll with the other substrate layer, and then ultraviolet radiation is used to cure through the clear film. Electron beam technology is used to cure the film combination if opaque films are involved. Of course, electron beams could also be used to cure clear films.

Cationic systems with rapid curing characteristics and high temperature resistance for bonding materials, such as metal to glass or other marked dissimilar materials, have been described [87]. These commercial products have excellent humidity, thermal aging, and thermal shock resistance, as well as other features desirable in adhesives. The new adhesives are said to be 100 times faster in cure, could be cured to twice the depth, were more flexible, and had shrinkage of one-half that of acrylates. Excellent adhesive point strength was obtained when glass was bonded to glass, glass to metal, and glass to plastic.

Laminated pressure sensitive adhesive labels can be much more complicated than the simplified label structure described earlier in this chapter. For example, the label structure and substrate for use on a polyethylene container for food might have the following configuration [88]:

Outer polypropylene surface film
Ultraviolet radiation-cured laminating adhesive
Water-based printed design and/or writing
Paper face stock carrying the above printing
Pressure sensitive adhesive (may be cured in various ways)
High density polyethylene container (i.e., substrate)
Food contact surface

Such labels with a polyolefin surface are used on juice, milk, soda, and water containers. The clear polypropylene surface film gives superior scuff and water resistance to these labels, in comparison to labels that merely have ultraviolet radiation-cured overprint coatings. The migration of components from the radiation-cured laminating adhesive through the polyethylene container was investigated on six laminating-adhesive products, and a number of them had no detected migration, whereas others did exhibit migration of acrylate monomer and photoinitiator [88]. Another study of laminating adhesives involved the investigation of different photoinitiators and lamp types on cure speed [89]. Even though the formulation was given in terms of commercial designations rather than chemical composition, the effort provides useful information and techniques for investigation. One conclusion from the study was that bis(2,4,6-trimethylbenzoyl)phenyl phosphine oxide gave four times the line speed of the other four photoinitiators used.

The adhesives described herein are cured by exposure to radiation. However, adhesives can be exposed to radiation after the curing process takes place if they are used in a medical device. Such devices or packages may be subjected to gamma ray or electron beam sterilization. In a study that involved exposure of various classes of adhesives including laminating, heat seal, and PSAs to gamma radiation, the effects ranged from nil through discoloration to moderate depolymerization [90]. The study concluded that the adhesives studied were sufficiently resistant to gamma radiation, but that caution is needed when adhesives are developed for the medical area wherein sterilization procedures may be used. Examples of medical devices that might be bonded with radiation-cured adhesives include needle assemblies, lancets, insulin syringes, and the like, wherein polymer may be bonded to polymer, polycarbonate to stainless steel, aluminum, and so on [91–93]. Both cationic [94] and free radical systems are used for such adhesives.

A compound suitable for use as a glass adhesive has been prepared by reacting hydroxypropyl acrylate with an unsaturated anhydride such as maleic anhydride [95]. Such unsaturated monomers were formulated with benzophenone and a synergist and optionally a trifunctional acrylate. The radiation-cured systems had adhesion if the glass surfaces were clean, and they had tensile strengths of about 500 psi (~35 kg/cm^2). There was some metal staining when the adhesive was used to bond metal and glass.

A description of some of the free-radical radiation-curable systems, including equipment configurations and the importance of factors such as stepless control of lamp output and the continuous measuring and control of radiation output, is published [96]. The information is for application or end-use fields, such as adhesive tapes, automotive uses, and labels. Related technology has also been described for cationic-curable systems [84]. Low intensity ultraviolet radiation equipment and systems can be used for curing an adhesive where equipment cost must be kept at a minimum [97]. An important factor is the user-friendly aspect of the technology, plus the fact that the units can be purchased as "off-the-shelf" items. The paper describes formulation, properties of cured film, and sources for low intensity lamps. The effect of formulation components on cure speed, surface cure, as well as physical and adhesive properties of systems cured with radiation has been reviewed with respect to curing with low intensity ultraviolet radiation sources [98]. Several formulations that would cure in a few seconds with a 35 mW/cm^2 source were developed. The use of low intensity ultraviolet radiation and visible light in the area of spot curing adhesives has been discussed in a number of publications [99–101].

References

[1] Steuben, K. C., "Radiation Curing of Pressure-Sensitive Adhesives: A Literature Review," *Adhesives Age*, Vol. 20, No. 6, June 1977, p.16.

[2] Dowbenko, R., "Radiation Curing," Ch. 30, *Handbook of Pressure-Sensitive Adhesive Technology*, D. Satas, Ed., Van Nostrand Reinhold Co., 1982.

[3] Zollner, S., "UV-Curable Acrylic Pressure Sensitive Adhesives Features, Mechanisms, Milestones," *Conference Proceedings of RadTech—Europe '99*, Berlin, Germany, 8–10 Nov. 1999, p. 543.

[4] Satas, D., "Radiation Curing of Adhesives," *RadTech Report*, Vol. 5, No. 4, July/August 1991, p. 10.

[5] Anon, "Photocurable Adhesives: 10 Tips for Best Performance," *Adhesives Age*, Vol. 40, No. 12, November 1997, p. 67.

[6] Dionne, C., "More Cooperative UV/EB Industry Focuses on Adhesive Users' Needs," *Adhesives Age*, Vol. 41, No. 1, January 1998, p. 34.

[7] Anon, "UV Curing Improves Production and Efficiency," *Adhesives Age*, Vol. 34, No. 4, April 1991, p. 21.

[8] Figov, M. and Levy, N., "UV/Heat/Sunlight Cured Adhesives Show High Bond Strengths," *Adhesives Age*, Vol. 35, No.13, December 1992, p. 19.

[9] Nuñez, C., McMinn, B., and Vitas, J., "Barriers to the Use of Radiation-Curable Adhesives in Manufacturing," *Adhesives Age*, Vol. 38, No. 1, January 1995, p. 33.

[10] Bluestein, C., "Radiant Energy Curable Adhesives," *Adhesives Age*, Vol. 25, No. 12, 1982 p. 19.

[11] Schlechter, M. M., "Specialty Adhesive Consumption to Grow 5.3% Yearly to 1991," *Adhesives Age*, Vol. 30, No. 4, 1987, p. 34.

[12] Bachmann, A. G., "Ultraviolet Light Curing 'Aerobic' Acrylic Adhesives," *Adhesives Age*, Vol. 25, No. 12, 1982, p. 31.

[13] Verschueren, K. and Wellens, W., "UV Curable Laminating Adhesives for Transparent and Nontransparent Materials," *Conference Proceedings of RadTech—Europe '99*, Berlin, Germany, 8–10 Nov. 1999, p. 565.

[14] Anon, "Rad-Curable Adhesives Get a Boost," *Adhesives Age*, Vol. 44, No. 6, June 2001, p. 8.

[15] Bloch, D. R., "Acrylic Monomers for Radiation-Cured Adhesives and Coatings," *Adhesives Age*, Vol. 37, No. 4, 1994 p. 30.

[16] Herze, P. Y., Philips, M., and Loutz, J. M., "Pressure Sensitive Adhesives Obtained by Irradiation," *Conference Proceedings of RadCure '86*, Baltimore, MD, 8–11 Sept. 1986, p. 12–29.

[17] Mizumachi, H., "Theory of Tack of Pressure Sensitive Adhesives," *Journal of Applied Polymer Science*, Vol. 30, No. 6, 1985, p. 2675.

[18] Barwich, J., Düsterwald, U., Meyer-Roscher, B., and Wüstefeld, R., "UV Curable Polyacrylate Hot Melt Polymers for PSAs," *Adhesives Age*, Vol. 40, No. 4, April 1997, p. 22.

[19] Ellerstein, S. M. and Palit, T. K., "Radiation Curable Pressure Sensitive Adhesives," Conference Papers, *RadTech '88—North America*, New Orleans, LA, 24–28 Apr. 1988, p. 461.

[20] Schlademan, J. A., "Tackifiers and Their Effect on Adhesive Curing," *Adhesives Age*, Vol. 40, No. 10, Sept. 1997, p.24.

[21] Hanrahan, M. J., and Milton M., "Radiation Curable Pressure Sensitive Adhesives," *Conference Proceedings of RadCure '86*, Baltimore, MD 8–11 Sept. 1986, p. 12–23.

[22] Schumacher, K. H., Düsterwald, U., and Fink, R., "UV-Acrylic Hot Melts," *Conference Proceedings of RadTech '99—Europe*, Berlin, Germany, 8–10 Nov. 1999, p. 559.

[23] Malik, R., "Advances in Radiation Curable PSAs," *End User Conference Proceedings of RadTech 2000*, Baltimore, MD, 9–12 April 2000, p. 130; *Adhesive and Sealants Industry*, Vol. 8 No. 4, May 2001, p. 54.

[24] Kardashian, R., "EB Curing/Crosslinking Aids PSA Product Makers," *Adhesives Age*, Vol. 30, No. 4, 1987, p. 24.

[25] Delaney, W. H. and Keough, A. H., "The Versatility of Electron Beam Contract Processing: Releases, PSA, Crosslinking, Grafting, and Cationic Cures," *Proceedings of RadTech '90—North America*, Chicago, IL, Vol. 2, 25–29 Mar. 1990, p. 103.

[26] Berejka, A. J., "Cost-Effective UV/EB PSAs Lead to Integrated Manufacturing," *Adhesives Age*, Vol. 36, No. 4, 1993, p. 26.

[27] Burger, P., Skinner, D., Field, P., and Draper, M., "Experiences with UV Curable PSAs on a Production Facility," *Conference Proceedings of RadTech '99—Europe*, Berlin, Germany, 8–10 Nov. 1999, p. 573.

[28] Fisher, R., "Tandem Coating of UV Curable Silicone Release and Water-Borne, Acrylic Pressure Sensitive Adhesives for the Tape & Label Industry," *Technical Conference Proceedings of RadTech 2000*, Baltimore, MD, 9–12 Apr. 2000, p. 1014.

[29] Patent: Hendricks, J. O., Minnesota Mining & Mfg. Co., Pressure Sensitive Adhesive Tapes, US 2,956,904. (1960).

[30] Patent: Baldwin, F. P. and Malatesta, Alberto, Exxon Research and Engineering Co., Conjugated Diene Butyl, US 3,965,213. (1976).

[31] Merrill, N. A., Gardner, I. J., and Hughes, V. L., "Using Conjugated Diene Butyl for Radiation-Cured PSAs," *Adhesives Age*, Vol. 35, No. 7, July 1992, p. 24.

[32] Berejka, A. J. and Looney, R. W., "Radiation Curing of Isoprene-Based PSAs," *Adhesives Age*, Vol. 34, No. 4, Apr. 1991, p. 30.

[33] Berejka, A. J., "Reengineered PSA Manufacture for VOC Reduction and Energy Savings," *Adhesives Age*, Vol. 40, No. 10, September 1997, p. 28.

[34] Bowtell, M., "Reactive Polymers for UV Curable PSAs," *Adhesives Age*, Vol. 40, No. 10, September 1997, p. 50.

[35] McElrath, K. O. and Robertson, M. H., "Heat Resistant Isobutylene Copolymers," *Adhesives Age*, Vol. 38, No. 10, Sept. 1995, p. 28.

[36] Dupont, M., Masse, M., and Schneider, J., "Rubber-based Radiation Curable Pressure Sensitive Adhesives," *Conference Proceedings of RadTech '99—Europe*, Berlin, Germany, 8–10 Nov. 1999, p. 551.

[37] Pappas, S. P. and Woods, J., "Using Dark Cure Processes to Improve UV Adhesives," *Adhesives Age*, Vol. 34, No. 4, April 1991 p. 24.

[38] Erickson, J. R., Zimmermann, E. M., Southwick, J. G., and Kiibler, Kathleen S., "Liquid Reactive Polymers for Radiation Curable High Performance PSAs," *Adhesives Age*, Vol. 38, No. 12, November 1995, p. 18.

[39] Schumacher, K. H. and Sanborn, T., "UV-Acrylic Hot Melts—Performance and Processing of an Attractive Raw Material for Self Adhesive Articles," *End User Conference Proceedings of RadTech 2000*, Baltimore, MD, 9–12 April 2000, p. 185; "UV-Curable Acrylic Hot Melts for PSAs," *Adhesives & Sealants Industry*, Vol. 8, No. 5, June/July 2001, p. 42.

[40] Patent: Fukukawa, S., Shimomura, T., Yoshikawa, N., and Murakami, T., Nitto Electric Industrial Co., Ltd., Process for the Production of Pressure-Sensitive Adhesive Materials, US 3,772,063. (1973).

[41] Patent: Christenson, R. M. and Anderson, Carl C., PPG Industries, Inc., Pressure-Sensitive Adhesive Articles and Method of Making Same, US 3,725,115. (1973).

[42] Kroll, M. and Fuller, H. B., "Radiation Curable Hot Melt PSAs with Excellent High Temperature Performance," *End User Conference Proceedings of RadTech 2000*, Baltimore, MD, April 9–12, 2000, p. 156.

[43] Patent: Brookman, R. S., Grib, S., and Pearson, D. S., The Firestone and Rubber Co., Process for the Preparation of Pressure Sensitive Adhesive, US 3,661,618. (1972).

[44] Patent: Kasper, A. A., Kendall Co., Irradiation of Pressure-Sensitive Adhesive Tape in an Oxygen-Free State, US 3,328,194. (1967).

[45] Stueben, K. C. and Patrylow, M. F., "Ultraviolet Cured Pressure Sensitive Adhesives, I. Acrylate Grafted Polyvinyl Alkyl Ethers," *Journal of Radiation Curing*, Vol. 9, No. 16, April 1982.

[46] Stueben, K. C., Patrylow, M. F., and Gibb, T. B., "Ultraviolet Cured Pressure Sensitive Adhesives, II. Monoacrylated Grafted Polyethers," *Journal of Radiation Curing*, Vol. 9, No. 20, April 1982.

[47] Patent: Stueben, K. C., Azrak, R. G., and Patrylow, Union Carbide Corp., Radiation Curable Pressure Sensitive Adhesive Compositions, US 4,165,266. (1979).

[48] Stueben, K. C. and Patrylow, M. F., "Ultraviolet Pressure Sensitive Adhesives, III. Correlations Between Curing/Formulating Parameters and Properties," *Journal of Radiation Curing*, Vol. 9, No. 4, April 1982.

[49] Lin, J. and Sun, B., "UV Curable, Screen Printable Pressure-Sensitive Adhesive," Technical Paper FC84-1019, *RadCure '84*, Atlanta, GA, 10–13 Sept. 1984.

[50] Takiguchi, R. and Uryu, T., "Polymeric Reaction of Polymer-Monomer System Irradiated by Low-Energy Electron Beam. IV. Adhesive Properties of Pressure Sensitive Adhesive," *Journal of Applied Polymer Science*, Vol. 40, No.1313, 1990.

[51] Miller, H. C., "Low Odor U.V. Pressure Sensitive Adhesives, *Workshop presentation, RadTech '92—North America*, Boston, MA, 26–30 Apr. 1992.

[52] Miller, H. C., "Formulating Low Odor UV-Curing PSAs," *Adhesives Age*, Vol. 36, No. 10, September 1993, p.32.

[53] Glotfelter, C., "UV Curable Monomers and Oligomers in PSA Applications," *Adhesives Age*, Vol. 40, No. 3, March 1997, p. 50.

[54] Johnson, M. A., "Some Recent Approaches to the Development of Acrylic UV-curable Pressure Sensitive Adhesives," *Conference Proceedings of RadCure '86*, Baltimore, MD, 8–11 Sept. 1986, p. 12–13.

[55] Kerr, III, Stuart R., Radl, M., Fitzgerald, T., Pinto, O., Frances, J.-M., "UV-Cure PSAs and Silicone Release," *End User Conference Proceedings of RadTech 2000*, Baltimore, MD, 9–12 April 2000, p. 165.

[56] Ramharack, R., Chandran, R., Shah, Smita, H., Deepak, O., John, and Foreman, P., "Ultra Low Voltage Curing of Acrylic Hot Melt PSAs," *Adhesives Age*, Vol. 39, No. 13, December 1996. p. 40.

[57] Huber, H. F. and Müller, H., "Pressure-Sensitive Adhesives Based on Radiation Curable

Polyesters," *Conference Proceedings of RadCure '86*, Baltimore, MD, 8–11 Sept. 1986, p. 12–1; "Radiation Curable Polyesters for the Formulation of Pressure-Sensitive Adhesives," *Rad-Cure Europe '87*, Munich, Germany, 4–7 May 1987, p. 8–35.

[58] Kauffman, T., Chappell, J., Acevedo, M., "The Use of Acrylated Polyesters in the Formulation of Radiation Curable Adhesives," *Proceedings of RadTech '98*, Chicago, IL, 1998, p. 310.

[59] Lutz, M. A., Scheibert, K. A., Gutek, B. I., and Peterson, A. L., "High Performance Ultraviolet Curable Silicones," *Proceedings of RadTech '90—North America*, Chicago, IL, Vol. 1, 25–29 Mar. 1990, p. 371.

[60] Kerr III, Stuart R., "UV Silicone Release Coatings: State of the Art," *Adhesives Age*, Vol. 39, No. 8, July 1996, p. 26; "Electron Beam Curing of Epoxy-Silicone Release Coatings," Vol. 41, No.11, November 1998, p. 27.

[61] Frances, J. M., "Formulating with UV/EB Epoxy Silicones and Adhesives," *Conference Proceedings of RadTech '99—Europe*, Berlin, Germany, 8–10 Nov. 1999, p. 581.

[62] Gray III, L. S., "Chemical and Cure Parameters That Influence Functional Coatings," Conference Papers, *RadTech '88—North America*, New Orleans, LA, 24–28 April 1988, p. 559.

[63] Bickford, Jr., R. H. and Jachman, J., "Innovative and Unique Applications of Radiation Cured Silicone Release Coatings," Conference Papers, *RadTech '88—North America*, New Orleans, LA, 24–28 Apr. 1988, p. 69.

[64] Eckberg, R. P., "Radiation Curable Silicones," *Conference Proceedings of Radcure '84*, Atlanta, GA, 10–13 Sept. 1984, p. 2-1; Conference Papers, *RadTech '88—North America*, New Orleans, LA, 24–28 Apr. 1988, p. 576.

[65] Gordon, G. V., Moore, P. A., Popa, P. J., Tonge, J. S., and Vincent, G. A., "Radiation-Cured Silicone Release Coatings: 'Sticking Lightly,'" *Technical Conference Proceedings of RadTech 2000*, Baltimore, MD, 9–12 Apr. 2000, p. 994.

[66] Riding, K. D., "Controlled Release Additives in UV Curable Epoxysilicone Chemistry," Vol. 1, *Proceedings of RadTech '90—North America*, Chicago, IL, 25–29 Mar. 1990, p. 377.

[67] Riding, K. D. and John, S. M., "UV Epoxysilicones Offer Better Release Against Acrylics," *Adhesives Age*, Vol. 33, No. 9, Sept. 1990, p. 20.

[68] Cyterski, D. J., "Radiation-Curable Release Coatings," *Conference Proceedings of RadCure '84*, Atlanta, GA, 10–13 Sept. 1984, p. 2–19.

[69] Varaprath, P. J., Wright, A. P., and Keryk, J. R., "Chemistry and Applications of Radiation Curable Silicones," *Conference Proceedings of RadCure '86*, Baltimore, MD, 8–11 Sept. 1986, p. 16–29.

[70] Bickford, R.H., Jr., "Applications–UV/EB Cured Silicone Release Coatings," *Proceedings of RadTech '90—North America*, Chicago, IL, 25–29 Mar. 1990, p. 109.

[71] Patent: Dougherty, J. A. and McKittrick, J., ISP Investments Inc., Release Coating Compositions Comprising an Acrylate-Functional Silicone Resin and a Vinyl Ether, US 6,011,079. (2000).

[72] Okamoto, Y., Crossan, D., and Ferrigno, K., "UV Curable Acrylate-Polysiloxanes," *RadTech '88—North America*, New Orleans, LA, 24–28 Apr. 1988, p. 167.

[73] Patent: Eckberg, R. P. and LaRochelle, R. W., General Electric Co., Ultraviolet Curable Epoxy Silicone Coating Compositions, US 4,279,717. (1981).

[74] Patent: Koerner, G., Kropac, V., and Weitemeyer, C., Th. Goldschmidt AG, "Process for the Manufacture of Organopolysiloxanes for Use in Adhesive Coating Materials," US 4,306,050. (1981).

[75] Patent: Hockemeyer, F., John, P., Muller, H., and Preiner, G., Wacker-Chemie GmbH, Crosslinkable Organopolysiloxanes and a Method for Preparing the Same, US 4,571,349. (1986).

[76] Patent: Varaprath, P. J. and Ziemelis, M. J., The Dow Chemical Corp., "Organopolysiloxane Compositions Curable by Ultraviolet Radiation," US 4,831,064. (1989).

[77] Eckberg, R., P., Riding, K. D., and Farley, D. E., "Novel Photocurable Organosilicon Compositions," *Proceedings of RadTech '90—North America*, Chicago, IL, 25–29 Mar. 1990, p. 358.

[78] Crivello, J. V. and Lee, J. L., "New Epoxy Functioal Silicone Monomers for Cationic UV Curing," *Proceedings of RadTech '90—North America*, Chicago, IL, 25–29 Mar. 1990, p. 432.
[79] Patent: Eckberg, R. P., General Electric Co., "Silicone Fluids Having Chloroalkyl and Epoxy Groups and Photocurable Silicone Coating Compositions," US 5,500,300. (1996).
[80] Patent: Kline, J. R., Monarch Marketing Systems, Inc., Silicone Release Coating Composition, US 6,022,050. (2000).
[81] Patent: Kline, J. R., Monarch Marketing Systems, Inc., Silicone Release Coating Composition. US 6,231,922. (2001).
[82] Shi, G. and DeMarco, J., "Light Cure Adhesives–Chemistry and Applications," *End User Conference Proceedings of RadTech 2000*, Baltimore, MD, 9–12 Apr. 2000, p. 122.
[83] Huber, H. F., Losensky, H. W., and Müller, H., "New Product Opportunities Through Specific Advantages of Radiation Cure," *Proceedings of RadTech '90—North America*, Chicago, IL, Vol. 1, 25–29 Mar. 1990, p. 131.
[84] Carter, W., "UV Laminating Opaque Substrates Using Cationic Dark Cure," *End User Conference Proceedings of RadTech 2000*, Baltimore, MD, 9–12 April 2000, p. 148.
[85] Ellerstein, S. M. and Lee, S. A., "Radiation Curable Laminating Adhesives," Conference Papers, *RadTech '88 North America*, New Orleans, LA, 24–28 Apr. 1988, p. 465.
[86] Carroy, A., "Factors Influencing the Adhesion Performance of Cationic UV-Cured Metal Coatings," Paper No. 4, *Proceedings of Aspect of Adhesion Conference, Paint Research Association*, Egham, England, 1992.
[87] Attarwala, S., Crossan, D., and Grabek, P., "High Temperature Resistant Adhesives for Glass Bonding," Conference Papers Addendum, *RadTech '88—North America*, New Orleans, LA, 24–28 Apr. 1988, p. 73.
[88] Lapin, S. C., "Migration of UV Curable Laminating Adhesive Components through HDPE Containers," *Proceedings of RadTech 2000*, Baltimore, MD, 9–12 Apr. 2000, p. 1007.
[89] Sciangola, D. A., "UV Curable Laminating Adhesives," *Proceedings of RadTech 2000*, Baltimore, MD, 9–12 Apr. 2000, p. 1023.
[90] Fries, J. A., "Radiation Curing and Sterilization of Adhesives," Conference Papers, *RadTech '88 North America*, New Orleans, LA, 24–28 Apr. 1988, p. 440.
[91] Perez, E. R., "Bonding Medical Devices with UV-Curable Adhesives," *Adhesives Age*, Vol. 37, No. 9, August 1994, p. 25.
[92] Beasley, John, "UV Curing Eliminates Solvents in Bonding Medical Plastics," *Adhesives Age*, Vol. 34, No. 4, April 1991.
[93] Dymax Corp., "We Create Solutions," Brochure, Medical Technology Group, Torrington, CT, 1991.
[94] Albright, L. D., Lamb, K. T., Carroy, A. C., and Carter, J. W., "Cationic Laminating Adhesives," *Proceedings of RadTech '98—North America*, Chicago, IL, 1998, p. 166.
[95] Mallik, A., "Radiation Curable Adhesive for Glass," *Proceedings of RadTech '90—North America*, Chicago, IL, Vol. 1, 25–29 Mar. 1990, p. 124.
[96] Müller, Rolf, "UV-Systems for Curing of Radiation-Curable UV-Acrylates," *End User Conference Proceedings of RadTech 2000*, Baltimore, MD, 9–12 Apr. 2000, p. 140; *Conference Proceedings of RadTech '99—Europe*, Berlin, Germany, 8–10 Nov. 1999, p. 591.
[97] Sutkaitis, D. and Crossan, D., "Low Intensity Ultraviolet Curing Adhesives," Conference Papers, *RadTech '88—North America*, New Orleans, LA, 24–28 Apr. 1988, p. 452.
[98] Gsuros, Z. and Crossan, D., "Designing Fast, Ultraviolet Curing, High Performance Adhesives for Low Intensity Light Sources," *Proceedings of RadTech '90—North America*, Chicago, IL, Vol. 2, 25–29 Mar. 1990, p. 21.
[99] Bachmann, C., "Expanding Capabilities with UV/Visible Light Curing Adhesives," *Adhesives Age*, Vol. 38, No. 4, April 1995, p. 14.
[100] EFOS USA Inc., "UV Spot Curing Advances Yield Greater Control," *Adhesives Age*, Vol. 39, No. 10, September 1996, p. 30.
[101] Beasley, J., "UV Spot Curing Facilitates Bonding in Critical Areas," *Adhesives Age*, Vol. 40, No. 4, April 1997, p. 20.

Additives and Miscellaneous

Antifoaming Agents

CERTAIN RADIATION CURABLE FORMULATIONS can develop foaming problems during the manufacturing process, and in certain instances, this foam will build up and become a difficulty. In other cases, the foam may only appear at particular times or with particular formulations. In either case, foam can be, at best, a nuisance and, at worst, can cause major losses in production. These additives, which usually are silicone fluids, modified silicones, or acrylics, can be obtained from several suppliers, and only a couple will be mentioned. The silicone fluids are different molecular weight dimethylpolysiloxanes,

$$CH_3-\underset{\displaystyle CH_3}{\overset{\displaystyle CH_3}{\underset{|}{\overset{|}{Si}}}}-O\left[\underset{\displaystyle CH_3}{\overset{\displaystyle CH_3}{\underset{|}{\overset{|}{Si}}}}-O\right]_n\underset{\displaystyle CH_3}{\overset{\displaystyle CH_3}{\underset{|}{\overset{|}{Si}}}}-CH_3$$

and the modified silicone compounds are variations of this structure, in which some or all of the methyl groups have been replaced with other groups.

A low viscosity silicone antifoam compound for inks and coatings, Dow Corning® 163 Additive was developed for use in radiation cure formulations [1]. It can be used as an additive to prevent buildup of foam or to destroy foam that has been generated. An additive that has been found to be effective in cycloaliphatic epoxide-based cationic formulations is an acrylic sold as PC-1344 Defoamer by Solutia, Inc. [2].

Expanding Monomers

As described in Chapter 5, most molecules shrink when they polymerize. Such shrinkage is a function of the polymerization mechanism (double bonds >> oxirane rings) and molecular weight (the lower the molecular weight, the greater the shrinkage). Other molecules will expand in volume when they polymerize. A common example of expansion that takes place when a phase change occurs is the transformation of water to ice. Most compounds shrink when they crystallize, but water is different. It expands. As water cools, it expands into all the microscopic interstices on a substrate, and this expansion reaches a maximum when it freezes. The result is extremely good adhesion as most people have experienced on cold winter mornings when they tried to scrape ice from the windshield after a freezing rain. Although not a polymerization, it points out in a common way why good adhesion is obtained from monomers that expand when they polymerize.

Expanding molecules were patented in the early 1970s [3]. The molecules involved polycyclic ring-opening monomers, such as the spiro-orthocarbonates, spiro-orthoesters, and bicyclic orthoesters [4]. These monomers have the following structure:

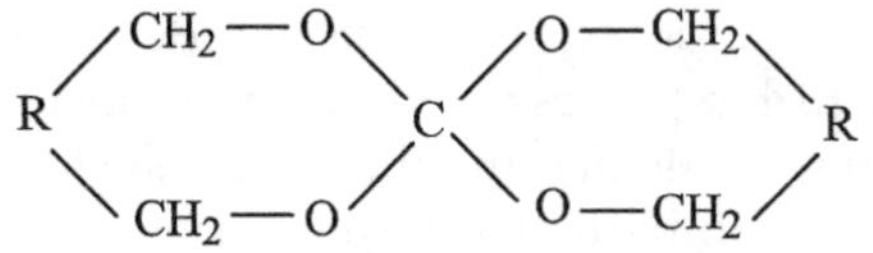

R(–CH_2–O–)(–CH_2–O–)C(–O–CH_2–)(–O–CH_2–)R

Generalized spiro-orthocarbonate

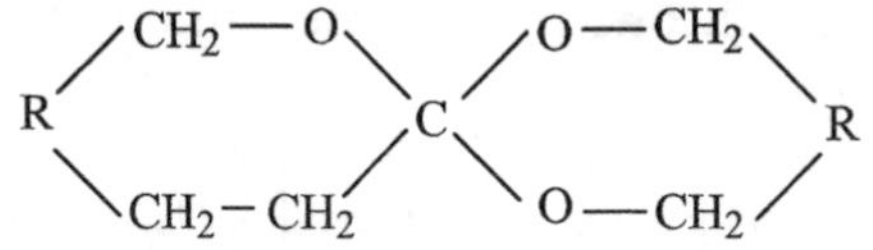

R(–CH_2–O–)(–CH_2–CH_2–)C(–O–CH_2–)(–O–CH_2–)R

Generalized spiro-orthoester

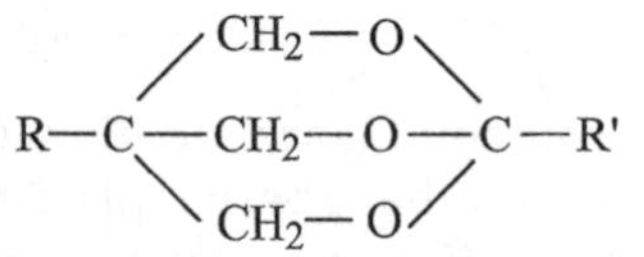

R–C(–CH_2–O–)(–CH_2–O–)(–CH_2–O–)C–R'

Generalized bicyclic orthoester

In these structures, R and R′ can be the same or different and can be norbornene, which will polymerize by cationic means with a Lewis or Bronsted acid, or —CH=CH_2, which will polymerize by free radical or cationic means. For example:

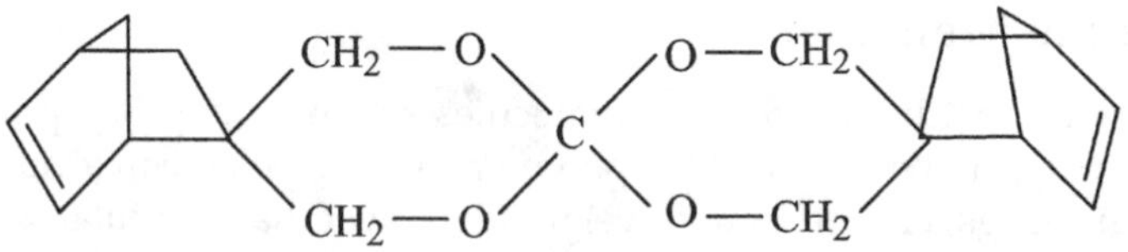

(norbornene)(–CH_2–O–)(–CH_2–O–)C(–O–CH_2–)(–O–CH_2–)(norbornene)

Norbornene spiro-orthocarbonate

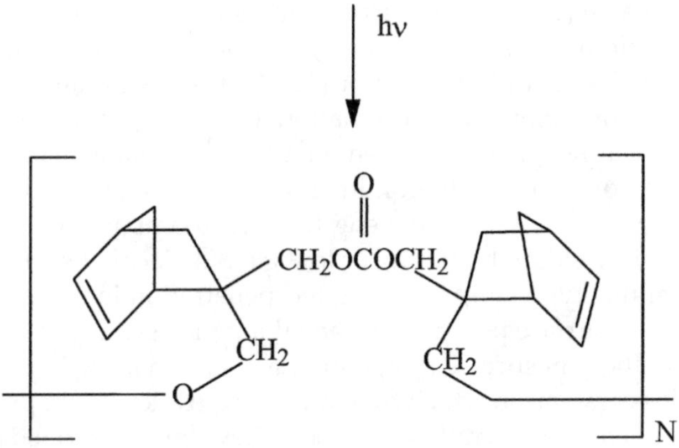

hν

O
CH_2OCOCH_2
CH_2 CH_2
O
N

Ring-opened unit

In this reaction, the monomer is ruptured at two points and a covalent bond is lost [5]. The net result is an expansion rather than contraction, due to polymerization. Monomers of this type are expected to have utility as coatings for aircraft and automobile head- and tail lights, plastics in general, optical fibers, photoresists, printed circuit assemblies, three-dimensional objects, and similar high quality end uses [6,7].

Gloss Control

Radiation cured coatings tend to have high gloss. The liquid, low viscosity systems are "instantly frozen" into the coating when they are exposed to radiation. As a result, unless special modifiers are used, the coating will have high gloss, a property that is aesthetically pleasing for many applications. However, there also are a number of end uses that require moderate to low gloss, and there is a need to control and modify gloss. (Also see **Textured Coatings** below, which is another way to obtain low gloss surfaces.)

One study investigated a number of commercial products as flatting agents for free-radical initiated, ultraviolet radiation cured coatings based on acrylated epoxides [8]. Diatomaceous earth, fumed silicas, and a proprietary reactive compound of undisclosed composition were used. Gloss could be varied with these compounds, and 60° gloss values of less than 10 could be obtained if a solvent were used for dilution purposes when high concentrations of the additives were used. With dual cure systems, in which radiation intensity is varied, i.e., increased, relatively low gloss urethane acrylate [9] and polyester/styrene [10] coatings can be achieved. Other ways to vary gloss include use of combinations of polydiene-based urethane acrylate with a standard urethane acrylate [11]; by exposure to radiation of different wavelengths in multiple steps, by adjusting the spectral distribution of radiation, the intensity, or the dose of the radiation, or by adjusting the time interval between the initial and subsequent irradiation steps [12]; and by use of a formulation containing an acrylated melamine, one or more silica, and a Norrish Type I and Type II photoinitiator [13]. Flat coatings can be obtained by first exposing a radiation-cure formulation to ionizing or actinic radiation in an oxygen-rich atmosphere and then, in a following step, exposing the system to ionizing or actinic radiation in an inert environment [14]. Irradiation of the surface with a 172 nm wavelength excimer lamp before the usual irradiation step will yield high matting or low gloss on lacquer systems [15]. The photons from this type lamp have limited penetration into the coating with only a thin-layer surface polymerization taking place. Microscopic shrinkage takes place in the thin surface film, and as a result, matting is produced.

Dual gloss systems in which two different gloss levels are produced on a substrate by a layering and radiation process have been described [16]. A combination of crystalline compounds or a crystalline compound and an amorphous compound in powder form can be used to make low gloss ultraviolet light curable powder coatings [17]. For example, a solid unsaturated polyester is combined with a vinyl ether terminated urethane oligomer, and a photoinitiator. The mixture is heated to form the coating, then allowed to cool and recrystallize to the low gloss finish, before curing with ultraviolet radiation.

Inorganic Glasses for Pigmented and Thick Section Cures

Particulate materials that are termed as radiation-transmissible materials or ultraviolet radiation transmissible (UVT™) materials were introduced to the market in 2000 [18]. These chemically and thermally stable spherical particles, which are marketed as UVT Sunspheres™ and UVT Sunspacers™, are solid and substantially noncrystalline in nature. When dispersed in radiation curable formulations, the spheres allow pigmented and/or thick, clear coatings to be cured readily with ultraviolet radiation [19]. The UVT materials appear to act as transmission paths that pipe radiation to the interior of coatings, inks, or sealants where it can activate the photoinitiator and cause complete cure throughout the film. The proprietary [20], engineered materials are available in average sizes of from 0.3 to 120 μm. They are easily dispersed in cationic and free radical curable formulations as well as in powder coating formulations at weight percentages up to 50% without a change in system viscosity. Concentrations of about 5–30% are recommended. Results indicate that expensive photoinitiator and pigment concentrations can be reduced when the UVT materials are incorporated into a formulation. Cured films have enhanced mechanical properties.

Odor

In the early days of cationic photocuring with onium salts, residual, malodorous compounds were evolved at times, particularly, it seemed, in partially or undercured coatings. Free radical photocuring had other odors as, for example, amine odors from synergists or odors from photoinitiator fragments associated with it. Today one does not hear much about this difficulty.

If organo-sulfurous odors should arise, they may be eliminated or minimized by the addition of nonaromatic carbon-carbon unsaturation to the formulation [21,22]. The compound can have one or more unsaturated groups; preferably, one of the carbon atoms of the unsaturation is bonded to a carbonyl group, and neither of these carbon atoms is bonded to more than one hydrogen atom. The general structure of such compounds is:

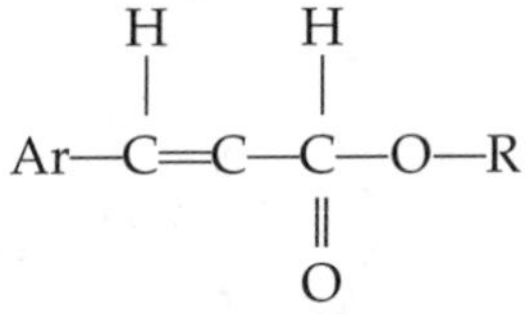

$$\mathrm{Ar{-}\overset{\displaystyle H}{\overset{|}{C}}{=}\overset{\displaystyle H}{\overset{|}{C}}{-}\underset{\displaystyle O}{\underset{\|}{C}}{-}O{-}R}\quad ,$$

wherein *Ar* is an aromatic radical such as phenyl, chlorophenyl, naphthyl, and the like, and *R* is hydrogen or lower alkyl, benzyl, chloroalkyl, etc. When 2.5 parts of trans-4-phenyl-3-butene-2-one was added to an onium salt/cycloaliphatic epoxide formulation that had a residual odor, after cure no odor was detectable.

Another study calls for addition of a compound that will generate a stable free radical to reduce or eliminate the odor [23]. Such compounds are organic hydrazyls, particularly triarylhydrazyls, with at least two and desirably three electron withdrawing groups on the aryl amino ring. The compound 2,2-diphenyl-1-picrylhydrazyl was found to be particularly effective. Radiation activated scavengers for the organic sulfur compounds include aromatic ketones, benzil compounds, benzoin compounds, aryl oxime compounds, and organic peroxide compounds [24]. Examples of specific compounds include α-benzoin oxime, 9-xanthone, 2-chlorothioxanthone, benzoyl peroxide, 4-chlorobenzil, benzoin isobutyl ether, acetophenone, α,α -diethoxyacetophenone, and the like. The latter class of compounds, i.e., the phenones, was found to be particularly effective.

Nonyellowing photoinitiators, that were an advance in low odor coatings for free radical-cured formulations, were based on a polymeric hydroxyalkylphenone structure [25]. Oligomeric and polymeric α-hydroxy-[4-(1-methylvinyl)] isobutyrophenones were found to be highly reactive photoinitiators for clear acrylate coating formulations. By starting with photoinitiators that have acrylate functionality, it was possible to make polymeric photoinitiators [26]. These photoinitiators, which were effective at low concentrations, became a part of the final coating, and thus, there are no photolysis fragments remaining in the cured coating.

Free radical generating photoinitiators can also have residual odors. Novel difunctional photoinitiators that have a low residual odor are of the form [27]:

$$A-C_6H_4-X-C_6H_4-B$$

where *A* and *B* are structures that will generate a free radical when irradiated with ultraviolet radiation and *X* is a group capable of altering the ultraviolet radiation absorption of the molecule remaining after free radical generation. A compound with a maximum absorption at 316 nm that has been particularly useful for these ends is [28]:

$$C_6H_5-C(=O)-C_6H_4-S-C_6H_4-C(=O)-C(CH_3)_2-SO_2-C_6H_4-CH_3$$

A ketosulphone benzophenone (LFC 1001, Lamberti S.p.A.)

This compound is organic solvent soluble and water insoluble, and it is easily incorporated into free radical monomer formulations. It is particularly useful in ink formulations.

Benzophenone/methyl diethanolamine and a commercially available acrylated benzophenone with reactive amine co-initiators were compared in formulations based on polyester acrylates, epoxy acrylates, and reactive diluents [29]. Photoinitiator and co-initiator concentrations, degree of cure, and reactivity of photoinitiator system were all found to have an effect on odor.

If such odors or others are a difficulty for a particular process, a methodology for testing odors [30] and for using fragrance additives [31] has been published. In regard to the fragrance additives, it should be kept in mind that such compounds are complex mixtures of aroma chemicals, and they merely mask and do not remove or destroy an annoying odor. A number of companies supply the compounds, which are generally termed natural or synthetic essential oils, aroma chemicals, masking compounds, and perfume compounds. Over 5000 compounds are available for these uses.

Scratch, Slip, and Abrasion Resistance

Clear, colored, or pigmented coatings that are used to decorate plastic, metal, or other substrates almost always have end uses that will subject them to scratching, staining, or abrasion attack. In addition to being decorative, many coatings are functional in nature. For example, clear automobile topcoats, floor coatings, and others are mainly functional, as are magnetic tape coatings. Scratch and abrasion resistance are important properties for these end uses. A coating with good scratch resistance will usually have good abrasion resistance, and slip can be a factor in improving scratch resistance.

Early studies of abrasion resistance [32] found that pentaerythritol triacrylate provided better abrasion resistance than the equivalent coatings based on trimethylolpropane triacrylate. When comonomers were studied, an 80/20 blend of pentaerythritol triacrylate and N-vinyl pyrrolidone had the best performance in the sense of abrasion resistance, optical clarity, and shrinkage. Shrinkage was measured by degree of curling. A patented coating [33], marketed as GAFGARD™ 233, was described as imparting super abrasion resistance to a variety of plastic substrates [34. 35]. For example, uncoated polycarbonate had a Taber Abrasion loss of 52.4 mg, but when coated with the novel coatings, the value dropped to 7.4 mg under comparable conditions. A recent study examined the wear resistance of a variety of radiation-cured acrylates by different test methods and concluded that test method does influence the results [36]. Test methods used by formulators should be in concert with those used by the end user. This was particularly true for abrasion resistance. A discussion of abrasion, scratch, and mar resistance and understanding the tests used to measure these properties is available [37].

Combinations of polyacrylic esters such as pentaerythritol tetraacrylate, dipentaerythritol hexaacrylate, or the like; carbamic compounds such as 1,3,5-tris(2-acryloyloxyethyl)isocyanurate; optionally, a diluent monomer such as 2-ethylhexyl acrylate, 2-phenoxyethyl acrylate, glycidyl acrylate, and the like; and a photoinitiator such as diethoxyacetophenone, benzoin isobutyl ether, and the like, have been found to have good abrasion resistance against steel wool [38]. The compositions have an inherent abrasion resistance to 000 steel wool and are useful as ophthalmic lenses.

When the abrasion resistance was investigated as a function of composition, cross-link density, and mechanical properties, it was found that when cross-linked density was increased by means of multifunctional monomers, abrasion resistance improved. However, it was found unexpectedly that other factors such as cure speed and photostability were decreased [39]. The study suggested that in the investigation of abrasion and scratch resistance, attention should be paid to the dependence of properties on free volume and cross-linking of the polymer matrix. The various tests that can be used in the development of scratch resistant topcoats have been described, and their exterior applications have been investigated [40].

Composite radiation-cured coating systems have been found to be highly abrasion resistant.[1] The systems are made up of a relatively soft base coating comprised of an aliphatic urethane acrylate combined with a multifunctional acrylate and a hard outer or surface coating. The hard abrasion resistant layer is comprised of a highly cross-linked acrylic copolymer that con-

[1] Patent: Hodnett, W. P., III, Thor Radiation Research, Inc., Protective Coating System for Imparting Resistance to Abrasion, Impact, and Solvents, US 5,254,395 (1993).

tains a minimum of 40 percent of a tri or higher functional acrylate monomer. In an example, the soft, inner coating had a Rockwell hardness ("M" scale)[2] of 35 and the hard surface coating a similar Rockwell hardness of 124. The two coatings are individually applied and the soft coating is radiation cured prior to application of the hard coating. The two coatings have excellent adhesion to each other. Ultraviolet radiation was used to effect the cure.

Other ways to improve abrasion resistance include the addition of silica acrylates, which are silica organosols in 1,6-hexanediol diacrylate, and tripropylene glycol diacrylate, which improve surface properties [41]. The silica particle size was 25 or 50 nm, and the optimum amount of silica to obtain optimum abrasion resistance was about 20%. Addition of fluoro or silicone acrylate monomers to the system has been used to modify surface properties [42]. The monomers had no effect on the cure rate of urethane acrylates, and the silicone acrylates were more efficient than the fluoro-acrylates in modifying the surface. Both considerably increased the hydrophobic character of the surfaces at concentrations as low as 1%. The results suggest that these chemicals can be used to modify surface chemical resistance, hydrophobicity, and slipperiness.

Cationic-cured, cycloaliphatic epoxide-based coatings were examined by five scratch tests, namely, pin test, ball test, rotating fiber test, ball mill abrasion test, and Taber abrader [43]. Scratch resistance improved with increasing cross-link density and increasing glass transition temperature by all test methods. Addition of wax to the coatings increased scratch resistance when scratch was measured by a single draw but decreased it when there was repeated movement. Addition of pigment decreased scratch resistance when the material removed remained between the device and the substrate, but it increased in the other tests.

Silane Coupling Agents

Silane coupling agents or silane adhesion promoters and their use in coatings have been described in a general sense by Plueddemann [44]. These compounds have the general structure:

$$R{-}Si(OCH_3)_3$$

wherein the group R—can be:

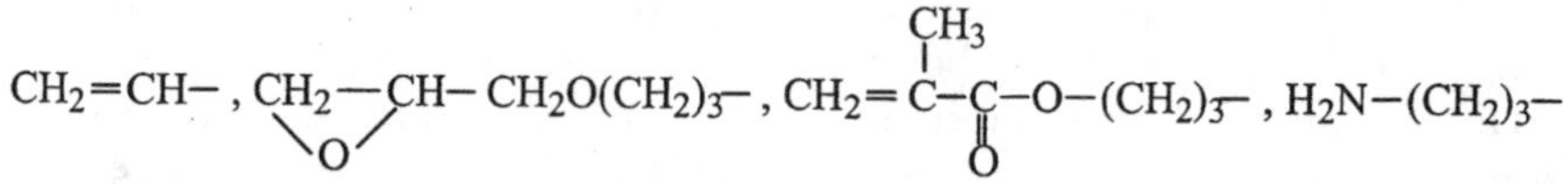

$CH_2{=}CH{-}$, $\underset{\backslash O /}{CH_2{-}CH}{-}CH_2O(CH_2)_3{-}$, $CH_2{=}\overset{CH_3}{\overset{|}{C}}{-}\underset{\overset{\|}{O}}{C}{-}O{-}(CH_2)_3{-}$, $H_2N{-}(CH_2)_3{-}$

Vinyl | Glycidyl epoxide | Methacrylate | Primary amine

[2] ASTM Test Method D 785, Test Method for Rockwell Hardness of Plastics and Electrical Insulating Materials.

or other groups such as chloropropyl, diamine, mercaptopropyl, cycloaliphatic, and others. The coupling agents provide adhesion by bonding to the matrix polymer system and to the substrate, which may also be a pigment particle. Judicious selection of the organofunctional group, and the system involved, is important to property improvement. For example, if an amine silane adhesion promoter were used in a cationic system catalyzed with an onium salt, the amine portion of the molecule would be expected to inhibit cure and, in fact, they do so. In such a system, one would expect better results with the cycloaliphatic epoxide or glycidyl epoxide functional molecules.

Silane coupling agents or adhesion promoters are tools that have enabled increased filler loadings, higher reinforcement level, and easier processing in plastic compounds, and many of these improvements can be carried over to coatings [45]. Procedures for using these formulating materials are available from suppliers [46]. A related article dealing with hyperdispersant technology is also useful to understanding how these compounds function [47].

Another way to improve adhesion is to employ a corona discharge treatment to the substrate [48]. The treatment, which places hydroxyl, carboxyl, and other functional groups on the surface, increases the surface energy of the substrate and promotes wettability of coatings on substrates and promotes adhesion between two materials that are to be bonded. The article [48] contains a listing of the surface tension for 33 polymers. The use of corona treatment on metallized paper, coated kraft paper, and metal foils is also discussed. Similar modification can be accomplished by flame or plasma treatment.

Surface modification of a variety of polymers has been accomplished by the use of only ultraviolet radiation [49]. The results indicated that such treatment belongs in the field of surface treatment, but that it only appears to work with some polymers and not others. Thus, it is necessary to test for any particular application. In another ultraviolet treatment technique, dilute aqueous- or solvent-based mixtures of hydrophilic acrylate monomers were applied and irradiated on polyolefins [50]. The aqueous system gave the best results, and the photo-grafted films had noticeably improved anti-fog performance in comparison with controls. The surface of both low- and high-density polyethylene has been modified by the ultraviolet radiation introduction of sulfonic acid groups through irradiation in the presence of oxygen and sulfur dioxide [51]. The surface of polystyrene has been modified with cyanide groups (CN) by irradiation with ultraviolet radiation of 254 nm in the presence of cyanogen bromide [52]. Both of these techniques demonstrate that it is possible to attach specific functional groups onto polymer surfaces and thus modify the surface properties and reactivity.

Although not related to silane coupling agents, plasma surface treatment is another way to alter the surface of substrates and improve wetting characteristics and adhesion [53]. Plasma treatment with inert gases, oxygen, ammonia, nitrous oxide, or other gases will allow a surface to become activated, and it will contain functional groups such as hydroxy, amino, carboxyl, etc. Plasma treatment allows surface properties to be modified for specific end uses, such as for devices used in the medical field [54]. Acrylic acid has been used to alter surface characteristics of polyethylene through electron beam irradiation [55]. Surface alteration was higher with high-density than low-density polyethylene. Improved flooring materials were manufactured when vinyl floor tiles were first treated with an aqueous acrylic acid solution and exposed to ultraviolet radiation [56]. When the treated tile was coated and cured in the usual manner with ultraviolet radiation, a tough, durable surface that had superior characteristics to other products was formed.

Surfactants

The radiation curing process is usually done at a very rapid rate, and coatings are expected to flow and level very quickly at ambient temperature. Surfactants added in small amounts of about 0.5% (0.1–1%) are effective in improving flow and leveling. In the early days of radiation curing, there were few surfactants designated as being useful in formulations developed for this technology. Today, many suppliers specify particular lines of surfactants or particular surfactants for radiation cure.

One exception was an early Dow Corning Corporation paper that summarized the silicone additives for radiation-curable coatings [57]. Organo-functional silicone fluids useful as surfactants are the polysiloxane-polyoxyalkylene copolymers. Typical of these are block copolymers of the ABA type:

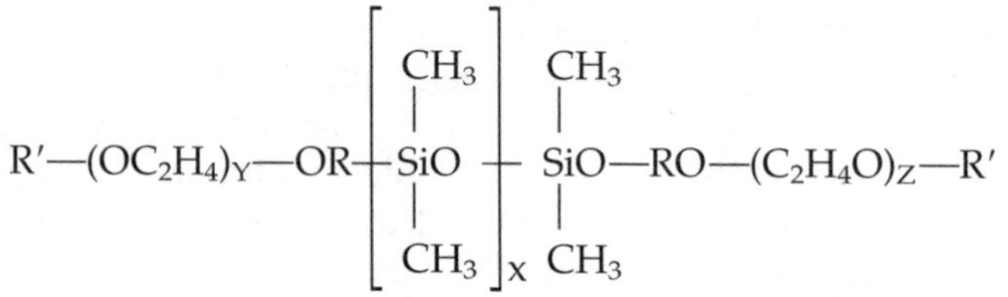

$$R'\text{—}(OC_2H_4)_Y\text{—}OR\text{—}\left[\begin{matrix} CH_3 \\ | \\ SiO \\ | \\ CH_3 \end{matrix}\right]_X\text{—}\begin{matrix} CH_3 \\ | \\ SiO \\ | \\ CH_3 \end{matrix}\text{—}RO\text{—}(C_2H_4O)_Z\text{—}R'$$

and graft copolymers similar to the following:

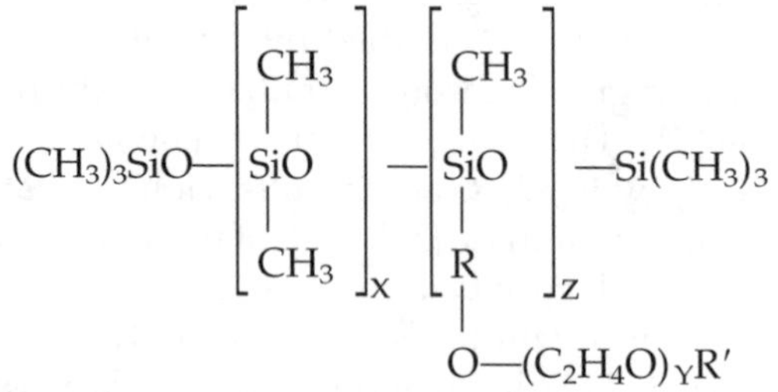

$$(CH_3)_3SiO\text{—}\left[\begin{matrix} CH_3 \\ | \\ SiO \\ | \\ CH_3 \end{matrix}\right]_X\text{—}\left[\begin{matrix} CH_3 \\ | \\ SiO \\ | \\ R \\ | \\ O\text{—}(C_2H_4O)_YR' \end{matrix}\right]_Z\text{—}Si(CH_3)_3$$

In these polysiloxane-polyoxyethylene structures, *R* is—O—$(CH_2)n$,—$(CH_2)_n$— and similar groups; *R′* is—H, —$(CH_2)_n$—CH_3,—CO—CH=CH_2, and similar groups; and *x*, *y*, and *z* usually have different average values, all of which are relatively low numbers. The resultant copolymers are liquids that are compatible with a number of the compounds used in formulating radiation curable coatings. In addition, these compounds do not affect the cured film's gloss or clarity and improve wetting, cure, adhesion, slip, and abrasion resistance on aluminum. Such additives modify the liquid coating without affecting the system cure or final physical properties [58].

A particularly useful graft copolymer of the polysiloxane-polyoxyethylene type had the following average structure:

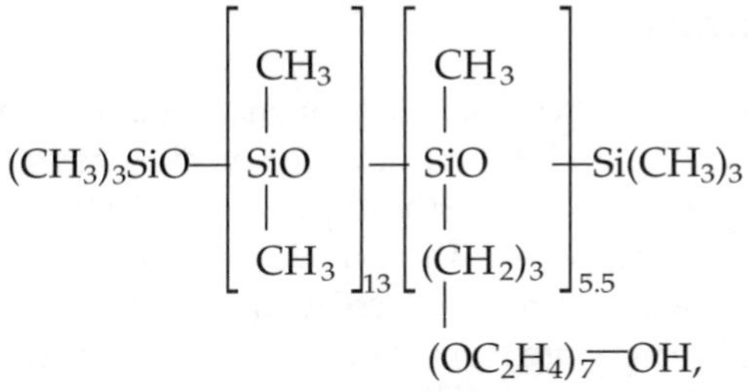

$$(CH_3)_3SiO-\left[\begin{array}{c} CH_3 \\ | \\ SiO \\ | \\ CH_3 \end{array}\right]_{13}-\left[\begin{array}{c} CH_3 \\ | \\ SiO \\ | \\ (CH_2)_3 \\ | \\ (OC_2H_4)_7-OH, \end{array}\right]_{5.5}-Si(CH_3)_3$$

and was marketed as SILWET™ L-5410 [59,60]. Such copolymers had a particular advantage, in that they would chain transfer with cycloaliphatic epoxide-based coatings through the hydroxyl group and become an integral part of the final coating. Another surfactant designed for radiation-cured overprint varnishes is SILWET™ RC-73, which improves flow, slip, and gloss [61].

Another class of surfactants that are useful in radiation-curable systems is the fluorosurfactants [62,63]. These surfactants are often derivatives of mixtures of primary alcohols that might have a structure such as the following:

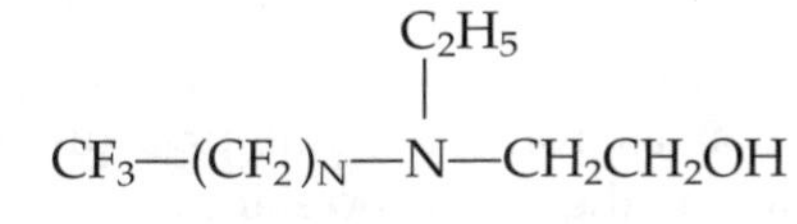

$$CF_3-(CF_2)_N-\underset{}{\overset{C_2H_5}{\overset{|}{N}}}-CH_2CH_2OH$$

For example, Fluorad® Fluorochemical Alcohol FC-10 is a compound with this structure in which *n* has an average value of 6.5 [64]. One particular fluorochemical surfactant recommended for use in radiation-cure formulations is FC-171 [65]. Fluorosurfactants are also marketed as Zonyl® products [66].

As would be expected, surface tension, γ, of the components and the final formulation are important to how well a coating will spread, flow, and level on a surface and to final properties. This parameter has been measured for a number of acrylates, formulations, and substrates by the pending drop and the du Nouy ring methods [67]. The formulation surface tensions were measured by an additive rule in which the formulation surface tension, γ, is equal to the summation of the surface tensions,γ_i of the individual components times their molar fraction, x_i. It was demonstrated that, as long as the components of the formulation are compatible, this additive method gives good results when compared to experimental values. Tape adhesion was measured, and its decrease as a function of increasing silicone surfactant additive concentration was demonstrated. A fluorinated surfactant additive could be used at a markedly lower level, and its effect was essentially nil on adhesion at these low levels. Methods for estimation of the surface tension of compounds by the group contribution method have been described and applied to photocure formulation [68]. Values of critical energy for various smooth polymeric surfaces are also given.

Textured Coatings

Textured is defined as a woven or interwoven appearance at a surface, and it is characterized by the number of strands or weaves per some unit of linear dimension. The net effect is to give the appearance of a flat coating. However, texturing can be achieved in different ways. What seems to be the first method for achieving a textured surface by means of ultraviolet radiation involves exposure of an acrylate formulation to short wave radiation produced by the 254 and 185 nm resonance lines of mercury from low pressure, "germicidal" mercury vapor lamps under inert conditions [69]. This radiation is preferentially absorbed at the surface of a responsive coating, causing cure and cross-linking in this region with accompanying shrinkage. The bulk of the coating remains uncured or cures to a low degree. In the process, the exposed coating remains for a controlled period of time in a dark or non-irradiation zone where the micro wrinkling takes place through the difference in shrinkage between the surface and the bulk causing texturing. The coating then enters the main curing region where the bulk of the coating is cured and the wrinkling is locked in place.

This process was improved later wherein the coating is exposed to broad range ultraviolet radiation, as from a medium pressure mercury vapor lamp (wavelength 180–400 nm), in the final processing step [70]. The final step may take place in an inert environment or in air. Another patent [71] provides for dark colored and light colored ultraviolet-radiation reflective and nonreflective backgrounds to be used during the texturing process, while another [72] is similar, except that it provides for dark and light colored infrared adsorbent and reflective backgrounds. Eximer lamps have been used to provide the short wavelength (172 nm) irradiation, as mentioned in the Gloss Control section above [15], and oxygen-rich control atmospheres produce altered surface appearances [14].

Surface texturing such as this has been commercially applied to clear or colored polycarbonate [73], poly(ethylene terephthalate), or other sheeting to provide a nonglare surface and a smooth reverse side. The reverse side can be printed and/or have other graphic arts applied. The artwork is protected from fading or other hostile environments by the cross-linked surface/film combination.

Texturing has also been accomplished by exposing a photocurable composition comprised of a multifunctional urea compound, an ethylenically unsaturated compound, and an aromatic ketone photoinitiator [74]. The oligomeric urea compound functioned as a synergist for a photoinitiator such as benzophenone. The unsaturated compound was trimethylolpropane triacrylate or another acrylate-functional compound.

Thick Section Curing

Thick section curing is of interest when sealants, some electronic packaging materials, reinforced fiber composites, and similar items are being considered. Thicknesses of up to about 0.200 in. (0.508 cm) were achieved when various cationic photoinitiators were used with cycloaliphatic and glycidyl epoxides [75]. Cure times were minutes and low levels of onium salt photoinitiator were used. No indication of mechanical properties for the cured articles was given. Others have considered the cure depth as a function of free-radical photoinitiator concentration and found that relatively low levels of photoinitiator are optimum for thick coatings [76]. The curing of thick film circuitry has also been studied [77]. The UVT Sunspacers™, which are inorganic glasses, mentioned in an above section will allow thick sections to be cured at greater than 25 ft/min (~7.6 m/min) [18–20].

Water-Based Systems

In an effort to combine the rapid cure and high quality of radiation-cure coatings and the ease of handling associated with conventional technology, there have been investigations into combining radiation cure and waterborne technology [78–80]. Isocyanate-terminated urethane oligomers have been prepared from poly(tetramethylene oxide) polyols, diisocyanates, and di- or trimethylol carboxylic acids such as 2,2-dimethylol propionic acid, 2,2,2-trimethylol acetic acid, and the like [81]. These urethanes were then capped with 2-hydroxyethyl acrylate, 2-hydroxypropyl acrylate, and the like to form urethane acrylates that could be made into aqueous emulsions. The compounds were useful as radiation-curable binders for textile printing [82]. Polyoxyethylene-modified urethane acrylates have also been prepared as radiation-curable emulsions and investigated for their coating properties [83]. Coating and mechanical properties were improved when a cross-linking agent and an acrylated epoxide were used in combination with the urethane acrylates. However, these compounds decreased emulsion stability. Acrylate oligomers have been dissolved in multifunctional acrylates, such as 1,6-hexanediol diacrylate and trimethylolpropane triacrylate, and then formulated with a surfactant [84]. Water was then added in successive amounts with tumbling to form emulsions. It appeared that water acted as a plasticizer, adhesion to steel was improved, and solvent resistance could be maintained if sufficient multifunctional monomer was present.

Several investigations of free radical-generating photoinitiators for waterborne systems have been made. These include a study of the excitation states of water-soluble photoinitiators derived from benzil, benzophenone, and thioxanthone [85]. Others have examined several commercial photoinitiators in self-emulsified unsaturated polyester, polyester acrylate, and urethane acrylates [86,87]. It appeared reactivity was increased after pre-drying to remove water, though if the photoinitiator was volatile, reactivity was decreased. A preliminary study of sodium benzoylmethylthiosulfate indicated the photoinitiator had high reactivity and produced high gloss, low yellowing coatings [88]. Other photoinitiators examined include bisacylphosphine oxides [89] and an oligomeric alpha-hydroxyketone [90].

Cationic systems based on cycloaliphatic epoxides have also been examined as aqueous radiation-cured systems [91,92]. These starting-point formulations involved adding hydroxyethyl cellulose and a surfactant to water. Then the epoxide, a polycaprolactone trifunctional polyol, and an onium salt photoinitiator were blended and added slowly to the aqueous mixture in a Cowles-type mixer. Emulsion stability varied from 3 to about 30 days. Cured properties were the same as for a control coating without water and hydroxyethyl cellulose.

The problem of high energy cost and potential for steam volatilization of photoinitiator when drying waterborne radiation cure systems was investigated [93]. The study indicated that a refrigerant-based dehumidification process would bring considerable energy savings, eliminate photoinitiator loss, and might be the only way to dry aqueous coatings on temperature sensitive substrates. Drying times with this process were about the same or less than with thermal ovens.

The effect of water contamination on the performance of urethane acrylate coatings has also been studied [94]. Both contamination and exposure to moisture after cure were examined. Such contamination affects physical performance and reduces coatings performance. This can be particularly important in the printed circuit assembly and fiber optic areas. The authors also describe methods for determining water content in raw materials and in cured coatings.

Weathering

Radiation-cured coatings had their beginning on substrates that were to be used indoors. However, with the strong success this technology has had over the past several years, exterior coatings are also being developed, used, and looked at as a new growth area.

In 1984, a series of formulations based on an aliphatic urethane acrylate, 1,6-hexanediol diacrylate, tetraethyleneglycol diacrylate, and 2-ethylhexyl acrylate were formulated with a wetting agent, photoinitiator, and various hindered amine light stabilizers. The cured panels were placed on test in Florida and also subjected to accelerated weathering in a Xenon Arc Weatherometer and a QUV weathering device [95]. The results indicated that significant improvements in weathering properties could be obtained when the stabilizers were added. In addition, the results indicated that the Xenon Arc device correlated better with Florida or natural weathering than the QUV device. About the same time, others were investigating the photodegradation of acrylated epoxides [96] and urethane acrylates [97], the effect of acrylate monomers on degradation [98], the influence of photoinitiators on color development [99], and the effect of titanium dioxide on durability [100].

A set of seven aliphatic urethane acrylates, a bisphenol A-based acrylated epoxide and a urethane diacrylate, and an aliphatic diacrylate and triacrylate was prepared by blending 80 parts of these compounds, 20 parts of 1,6-hexanediol diacrylate, and 5 parts of an alpha-hydroxy acetophenone photoinitiator [101]. The polyol portion of the urethane acrylates was either polyether, polyether/polyester, or branched polyester. In the latter category, different molecular weight polyols, ranging from 500 to 5000, were used. After radiation cure, cured panels were subjected to accelerated weathering in a QUV weathering device for 3000 h. The aliphatic urethane acrylates had good weathering characteristics, and the mechanical properties were also well conserved. Another study investigated both the free radical scavenging hindered-amine light stabilizers (HALS) and ultraviolet light absorbers that filter out harmful radiation in the 290–380 nm range [102,103].

New high quality clear topcoats have been tested against standard two-package topcoats used in the automotive industry [40]. The ultraviolet radiation cured coating started at a gloss level and ended at a higher gloss level after 6000 h accelerated weathering. It was pointed out that the higher gloss level does not mean an optical quality difference and is merely related to refractive index differences. The radiation-cured coating underwent cracking after 2000 h exposure, and it is thought this was due to poor elasticity in the coating.

A number of clear acrylic coatings were cured and subjected to accelerated weathering in a SUN-test for 1000 h, wherein light, temperature, and humidity influence the coating [104]. Acrylated epoxides exhibited the most yellowing. Conventional and amine-modified polyester acrylates had only slight yellowing, and polyether acrylates and urethane acrylates were essentially unchanged by the SUN-test exposure. In tests with coatings that contained light stabilizers, there was improved stability over those without the additives. The use of light stabilizers also improved mechanical properties as evidenced by improved cracking resistance.

References

[1] Materials News, Dow Corning, "New Foam Control Additive Introduced for UV-Cure and EB-Cure Inks and Coatings," January/February 1994, p. 3.
[2] Monsanto Co. (now Solutia, Inc.), "PC-1344 Defoamer," Product Data Sheet, 1985.
[3] Patent: Bailey, W. J., Polycyclic Ring-Opened Polymers, US 4, 387,215. (1983).
[4] Cohen, M. S., Bluestein, C., and Dunkel, M., "Monomers Which Expand on Polymerization," *Proceedings, RadCure* '84, Atlanta, GA, 10–13 Sept. 1984, p. 11-1.
[5] Prane, J. W., "Expanding Polymers Reduce Shrinkage of Cured Adhesive," *Adhesives Age*, Vol. 30, No. 3, March 1987, p. 48.
[6] Bluestein, C., Cohen, M., and Mehta, R., "Benefits of Expanding Monomers in UV Cure Epoxy Cationic Coatings," *Proceedings of RadTech '88—North America*, New Orleans, LA, 24–28 Apr. 1988, p. 389.
[7] Bluestein, C. and Cohen, M. S., "Expanding Monomers—Recent Advances in Light Cured Compositions," Conference Paper, *RadTech—Europe '89*, Florence, Italy, 9–11 Oct. 1989.
[8] Kosnik, F. J., "Gloss Reduction of Radiation Curable Coatings," Conference Papers, *RadTech '88—North America*, New Orleans, LA, 24–28 Apr. 1988, p. 492.
[9] Patent: McDowell, J. R., Lord Corp., Low Gloss Finishes by Gradient Intensity Cure, US 4,169,167. (1979).
[10] Garratt, P. G., "Ultraviolet Radiation Dual-Cure Processes for the Production of Low Gloss Films," *Conference Proceedings of RadCure—Europe '87*, Munich, Germany, 4–7 May 1987, p. 10–23.
[11] Patent: Kurpiewski, T. and Heinze, R. E., Radiation Curable Coating Compositions, US 4,780,487. (1988).
[12] Patent: Matthews, J. C. and Couch, R. W., Fusion Systems Corp., Method and Apparatus for Providing Low Gloss and Gloss Controlled Radiation-Cured Coatings, US 4,313,969, 1982.
[13] Patent: Vincent, K. D., Hewlett-Packard Co., Radiation Curable Barrier Coating Having Flexibility and Selective Gloss. US 4,230,550. (1980).
[14] Patent: Hahn, E. A., PPG Industries, Inc., Method of Producing Flat (Non-Glossy) Films. US 3,918,393. (1975).
[15] Roth, A., "High Matting Without Matting Additives," *Conference Proceedings of RadTech—Europe '99*, Berlin, Germany, 8–10 Nov. 1999, p. 661.
[16] Patent: Sachs, P. R., GAF Corp., Dual Gloss Coating and Process Therefor. US 4,309,452. (1982).
[17] Patent: Daly, A. T., Haley, R. P., Reinheimer, E. P., and Mill, G. R., Morton International, Inc., Method for Producing Low Gloss Appearance with UV Curable Powder Coatings. US 6,017,593. (2000).
[18] Koleske, J. V., "Radiation-Curable Coatings on the Upswing," *Paint & Coatings*, XVI, Vol. 4, No. 34, June 2000.
[19] "UVT Sunspacers and UVT Sunspheres," Brochure, Suncolor Corp., North Canton, Ohio.
[20] Patent: Smetana, D. A. and Koleske, J. V., Suncolor Corporation, Radiation-Curable Compositions and Cured Articles, US 6,350,792. (2002–02).
[21] Patent: Carlson, R. C., Minnesota Mining and Manufacturing Co., Addition of Ethylenically Unsaturated Materials to Control Odor in Photopolymerizable Epoxy Compositions, US 4,218,531. (1980).
[22] Patent: Carlson, R. C., Minnesota Mining and Manufacturing Co., Controlling Odor in Photopolymerization, US 4,324,679. (1982).
[23] Patent: Schlesinger, S., I., American Can Co., Cationically Polymerizable Compositions Containing Sulfonium Salt Photoinitiators and Stable Free Radicals as Odor Suppressants and Method of Polymerization Using Same. US 4,306,953, (1981).
[24] Patent: Schlesinger, S. I. and Kester, D. E., American Can Co., Cationically Polymerizable

Compositions Containing Sulfonium Salt Photoinitiators and Odor Suppressants and Method of Polymerization Using Same. US 4,250,203. (1981).

[25] Bassi, G., Li, C., Luciano, and Broggi, F., "Advance in Low-Odor Coatings: A New Class of Polymeric Nonyellowing Photoinitiators," *Proceedings of RadTech—Europe '87*, Munich, Germany, 4–7 May 1987, p. 3–15.

[26] Koehler, M. and Ohngemach, J., "Polymers with Photoinitiator Properties," *Proceedings of RadTech Europe '87*, Munich, Germany, 4–7 May, 1987, p. 3–1.

[27] Visconti, M., Cattaneo, M., Meneguzzo, E., and Bassi, G., "A Novel Low Odor Photoinitiator for Dark Pigmented Systems," *Proceedings of RadTech—Europe* '99, Berlin, Germany, 1999, p. 117.

[28] Visconti, M., Cattaneo, M., Casiraghi, A., Norcini, G., and Bassi, G., et al., "LFC 1001, A Novel, Low Odor Photoinitiator For Dark Pigmented Systems," *Proceedings of RadTech 2000*, Baltimore, MD, 9–12 Apr. 2000, p. 414.

[29] Guarino, J. P. and Ravijst, J. P., "Low Odor UV Cure Coatings," Conference Papers, *RadTech '88—North America*, New Orleans, LA, 24–28 Apr. 1988, p. 523.

[30] Fales, N. J., Stover, L. C., and Lamm, J.D., *Polymer-Plastic Technology and Engineering*, Vol. 21, No. 2, 1983, p. 111.

[31] Suran, V. M., "How to Use Fragrance Additives," *Plastics Technology*, June 1984, p. 65.

[32] Tu, R. S., "Radiation-Curable Abrasion-Resistant Coatings Based on Pentaerythritol and Trimethylolpropane Triacrylates," *Conference Proceedings of Radiation Curing VI*, Chicago, IL, 20–23 Sept. 1982, p. 10-1.

[33] Patent: Tu, S. and Lorenz, D. H., GAF Corp., Abrasion Resistance Radiation Curable Coating, US 4,319,811 (1982).

[34] Kushner, L. and Tu, R. S., "Imparting Super Abrasion Resistance to Plastic Substrates," *Conference Proceedings of Radiation Curing VI*, Chicago, IL, 20–23 Sept. 1982, p. 8-1.

[35] Kushner, L. and Tu, R. S., "New Coatings are Radiation Curable and Abrasion Resistant," *Modern Plastics*, April 1983, p. 87.

[36] Dulany, L., "Energy Curable Resins for Increased Scratch and Abrasion Resistant Coatings," *Proceedings of RadTech 2000*, Baltimore, MD, 9–12 Apr. 2000, p.103.

[37] Muzeau, E., von Stebut, J., and Magny, B., "The Scratch Resistance of Radiation Curable Coatings," *Proceedings of RadTech 2000*, Baltimore, MD, 9–12 Apr. 2000, p. 314.

[38] Patent: Hegel, R. F., Minnesota Mining and Manufacturing Co., Ultraviolet Light Curable Compositions for Abrasion Resistant Articles. US 4,650,845. (1987).

[39] Krongauz, V., "Abrasion Resistance and Molecular Properties of UV-Cured Polymer Films," *Proceedings of RadTech—Europe '99*, Berlin, Germany, 8–10 Nov. 1999, p. 435.

[40] Rekowski, V. and Frigge, E., "Progress in UV Clear Coats for Exterior Use," *Proceedings of RadTech Europe '99*, Berlin, Germany, 8–10 Nov. 1999, p. 89.

[41] Eranian, C. V., Faurent, C., Noireaux, P, and Deveau, J., "Colloidal Silica Acrylates for High Performance in Radiation-Curable Coatings," *Proceedings of RadTech—Europe '99*, Berlin, Germany, 8–10 Nov. 1999, p. 523.

[42] Zahouily, K., Boukaftane, C., Decker, C., "Surface modification of UV-Cured Urethane Acrylate Coatings by Addition of Fluoro or Silicone Acrylic Monomers," *Conference Proceedings of RadTech '99—Europe*, Berlin, Germany, 8–10 Nov. 1999, p. 803.

[43] Lange, J., Luisier, A., and Hult, A., "Influence of Crosslink Density, Glass Transition Temperature and Addition of Pigment and Wax on the Scratch Resistance of an Epoxy Coating," *Journal of Coatings Technology*, Vol. 69, No. 872, Sept. 1997, p. 77.

[44] Puddleman, E.P., "Silane Adhesion Promoters in Coatings," *Progress in Organic Coatings*, Vol. 11, 1983, p. 297.

[45] Marsden, J. C., "Functions, Applications, and Advantages of Silane Coupling Agents," *Plastics Compounding*, July/August1978. p. 5.

[46] Union Carbide Corp., "Organofunctional Silanes," *Adhesives and Sealants: Application Techniques*, Product Application Information.

[47] Hampton, J. S., "Hyperdispersant Technology for Non-Aqueous Coatings," *Modern Paint and Coatings*, Vol. 75, No. 6, June 1985, p. 46.
[48] Lekan, S. F., "Corona Treatment as an Adhesion Promoter for UV/EB Curable Coatings," Conference Papers, *RadTech '88—North America*, New Orleans, LA, 24–28 Apr. 1988, p. 334.
[49] Burger, P., Weidenhammer, P., Armbruster, K., Rekowski, V., and Osterhold, M., "Surface Pretreatment of Polymers Using UV Light," *Conference Proceeding of RadTech '99—Europe*, Berlin, Germany, 8–10 Nov. 1999, p. 665.
[50] Kyle, D.R., "Photochemical Surface Modification of Polyolefins Using Ultraviolet (UV) Light," *Proceedings of RadTech 2000*, Baltimore, MD, 9–12 Apr. 2000, p. 478.
[51] Kavac, T., Kern, W., Ebel, M. F., Svagera, R., Surface "Modification of Polyethylene by Photoinitiated Introduction of Sulfonic Acid Groups," *Conference Proceeding of RadTech '99—Europe*, Berlin, Germany, 8–10 Nov. 1999, p. 725.
[52] Meyer, U., Hoiser, E. M., Kern, W., Ebel, M. F., Svagera, R., "Photomodification of Polymer Surfaces: Introduction of CN Groups," *Conference Proceeding of RadTech '99—Europe*, Berlin, Germany, 8–10 Nov. 1999, p. 731.
[53] Cheremisinoff, P. N., Farah, O. G., and Ouellette, R. P., *Radio Frequency/Radiation and Plasma Processing*, Technomic Publishing Co., Lancaster, PA, 1985, p. 213.
[54] Kaplan, S. L. and Gehrke-Neumann, C., "Applications for Plasma Surface Treatment in the Medical Industry," *Adhesive and Sealants Industry*, Vol. 7, No. 4, Apr. 2000, p. 36.
[55] Harada, J., Chern, R. T., and Stannet, V. T., "Grafting of Acrylic Acid on Polyethylene Pre-Irradiation with Electron Beam," *Proceedings of RadTech '90—North America*, Chicago, IL, Vol. 1, 25–29 Mar. 1990, p. 493.
[56] Patent: Bolgiano, N. C. and Sigman, W. T., Armstrong World Industries, Inc., Process for Providing Improved Radiation-Curable Surface Coverings and Products Produced Thereby, US 4,421,782. (1983).
[57] Gordon, D. J., "New Silicone Additives for Radiation Curable Coatings," Technical Paper FC76-505, The Association for Finishing Processes of SME, Dearborn, MI, 1976.
[58] Tonge, J., "Silicone and UV/EB Boost Compliance and Productivity," *Modern Paint and Coatings*, Vol. 84, No. 7, July 1994, p. 44.
[59] Patent: Koleske, Joseph V., Union Carbide Corp., "Conformal Coatings Cured with Actinic Radiation," US 5,155,143. (1992).
[60] Surface Active Copolymers, Brochure, Union Carbide Corp., 1983.
[61] Anon, "Additives Helping to Meet the Low-VOC Challenge," *Modern Paint and Coatings*, Vol. 83, No. 7, July 1993, p. 42.
[62] Gehlhoff, L., "Fluorosurfactants as Performance Additives," *Modern Paint and Coatings*, Vol. 88, No. 8, August 1998, p. 22.
[63] Allison, M. C., "Fluorochemicals for Aqueous and Non-Aqueous Paint Systems," *Polymers Paint Colour Journal*, Vol. 176, No. 4174, September 1986, p.688.
[64] Fluorad® Fluorochemical Alcohol FC-10, 3M Co., Product Information Brochure, 1984.
[65] "3M Brand Photosensitizer FC-510," Technical Data Sheets, Commercial Chemicals Div., 3M Co., 1979.
[66] Zonyl® Fluorosurfactants, Technical Data Brochure, Dupont Co., 1992.
[67] Magny, B. and Cavalié, H., "Importance of Surface Tension for Radiation Curable Coatings Properties," *Conference Proceedings of, RadTech Europe '99*, Berlin, Germany, 8–10 Nov. 1999, p. 71.
[68] Wolinski, L. E., "The Wetting of Low-Energy Surfaces by Ultraviolet Light Cured Coatings," *Proceedings of Radcure '86*, Baltimore, MD, 8–11 Sept. 1986, p. 8–17.
[69] Patent: Osborn, C.L. and Troue, H. H., Union Carbide Corp., US 3,840,448. (1974).
[70] Patent: Troue, H. H., Process for Producing Textured Coatings, Union Carbide Corp., US 4,421,784. (1983).
[71] Patent: Troue, H. H., Process for Producing Textured Coatings, Union Carbide Corp., US 4,483,884. (1984).

[72] Patent: Troue, H. H., Union Carbide Corp., Process for Producing Textured Coatings, US 4,485,123. (1984).

[73] Mais, M. G., "Surface Texturing of UV Curable Coatings," *Conference Proceedings of Rad-Cure '84*, AFP/SME, Atlanta, GA, 10–13 Sept. 1984, p. 7–1; *Proceedings of Conference on Radiation Curing—Asia*, Tokyo, Japan, 20–22 Oct. 1986, p. 285.

[74] Patent: Beckett, A. D., Koleske, J. V., and Gerkin, R. M., Process for Obtaining Textured Coatings from Photo-Curable Urea-Containing Compositions, Texaco Chemical Co., US 5,212,271. (1993).

[75] Patent: Smith, G. H. and Olofson, P. M., Mining and Manufacturing Company, Complex Salt Photoinitiators, Minnesota, US 4,231,951. (1980).

[76] Van Landuyt, D. C., "Practical Aspects of Thick Film UV Curing," *Journal of Radiation Curing*, Vol. 11, No. 4, July 1984, p. 4.

[77] Green, William J., "Processing Polymer Thick Film Circuitry with Radiation Curing," *Journal of Radiation Curing*, Vol. 10, No. 2, April 1983, p. 4.

[78] Loutz, J. M., Demarteau, W., and Herze, P. Y., "Water-Based UV-EB Coating Systems," *Proceedings of RadCure—Europe '87*, Munich, Germany, 4–7 May 1987, p. 5–37.

[79] Kosnik, Frank J., "Water Offers Possibilities in Radiation-Cure Coatings," *Modern Paint and Coatings*, Vol. 79, No. 6, June 1989, p. 42.

[80] Gerst, D. D., "Water Borne Radiation Curable Oligomers," Radiation Curing, Vol. 9 No. 11, November 1982, p.27; p. 9–21, *Proceedings of Radiation Curing VI*, Chicago, IL, 20–23 Sept. 1982.

[81] Patent: Park, K., Bryant, G. M., and Carr, F. G., Union Carbide Corp., Acrylyl Capped Urethane Oligomers, US 4,153,778. (1979).

[82] Park, K., Frame, R. L., and Bryant, G. M., "Elastomeric Radiation-Curable Binders for Textile Printing," *Textile Chemist and Colorist*, Vol. 2, No. 5, May 1979, p. 107.

[83] Song, M. E., Kim, J. Y., and Suh, K. D., "Preparation of UV Curable Emulsions Using PEG-Modified Urethane Acrylates and Their Coating Properties," *Journal of Coatings Technology*, Vol. 68, No. 862, November 1996, p. 43.

[84] Jesseph, S. P. and Costanza, J.R., "Multifunctional Monomers in Water Borne Radiation Curable Coatings," *Radiation Curing*, Vol. 11, No. 3, Aug. 1984, p. 10.

[85] Fouassier, J. P., and Lougnot, D. J., "Reactivity of Water Soluble Photoinitiators," *Proceedings of RadCure* '86, Baltimore, MD, 8–11 Sept. 1986, p. 4–1.

[86] Pietschmann, N., "Photoinitiator Efficiency in Water-Borne UV Curable Coatings," *Proceedings of RadTech—Europe '99*, Berlin, Germany, 8–10 Nov. 1999, p. 785.

[87] Koehler, M. and Ohngemach, J., "Water Soluble and Hydrophilic Alpha-Type Photoinitiators," *Proceedings of RadTech '88—North America*, New Orleans, LA, 24–28 Apr. 1988, p. 150.

[88] Bassi, G. Li, Broggi, F., and Revelli, A., "Water Soluble Photoinitiators: Unimolecular Splitting Compounds for Aqueous Solution Systems," in *Proceedings of RadTech '88—North America*, New Orleans, LA, 24–28 April 1988, p. 160.

[89] Burglin, M, Köhler, M., Dietliker, K., and Wolf, "Curing of Outdoor Water-Borne Coatings—A Novel Application of Bisacylphosphine Oxide Photoinitiators," *Proceedings of RadTech 2000*, Baltimore, MD, 9–12 Apr. 2000, p. 577.

[90] Cattaneo, L., "A Highly Efficient Photoinitiator for Water-Borne UV Curable Systems," *Proceedings of RadTech 2000*, Baltimore, MD, 9–12 Apr. 2000, p. 1057.

[91] Eaton, R. F., Hanrahan, B. D., and Braddock, J. K., "Formulation Concepts for Both High Solids and Water Based UV Cured Cycloaliphatic Epoxy Wood Coatings," *Proceedings of RadTech '90—North America*, Chicago, IL, Vol. 1, 25–29 Mar. 1990, p. 384.

[92] Union Carbide Chemical and Plastics Company, "Water Based Emulsions for CYRACURE® Formulations," Technical Bulletin No. 2, 14 May 1990.

[93] Rajasinghe, N., "Drying Water-Based Radiation Curable Coatings Using Dehumidification," *Proceedings of RadTech 2000*, Baltimore, MD, 9–12 Apr. 2000, p. 973.

[94] Kallendorf, C. J. and Woodruff, R. T., "The Effect of Water Contamination on the Performance of Urethane Acrylate Coatings," *Proceedings of RadCure '86*, Baltimore, MD, 8–11 Sept. 1986, p. 9-1.

[95] Gatechair, L. R. and Evers III, H. J., "Improving the Weatherability of UV Cured Coatings," *Proceedings of RadCure '84*, Atlanta, GA, 10–13 Sept. 1984, p. 6–17.

[96] Bendaikha, T. and Decker, C., "Photodegradation of UV Cured Coatings: I. Epoxy-Acrylate Networks," *Journal of Radiation Curing*, Vol. 11, No. 2, Apr. 1984, p. 6.

[97] Decker, C., Moussa, K., Bendaikha, "Photodegradation of UV-Cured Coatings. II. Polyurethane-Acrylate Networks," *Journal of Polymer Science*, A: Polymer Chemistry, Vol. 29, 1991, p. 739.

[98] Gismondi, T. E., "The Influence of Acrylate Monomers on the Resistance of UV-Cure Coatings to UV-Induced Degradation," *Journal of Radiation Curing*, Vol. 11, No. 2, April 1984, p. 14.

[99] Schmid, S. R., "Photoinitiators and Their Influence on Color Development in UV Cured Films," *Journal of Radiation Curing*, Vol. 11, No. 2, April 1984, p. 19.

[100] Simpson, L. A., "Influence of Titanium Dioxide Pigment on Durability of Paint Films," *Polymers, Paint, Colour Journal*, Vol. 176, No. 4168, May 1986, p. 408.

[101] Yang, B., "Weathering Resistant Oligomers for Long-Term Applications," *Modern Paint and Coatings*," Vol. 86, No. 5, May 1996, p. 40.

[102] Valet, A. and Wostratzky, D., "Light Stabilization of UV Cured Coatings: A Progress Report," *RADTECH Report*, Vol. 10, No. 6, November/December 1996, p.18.

[103] Valet, A., "Light Stabilization of Radiation Cured Coatings," *Polymer, Paint Colour Journal*, Vol. 185, No. 31, 1995.

[104] Beck, E., Enenkel, P., Königer, R., Lokai, M., Menzel, K., Schwalm, R., "Weathering of Radiation Curable Coatings," *Proceedings of RadTech—Europe '99*, Berlin, Germany, 8–10 Nov. 1999, p. 531; *Proceedings of RadTech 2000*, Baltimore, MD, 9–12 Apr. 2000, p. 619.

Safety and Health Considerations

SAFETY CAN BE SUMMARIZED IN TWO WORDS, "BE CAREFUL." No matter which words are used, if an investigator does not regard safety with the most respect possible and does not take it on herself or himself as a responsibility, little can be done. In this day and age, there is a great deal of information about chemical compounds and about process equipment, so all workers should be sure they are informed about toxicity and safe handling procedures. Ask a supplier of materials what they are and how they should be handled, obtain and read Material Safety Data Sheets, and ask questions if you have even the slightest wondering about any facet of what is supplied to you. A listing of the various health and safety items that should be considered including chemical hazards, storage and handling, personal protective equipment and hygiene, housekeeping, first aid, as well as ultraviolet and electron beam radiation process equipment points of concern have been tabulated by RadTech and can be found on its web site. Keep in mind that "safety" and "toxicity" do not necessarily refer to the same thing. Toxicity has to do with a chemical compound's hazard relative to some exposure condition and is usually specified as either the oral or dermal LD-50. Safety has to do with risk assessment, the chance of suffering an adverse effect from exposure, the aspects of taking care not to be exposed by inhalation, skin absorption, or other means, and ensuring that others are not exposed through entry of compounds into the environment. In a similar vein, "toxicity" doesn't necessary mean the same as "irritation." Dufour [1] has discussed the irritation of acrylate monomers and oligomers, the fact that contact often causes irritation but can be prevented, and made clear the distinction between irritation and toxicity.

Frequently asked questions and their answers about radiation in the workplace have been compiled and are useful for those working with or contemplating working with the technology [2]. An air sampling method for monitoring exposure was developed some years ago [3], and safe-handling information for the industry can be found in various places [4–7]. The air sampling technique [3] involves the use of silica gel tubes and pumps capable of sampling at 500 cc/min. The gel is desorbed with acetone and gas chromatography is used to analyze the desorbate. It is pointed out [5] that one should be cautious of early data taken with acrylates, since little was really understood about their toxicology in, say, about the 1970s and early 1980s.

Is radiation dangerous? Surely it is if not controlled and treated with proper respect [8–14]. Keep in mind that we are bombarded every second of our lives with some type of radiation that reaches us from outer space, from the sun that sends infrared, visible, and ultraviolet radiation to the earth, at the dentist, doctor, or hospital where X-rays are used, and so on. Too much time on the beach can result in sunburn, which certainly is not good for anyone, and if repeated too often can have far reaching consequences. It is the same with ultraviolet curing equipment if one uses equipment that is not properly shielded, does not wear proper eye protection if and when it is needed, and so on. Details about the effect of ultraviolet radiation on skin and eyes are available [15]. When properly designed, installed, and operated, ultraviolet radiation equipment for three-dimensional curing can be safe for personnel and the environment [16]. In a general sense, today suppliers know and understand the problem and have taken precautions to ensure the safety of workers. But, individuals must read the information supplied, and if it isn't supplied ask for it. This message was given in the first paragraph also, and it cannot be repeated enough. It is worth pointing out that one company has offered a tailored insurance program and a premium credit when manufacturers adopt radiation-curing technology [17].

Federal, state, and local regulations are a part of safety. These regulations, cumbersome as they may seem at times, protect you, the worker, the people around you, and the environment, which in itself protects everyone. The topic of regulation is too broad to cover in this section, except for a brief mention [18]. Radiation curing with ultraviolet or electron beam radiation can eliminate or drop to almost nothing air pollution—volatile organic content, hazardous air pollutants, and oxides of nitrogen—from a coating line. Since these technologies reduced VOC to essentially nothing, they have reduced their ozone output, the actual air pollution culprit, to zero [19]. In effect, all users of electron beam or ultraviolet radiation technology in the coating field have been able to increase production and keep plant emissions well below the standards of the U. S. Environmental Protection Agency (EPA). This has been done without slowing industrial growth, and in fact, has allowed manufacturers to broaden their product lines. Radiation technology has allowed companies to improve safety and meet environmental standards and to do so while increasing profitability through increased productivity, improved coating performance, and other economic savings from oven elimination and solvent recovery or disposal elimination. The Canadian Environmental Protection Act, which took effect on February 1, 1991, has been summarized, and comments about various aspects have been made [20].

A summary of the regulations regarding industrial chemicals in the United States, Europe, and Japan in 1987 can be found in the literature [21], though there certainly have been changes or modifications since then. In 1988, a review of the key elements of the Toxic Substances Control Act (TSCA) [22] and related TSCA information [23] were published along with related statutes from Europe (the EEC "Sixth" Amendment), Japan, and Australia. The review has particular emphasis on the manufacture, distribution, and use of chemicals used in the radiation-cure industry. TSCA imposes regulatory restrictions on the commercialization of new chemicals, and these restrictions need to be addressed when new products are introduced [24]. Such restrictions may change and manufacturers should maintain awareness of such changes. In fact, compliance with regulations begins with awareness [25].

The RadTech website (*www.radtech.org*) has a number of excellent articles that deal with safety in the radiation curing work environment. Their Product Stewardship Program is available there and elsewhere [26]. The program is voluntarily implemented by facilities that manufacture or use ultraviolet or electron beam radiation products and technology. Its purpose is to establish health, safety, and environmental considerations early in product formulation and to develop a way to communicate the appropriate safeguards for product use and disposal to customers. It is inherent in this program that product stewardship is a shared responsibility. In addition, this factor of stewardship is understood by all those responsible for the various facets of an operation, i.e., by formulation, manufacturing, marketing, and customer support personnel.

Articles on the RadTech website include a description of the results of permeation testing of five acrylates (2-phenoxyethyl acrylate, 1,6-hexanediol diacrylate, tripropylene glycol diacrylate, trimethylolpropane triacrylate, an ethoxylated trimethylolpropane triacrylate) for three different handling glove materials (butyl rubber, natural rubber latex, and nitrile rubber) [27]. This eleven-page study examined thin gloves (0.1 mm), medium thickness gloves (0.45 mm), and thick gloves (0.56 mm), each of which was considered for different time periods and work environments. Included in the study were the gloves used in clean-up situations where solvents were added to the acrylates. Two different solvent mixtures were studied. In addition to the thickness and polymer type for the gloves, information about trade names, suppliers, and price are included. A slightly modified version of an ASTM standard test method was used as a protocol [28]. Migration of six acrylate monomers through the packaging materials high density polyethylene and poly(ethylene terephthalate) have been studied [29,30]. The studies were done at 25° C and 45° C for 10 days and for 29 days. The results indicated that with hexanediol diacrylate, no migration took place at 25° C for either packaging material after 10 days, and in the case of poly(ethylene terephthalate), the time was extended to 29 days at this temperature. Results for the other times and temperatures varied from no detectable monomer migration to various amounts. There is a list of chemicals (acrylates) used in the radiation curing industry that the United Kingdom union workers are reluctant to use [31]. With improved training and handling techniques, the list had been reduced from the first one devised, when dermatitis and skin sensitization were definite problems, but the list had not been eliminated in 1987. A discussion of the various facets of safe handling of radiation curable materials, their hazards, and guidelines for safe and efficient use has been detailed [32,33]. Regulatory concerns when packaging children's products and food have been recently discussed [34–36]. A "Specialty Acrylates and Methacrylates Panel" known as SAM has been set up for negotiating with government agencies regarding regulatory restrictions regarding manufacturing under TSCA [37].

Draize scores [38], skin and eye irritation ratings, and the acute dermal and oral toxicity of several commonly used acrylates and the commonly used cycloaliphatic epoxides have been tabulated [39]. For comparison purposes, the same parameters were tabulated for several commonly used solvents, and the comparison was carried to other parameters, such as flash point, systemic toxicity, mutagenicity, and effect on reproductive system. In 1979, a compilation was made of particular cases dealing with contact dermatitis and acrylates [40], as well as skin irritation due to hexanediol diacrylate [41], delayed irritation due to hexanediol diacrylate and butanediol diacrylate [42], allergenicity of trimethylolpropane triacrylate [43], and carcinogenicity of acrylates including oligomers [44,45]. Allegeric contact dermatitis from urethane acrylate in inks was also reported [46]. Toxicity data and other health, safety, and environmental factors dealing with glycidyl and cycloaliphatic epoxides also have been studied [47–50]. An interesting article deals with the coatings chemistry and toxicology, as well as some simple definitions of terms used in this area and comparison risks [51,52]. The National Institute of Occupational Safety and Health (NIOSH) and the EPA have assessed the potential for exposure to acrylates in radiation-cured products [53]. The inhalation hazards of low vapor pressure chemicals and the potential for respiratory danger if acrylates are inhaled have been discussed [54]. A safety assessment of triethylene glycol divinyl ether has been made [55].

The RadTech website has information about radiation-cured coatings and food contact applications [56,57]. The U. S. Food and Drug Administration (FDA) allows the use of such coatings as components of food packaging under certain conditions and in compliance with certain regulations. Quotes dealing with FDA regulations can be found for packaging. Details are left for the interested investigator to peruse. Also found at this site is a letter from particular law offices containing advice regarding the use of radiation-cured adhesives in food packaging applications [58]. Again, perusal and interpretation is left to those interested in the topic. Information about the use of cationic ultraviolet radiation cured coatings in food packaging in Germany indicates that there is a potential for full compliance with the stringent safety demands of this area [59]. The investigators obtained analytical results from extraction and migration tests for specific residual coating components. The study includes experimental details and methods of analysis.

A study dealing with the migration of cationic-cured inks and a safety assessment has been made [60]. In addition, the extractables [61,62], volatiles [63,64], and retortability [65] of cationic coatings and inks have been investigated. Other factors, such as moisture vapor transmission and vacuum outgassing, have been studied in terms of conformal coating usage and military specification approval [64].

References

[1] Dufour, P., "Irritation of Acrylic Prepolymers and Monomers," *Conference Proceeding of RadTech—Europe '87*, Munich, Germany, 4–7 May 1987, pp. 2–27.

[2] "UV/EB Health and Safety: Answers to Frequently Asked Questions," *RADTECH Report*, Vol. 13, No. 5, September/October 1999, p. 60.

[3] Bosserman, M. W. and Ketcham, N. H., "An Air Sampling and Analysis Method for Monitoring Personal Exposure to Vapors of Acrylate Monomers," *American Industrial Hygiene Association Journal*, No. 20, January 1980, p. 41.

[4] Burak, L., "Advantages of the UV/EB Curing Process and Recommended Safety Practices," *Paint and Coating Industry*, Vol. XVII, No. 4, 2001, p. 62.

[5] Golden, R., "Safety and Handling of UV/EB Curing Materials," *Journal of Coatings Technology*, Vol. 69, No. 871, August 1997, p. 83.

[6] Ansell, J. M., "Evaluating the Safety of Radiation-Curable Coatings," *Journal of Radiation Curing/Radiation Curing*, Vol. 18, No. 12, Summer 1991.

[7] NPCA Subcommittee on Radiation Cured Coatings, "Safe Handling and Use of Ultraviolet/Electron Beam (UV/EB) Curable Coatings, *National Paint & Coatings Association*, August 1980.

[8] Rudolph, A. C., "Radiation Hazards, A Rational Approach," *Proceedings of RadTech '90—North America*, Chicago, IL, Vol. 2, Mar 25–29 1990, p. 100.

[9] Moss, C. E., "Radiation Hazards Associated with Ultraviolet Radiation Curing Processes—An Update," *Radiation Curing*, Vol. 12, No. 10, May 1985.

[10] Kovachik, C., "Radiation Curing® Safety Reference Guide," *Radiation Curing*, Vol. 14, May 1987, p. 6.

[11] Baer, G. F., "Safety and Handling Considerations of UV Equipment," *Conference Proceedings of RadCure '84*, Atlanta, GA, 10–13 Sept. 1984, pp. 4–5.

[12] Nablo, S. V., Fishel, M., and Quintal, B. S., "Developments in Product Handling for Self-Shielded Electron Processing," Conference Papers Addendum, *RadTech '88—North America*, New Orleans, LA, 24–28 Apr. 1988, p. 7.

[13] Siegel, S. B., "Ultraviolet Curing System and Process," Conference Papers Addendum, *RadTech '88—North America*, New Orleans, LA, 24–28 Apr. 1988, p. 19.

[14] Ramler, W. J., "Performance Characteristics of a WIP Electron Beam System," Conference Papers Addendum, *RadTech '88—North America*, New Orleans, LA, 24–28 Apr. 1988, p. 23.

[15] Schaper, K. L., "Ultraviolet Radiation: Effects on Skin and Eyes," Conference Papers Addendum, *RadTech '88—North America*, New Orleans, LA, 24–28 Apr. 1988, p. 34.

[16] Arnold, H. S., "Health and Safety Requirements for 3D UV Processing," Conference Papers Addendum, *RadTech '88—North America*, New Orleans, LA, 24–28 Apr. 1988, p. 37.

[17] West, B., "Keep the Presses Running with the Hartford: A Powerful Partnership for Your Printing Operation," Brochure, *The Hartford*, Summer 2000.

[18] Brezinski, J. J., "Regulation of Volatile Organic Compound Emissions from Paints and Coatings," Ch. 1, *Paint and Coating Testing Manual*, Gardner Sward Handbook, J. V. Koleske, Ed., 14th ed., ASTM International, West Conshohocken, PA, 1995.

[19] Ross, A., "Eliminating Air Pollution (VOC & HAP) at the Source Through Use of Ultraviolet or Electron Beam Polymerization," Internet, *www.radtech.org/regulatory*, Dec. 11, 2000.

[20] Havery, M. J., "The Canadian Environmental Protection Act," *Proceedings of RadTech '90—North America*, Chicago, IL, Vol. 2, 25–29 Mar. 1990, p. 143.

[21] Keener, R. J., Plamondon, J. E., and West, A. S., "Recent Developments in the Regulation of Industrial Chemicals in the United States, Europe and Japan," *Conference Proceedings of RadCure '86*, Baltimore, MD, 8–11 Sept. 1986, p. 11-1; *Proceedings of RadTech '90—North America*, Chicago, IL, Vol. 2, 25–29 Mar. 1990, p. 115.

[22] West, A. S., "The Regulatory Framework: TSCA and Worldwide Analogs," Conference Papers, *RadTech '88—North America*, New Orleans, LA, 24–28 Apr. 1988, p. 91.

[23] Hayes, D. J., "TSCA Pressure Points," Conference Papers, *RadTech '88—North America*, New Orleans, LA, 24–28 Apr. 1988, p. 113.

[24] Nelson, P. C. and Moran, E. J., "Regulatory Restrictions on Commercializing New Chemicals," Conference Papers, *RadTech '88—North America*, New Orleans, LA, Apr. 24–28 1988, p. 120.

[25] Rauscher, G., "Chemical Regulations—Compliance Begins with Awareness," *Proceedings of RadTech '90—North America*, Chicago, IL, Vol. 2, 25–29 Mar. 1990, p. 134.

[26] "Product Stewardship Program Guide," *Proceedings of RadTech 2000*, Baltimore, MD, 9–12 Apr. 2000, p. 336.

[27] Zwanenburg, R., "Adequate Protective Gloves for Working with UV/EB-Curing Acrylates," European Association for the Promotion of UV and EB Curing, Internet *www.radtech.org*, Nov. 26, 1999; *Proceedings of RadTech—Europe '99*, Berlin, Germany, 8–10 Nov. 1999, p. 375.

[28] ASTM F 739: Standard Test Method for Resistance of Protective Clothing Materials to Penetration by Hazardous Liquid Chemicals, ASTM International, West Conshohocken, PA.

[29] Park, M. and Buehner, R., "An Effective Cure," *Adhesive Age*, Vol. 43, No. 9, September 2000, p. 27.

[30] Lapin, S. C., "Migration of UV Curable Laminating Adhesive Components Through HDPE Containers," *Proceedings of RadTech 2000*, Baltimore, MD, 9–12 Apr. 2000, p. 1007.

[31] Horwood, E., "Safe Handling of UV and EB Resins—A UK Union View," *Conference Proceeding of RadTech—Europe '87*, Munich, Germany, 4–7 May 1987, p. 2-1.

[32] Lawson, K., "Safe Handling of UV/EB Curable Materials," *Conference Proceeding of RadTech—Europe '87*, Munich, Germany, 4–7 May 1987, pp. 2–11.

[33] Robinson, C. J., "Control of Exposure to Acrylate Based Radiation Curable Coatings," *Conference Papers Addendum, RadTech '88—North America*, New Orleans, LA, 24–28 Apr. 1988, p. 30.

[34] Belilos, E., "Regulatory Implications for UV/EB-Cured Coatings Used in Packaging Children's Products," *RADTECH Report*, Vol. 13, No. 5, September/October 1999, p. 51.

[35] Hurd, P., "Cost and Procedures for Obtaining Suitable Status of UV/EB Coatings in Food Packaging Materials," *RADTECH Report*, Vol. 13, No. 5, September/October 1999, p. 44.

[36] Kinter, M. Y., "Environment, Safety and Health, UV Inks and You," *RADTECH Report*, Vol. 13, No. 5, September/October 1999, p. 1.

[37] Adams, G.L., "Specialty Acrylates and Methacrylates Panel Status Report," *Proceedings of RadTech '90—North America*, Chicago, IL, Vol. 2, 25–29 Mar. 1990, p. 139.

[38] Draize, J. H., "Appraisal of the Safety of Chemical in Foods, Drugs, and Cosmetics, Dermal Toxicity," Association of Food and Drug Officials of the United States, Texas State Department of Health, Austin, TX, 1959.

[39] Cantrell, "Safety and Handling of UV/EB Curing Materials," *RADTECH Report*, Vol. 13, No. 5, September/October 1999, p. 23.

[40] Schooley, Willard A., "Contact Dermatitis and UV Curables," *Radiation Curing*, Vol. 6, No. 4, May 1979.

[41] Gelbke, H. P. and Zeller, H., "Skin Irritation Due to Hexanediol Diacrylate; Problems in the Evaluation of Results From Animal Studies," *Journal of Oil Colour Chemists Association*, Vol. 64, 1981, p.186

[42] Malten, K. E., den Arend, J. A. C. J., and Wiggers, R. E., "Delayed Irritation: Hexanediol Diacrylate and Butanediol Diacrylate," *Contact Dermatitis*, Vol. 5, 1979, p. 178.
[43] Björkner, B., "Allergenicity of Trimethylol Propane Triacrylate in Ultraviolet Curing Inks in the Guinea Pig," *Acta Dermatovener*, Stockholm, Vol. 60, No. 6, 528, 1980.
[44] Björkner, B., "Sensitization Capacity of Acrylated Prepolymers in Ultraviolet Curing Inks Tested in Guinea Pig," *Acta Dermatovener*, Stockholm, Vol. 7, p. 61, 1981.
[45] DePass, L. R., "Carcinogenicity Testing of Photocurable Coatings," *Radiation Curing*, Vol. 9, No. 18, August 1982.
[46] Nethercott, J. R., "Allergic Contact Dermatitis Due to Urethane Acrylate in Ultraviolet Cured Inks," *British Journal of Industrial Medicine*, Vol. 40, No. 3, Aug. 1983, p. 241.
[47] Weil, C. S., Condra, N., Haun, C., and Striegel, J.A., "Experimental Carcinogenicity and Acute Toxicity of Representative Epoxides," *Industrial Hygiene Journal*, Vol. 24, July/ Aug. 1963, p. 305
[48] Berger, J. E., Darmer, K. I., and Phillips, C. F., "Health, Safety, and Environmental Factors Affecting Epoxy Resins," *High Solids Coatings*, Vol. 3, No. 6, December 1980, p. 16.
[49] Van Duuren, B. L., Langseth, L., and Goldschmid, B. M., "Carcinogenicity of Epoxides, Lactones and Peroxy Compounds. VI. Structure and Carcinogenic Activity," *Journal of the National Cancer Institute*, Vol. 39, No. 6, Dec. 1967, p. 1217.
[50] Van Duuren, B. L., Orris, L., and Nelson, N., "Carcinogenicity of Epoxides, Lactones, and Peroxy Compounds. Part II," *Journal of the National Cancer Institute*, Vol. 35, No. 4, Oct. 1965, p. 709.
[51] Myer, H. Everett, "Toxicology, the Law, and the Coatings Chemist," *Journal of Coatings Technology*, Vol. 56, No. 710, March 1984, p. 51.
[52] Leimgruber, R. A., "Safety Aspects of Varnishes, Printing Inks and Adhesives," Conference Reading, *RadCure—Europe '87*, Munich, Germany, 4–7 May 1987.
[53] McCammon, C. S., "An Assessment of the Potential for Exposure to Multifunctional Acrylates in Radiation Cured Inks, Coatings, and Adhesives," Conference Papers, *RadTech '88—North America*, New Orleans, LA, 24–28 Apr. 1988, p. 102.
[54] Zeliger, H. I., "Inhalation Hazard of Low Vapor Pressure Chemicals," *Proceeding of RadTech '90—North America*, Chicago, IL, Vol. 1, 25–29 Mar., 1990, p. 352.
[55] Ansell, J. M., "Safety Assessment of a New Reactive Diluent: Triethylene Glycol Divinyl Ether," Vol. 1, *Proceedings of RadTech '90—North America*, Chicago, IL, 25–29 Mar. 1990, p. 354.
[56] "UV/EB Coatings and Food Contact Applications," Internet, *www.radtech.org/regulatory* (Dec. 11, 2000).
[57] Hurd, P., "Customer Assurance Solutions for UV/EB in Food Packaging Markets," *RADTECH Report*, Vol. 13, No. 5, September/October 1999, p. 47.
[58] Nielsen, C. R., Keller and Heckman, "UV/EB Cured Adhesives; FDA Status for Food Packaging Applications," letter to Gary M. Cohen, Executive Director, *RadTech International, North America*, Washington, D.C., 10 Apr. 2000.
[59] Gaube, H. G. and Ohlemacher, J., "Cationic UV Curable Coatings: Aspects of Consumer Safety in Food Packaging," Conference Paper, *RadTech—Europe '89*, Florence, Italy, Oct. 9–11, 1989.
[60] Carter, J. W. and Jupina, M. J., "Cationic UV Ink Migration and Safety Assessment," *Proceedings of RadTech—Europe '97*, Lyon, France, June 16–18, 1997, p. 250; *Proceedings of RadTech—Asia '97* Conference, Yokohama, Japan, 4–7 Nov. 1997, p. 514.
[61] Carter, J. W., Sachs, W. H. and Braddock, J. K., "Formulating Variable Affecting Extractables in Cationic UV-Cured Epoxide Coatings," *Proceedings of 19th Water-Borne, Higher-Solids, and Powder Coating Symposium*, New Orleans, LA, 26–28 Feb. 1992, p. 421.
[62] Carter, J. W., Davis, M. S. and Jupina, M. J. "Cationic UV Coating Extractables," *RadTech '96—North America, Nashville, TN*, 28 Apr.–2 May, 1996, p. 29.
[63] Carter, J. W., Bandekar, J., Jupina, M. J., Kosensky, L. A., and Perry, J. W., "Volatile Contents

of UV Cationically Curable Epoxy Coatings," Technical Paper, *Proceedings of 2nd Biennial Conference: Low- and No-VOC Coating Technologies*, Durham, NC, 13–15 Mar. 1995.

[64] Union Carbide Corp., Data Sheet, "ENVIBAR UV1244 and UV1244T Environmental Barrier Coatings," 1988.

[65] Carter, J. W. and Braddock, J. K., "Effects of Key Variables on Retortability, Flexibility, and Other Physical Properties of Cationic Coatings. I.," *Proceedings of RadTech—Asia '93*, Tokyo, Japan, 10–13 Nov. 1993, p. 314; "II." *Proceedings of RadTech '94—North America*, Orlando, FL, 1–5 May 1994, p. 246; "III." *Proceedings of RadTech—Europe '95*, Maastricht, The Netherlands, 25 Sept. 1995, p. 499.

13

End Uses

THERE ARE NUMEROUS END USES for radiation-cured materials. In a generalized sense, the end uses are coatings, adhesives, sealants, and three-dimensional objects or stereolithography. Coatings are the general subject of this book, and adhesives have been treated in a separate chapter. Within each of these general categories, there are a number of specific end uses, and it is beyond the scope of this book to treat each of these separately. However, some of them are sufficiently developed so that a small section about each can be included. These are given below and referenced where possible. The final section is mainly a table of end use categories with references.

Automotive

Although radiation curing has been used in the automotive area for over 20 years [1,2], it has really been in the past few years that it has taken a firm hold on a portion of the market [3] as evidenced by the strong showing through papers presented at recent radiation curing meetings. In addition to the usual energy, space, time, productivity increase, etc., savings attributed to radiation-cured coatings, the durability of the automotive coatings has been demonstrated by 5–7 year Florida exposure, coupled with major improvements in scratch and mar resistance, and there have been marked reductions in VOC from 40–60% to 0–15%, as well as the potential to eliminate ovens, etc. In addition, the technology has found many commercially successful applications that include [4–6]:

- Electrical and electronic applications such as coil terminators, component marking inks, conductive inks for rear window defrosters, motor balancing compounds, potting compounds, printed circuitry through solder masks and conformal coatings, screen printed membrane switches, sensor switch encapsulants, tacking adhesives, and windshield wiper motor sealing adhesives.
- Headlamp and tail lens assembly applications such as abrasion resistant metalization topcoats, electrical connector potting compounds, glass and plastic lens and reflector adhesives, and primers for metalization.

- Graphics and identification uses such as battery labels, dashboard instrument screen printing and gloss control topcoats, logos on glass substrates, nameplates that are deep cured, and screen-printed oil filter housings, instructions on windshield washers, and fleet markings.
- Functional applications including solid or foamed gasketing compounds, primers and topcoats for plastic parts that are metallized, release coatings for interior simulated leather products, sealants for airbag explosive cartridges, side body molding coatings, wheel cover coatings, wiring harness potting, wood grain printing and topcoats for interior laminates including sealers and topcoat for customized van and RV wood components, and others.

Studies are continuing as technologists strive for improved abrasion, scratch, and mar resistance [7–9] and other physical properties [10]; for new clearcoat technology based on urethane acrylates [11], weathering studies [12], and methods for use of three-dimensional curing on automobile bodies [13]; and for use of new curing technology such as dual-cure systems [14].

Automobile refinishing, where time saved in a drying or curing operation is particularly important, presents the area of ultraviolet radiation-cured coatings an excellent opportunity [15]. Room temperature drying of a solvent-based, two-package polyurethane, high solids system requires about 16 h for the primer and 15 min for the basecoat. If the coating is cured in an oven at 60° C, primer or basecoat drying time is 30 min and the clearcoat is 30 min. If an infrared radiator is employed where the temperature is 70–90° C, these times can be decreased to 10 min each. If an ultraviolet radiation plus infrared radiation system is employed, the plan is to decrease the curing time to a total of 10 min.

Electrical/Electronics

The electrical and electronics areas involve a number of heat sensitive substrates, and radiation curable coatings, adhesives, and sealants are ideally suited for these areas. Particular molecular design is needed to meet the high-technology specifications for this industry [16]. Some of the early end uses for radiation-cured coatings were as soldering masks, photoresists, conductive inks, and other items used to develop flat wires for printed circuitry [17–27]. Such coatings can be applied to rigid printed circuit boards without a mask when screen printing techniques are used [28], or they may be screen printed onto flexible printed wiring circuitry as a top or cover coating.

Conformal coatings, which are used to encapsulate assemblies of chips, capacitors, and other components, on printed circuit boards are radiation-cured with cationic and free radical systems [29–32]. Conformal coatings protect printed circuit assemblies from hostile environments such as moisture, dust, body fluids, and the like, and make performance of many electronic items more uniform and reliable.

Light emitting diodes, as are used in many applications, can be protected and have improved life [33], and radiation-curable adhesives increase productivity for pin connections on liquid crystal displays [34]. The electronic ballast sealants used in fluorescent light systems are decreased in weight (from 8 lbs to 2 lbs) improving safety consideration and ease of installation, require one-third less power wattage, and remove flickering of the light when sealed with a radiation cure sealant [35]. The ballast system is the transformer that powers fluorescent lighting.

Surface component mounting adhesives cured with radiation can improve thermal expansion mismatching [36] and can be conductive [37–39] or not conductive in nature. "Multiwire" products are related to printed circuit boards in that they are built on an epoxy/fiberglass composite substrate [40]. However, rather than flat wires for the circuitry and connections in printed circuit boards, Multiwire uses discrete insulated wires for the circuitry and connections. The round wires are usually insulated with a polyimide coating and may cross over one another on the same level. Such boards are also amenable to radiation-cured coatings or adherents.

The list of electrical/electronic end uses also comprises tacking of wires and parts, electromagnetic interference shielding reduction, tamper proofing, glob topping and encapsulation, potting, use of laser curing for imaging microcircuits [41], inks for component marking [42], and the like. Radiation curing has been investigated for use in polymer-thick-film technology, an additive process for making electronic circuitry [43].

Magnetic Media

Valdemer Poulsen of Denmark in 1898 invented magnetic recording when he developed a method for recording sound [44]. The technology didn't grow much until the 1950s. Today, however, this technology has an important place in the electronics industry, with products such as magnetic tape, program cards of various types, computer disks, audio materials for pleasure and education, and so on. High loadings of particulate, magnetic gamma-ferric oxide, ferric oxide with Cr^{+2}, chromium oxide, and iron-cobalt alloys are dispersed in a solution of a binder. A number of binders have been used to hold the particles on the film or other substrate. These include polyurethanes, vinyl chloride/vinyl acetate copolymers, vinyl chloride/vinyl acetate/vinyl alcohol copolymers, polyacrylates, nitrocellulose, and many others. Particularly important are blends of other polymers with polyurethanes [45].

The substrates used for tapes and other end uses are poly(vinyl chloride), poly(ethylene terephthalate), polypropylene, and other film forming polymers. Poly(ethylene terephthalate) is the most widely used substrate because of its durability, strength, dimensional stability, cost, and versatility [46]. Because of the high loadings of magnetic particles and the high molecular weight binders, large amounts of solvent were needed to obtain application viscosities. A useful study of pigment volume concentrations of magnetic pigments is available in the literature [47]. The technology of particle density, which determines the packing of the pigment in the dried coating, also has an effect on magnetic tape performance.

With emission difficulties, a desire to speed up the manufacturing process, and a search for improved quality products, the area was a natural one to be investigated by those involved in radiation curing and, particularly, in curing with electron beams. The market was large.

By the 1980s, it was said that magnetic media was the driving force for electron beam technology growth in the radiation curing field [48].[1] In one study, a number of different products that contained radiation curable, ethylenic unsaturation was synthesized and formulated with ferric oxide and other ingredients, including relatively large amounts of solvent [49]. After drying, the coatings were exposed to ionizing radiation and cured to form useful magnetic tape materials. A summary of the types of magnetic oxides for audio, video, and computer tapes that were in vogue in the early 1980s was published [50]. This information can provide investigators with information to compare with today's products that are available from various suppliers.

Low or moderate molecular weight urethane acrylates that were difunctional, trifunctional, or comb-branched were blended with each other or with a vinyl polymer or Phenoxy, magnetic iron oxide, and solvent. The blends were coated on poly(ethylene terephthalate) and cured with electron beam radiation in a statistical study [45]. Parameters investigated included hardness, tensile properties, gel fraction, swelling, gloss, and abrasion loss. The study indicated that optimum properties were obtained with blends rather than with single polymers, that glass transition temperature was a more important parameter than cross-link density, and that the statistical approach used was an efficient way to optimize formulation studies. An earlier patent had indicated that polyurethanes with unsaturated end groups combined with Phenoxy and cured by irradiation yielded useful magnetic media materials [51]. Blends of polyurethanes with urethane acrylates and cure with electron beam radiation indicated that good films could be obtained initially, and the added advantage of adding semirigid, elastic properties was ob-

[1] Private Communication. In the early 1980s, moisture-cure urethanes were the standard binders used for magnetic media. Electron beam technology afforded formulations with improved stability coupled with rapid and consistent curing. At the time this was a high value added market, but the market was soon flooded with low cost binder systems. This discouraged capital investment in equipment and the use of relatively high-cost materials.

tained by the subsequent electron beam irradiation [52]. Urethane acrylates based on a caprolactone acrylate or a caprolactone polyol and different isocyanates were prepared in 2-ethylhexyl acrylate, compounded with a magnetic iron oxide, and cured with an electron beam to form useful magnetic media coatings [53].

Blend systems based on vinyl chloride/vinyl acetate/vinyl alcohol copolymer, a 2-mole caprolactone acrylate, and a commercial polyurethane were formulated with a free radical photoinitiator and cured with ultraviolet light to make binder systems that were suitable for magnetic media [54].

There are many studies in the literature dealing with radiation-cured magnetic media, including studies of the energy deposition and backscattered electrons on magnetic properties [55], of how binder molecular weight affects slurry milling efficiency and magnetic tape performance [56], wettability characteristic of electron beam curable coatings in comparison to those of conventional magnetic media coatings [57], the effect of electron beam irradiation on the magnetic media lubricant poly(perfluoroproplyene oxide) [58], performance comparison of magnetic tapes made by irradiation and by thermal means [59], methods for forming a magnetic recording layer on a base plate [60], the use of carboxylic acid-functional polyurethanes that have a strong affinity for magnetic pigments [61], and the use of a radiation curable binder with α-methyl styrene [62].

Optical Components and Materials

Radiation curing has played an important role in the growth of optical media. For example, all digital video discs (DVD) are cemented together with a clear, ultraviolet radiation curable adhesive. The importance of optical fibers, their relationship to the telecommunications industry, and the significance of coatings in their development have been acknowledged facts. This industry, that uses the optical fibers to send streams of digital pulses of voice, video, and data, came into being in the 1970s and has played a role in almost everyone's life since it became a commercial reality in the early 1980s [63]. Some early papers dealt with products and adhesives for optical components [27, 64]. Particular formulations for coatings and adhesives and the assembly of components have been studied [65–70], including studies in special areas such as buffer coatings [71, 72], cabling [73], polyfunctional core coatings [74], and interaction of ultraviolet radiation-cured coating and fiber transmission properties [75].

Thermal and aging properties must be considered very important. Optical fibers are expected to maintain mechanical properties for more than 20 years. Accelerated testing has indicated radiation-cured coatings would be mechanically stable for up to 100 years under ambient conditions [76]. Radiation-cured adhesives can also be used for end-to-end splicing, termination of fiber bundles, construction of optical sensors, and other areas in the optical area [77]. Through the years, many innovations involving radiation curing and optical devices have been made, as is evidenced by the numerous patents in the area. Some recent studies include manufacturing methods for optical recording media [78–80] and free radical curable compositions for optical components [81].

Printing Inks and Graphic Arts

Both actinic and ionizing [82] radiation are used for curing printing inks, as well as both cationic [83,84] and free radical [85,86] photoinitiation with epoxides, vinyl ethers, and ethylenically unsaturated monomers as formulating tools. Safety and environmental concerns [87,88] and process control with radiometry have been considered [89].

Radiation curing has found use in every type of printing process [90–92], including intaglio [93], electrophotography [94], narrow and wide web flexography [95–101], sheet fed or web offset/lithograph [102–110, 113], rotary letterpress [111], screen [111,112], direct and offset gravure, ink jet, and pad printing. Particular area studies include bank notes [114,115], food packaging [116,117], high-speed aluminum can decoration [118], postage stamps [119], coatings that prevent ink fading without loss of adhesion [120], decorative and protective packaging [121,122], coatings for plastics such as polycarbonate [123,124], use of commercial products in inks [125,126], and the use of excimer radiation for ink curing [127].

Stereolithography or Three-Dimensional Object Curing

Stereolithography may also be defined as a photochemical process that enables rapid preparation of solid, three-dimensional models, masters, or patterns that can have any shape. Other names for the technology include "Rapid Prototyping," "Desktop Manufacturing," "Solid Freeform Fabrication," and "Layered Manufacturing." It was first introduced in 1987 as rapid prototyping [128,129]. The essential components of the process are [130,131]:

- A computer with a Computer-Aided Design (CAD) or Magnetic Resonance Image (MRI) software package,
- A laser to provide light energy,
- A mirror system for directing and/or focusing light energy,
- A movable platform located in a reservoir (bath),
- A reservoir containing a photoreactive chemical formulation.

The software sends location information to the laser/mirror system that then directs light energy to specific regions of a thin layer of photoreactive chemicals (i.e., a radiation-curable formulation) present on a movable platform that is located in a reservoir of the chemicals. Simultaneously, the information package directs the system to lower the platform slowly with the cured "slice" of the object into the reservoir. In this manner, complex objects are built up in thin layers of about 0.0005 in. (~0.013 mm) to 0.020 in. (~0.51 mm) on the platform.

Variations on this method exist, but they all deal with methods for the accurate and rapid production of solid models. For example, a computer operator may create a three-dimensional virtual image of an object to be manufactured [132]. Software then slices the composition into layers and generates a precise image of each layer. Then by using a unique masking system generated on a solid base, complex shapes are built up layer by layer. As the object is built up, material under the mask and uncured after the exposure process is removed and replaced with wax before the next curable layer is applied. When the process is completed, the wax is removed by warming or rinsing.

Both free radical [133–135] and cationic systems [129,136] are used. Acrylate formulations and the factors that must be considered in formulation control have been investigated for this end use [137]. These include nature of the components, laser wavelength and intensity, focused spot size and trace speed, and number of scans or dosage per polymerized layer. Strength characteristics, extractables, distortion ratios, solidification depth, and polymerization efficiency were measured properties. Actual and imaginary requirements for the process have also been delineated [138]. These chemical systems produce a variety of items, including models or molds for intake manifold, turbine blades, automobile parts, tools for use in injection-molding, and so on.

The chemical and physical changes that take place in the cured/uncured mixture of monomers and polymers directly under the laser radiation have been studied [139]. A mathematical model based on coupled partial differential equations has been solved for 1,6-hexanediol diacrylate. Although the model's predictions followed expectations, the model was not experimentally tested.

Since these objects are produced rapidly and precisely, it is easy to see where distortion, which might be caused from shrinkage, could lead to difficulties. Distortion control additives for free radical-cured systems that are based on a multifunctional isocyanate, one or more multifunctional polyols, a diol chain extender, and an acrylate with from 1 to 5 acrylate groups per molecule have been developed [140]. It is claimed that these additives will provide a reduction in distortion of up to about 20%.

There are a number of sites on the Internet where one can find information about stereolithography. The field is a rapidly changing one, and a site for the latest updates is *www.cc.utah.edu/~asn8200/rapid.html*. Simply inserting the word "stereolithography" into a search engine returned 58 hits.

Wood Coatings

Wood is a renewable resource that is complex in nature. It is used in a broad variety of end uses, and in many of them it is coated. Wood varies by

- type, hardwood and softwood,
- species that are far too numerous to mention,
- springwood and summerwood,
- growth history—virgin, second growth, third growth—with the wood having a more flat grain as the number of growth cycles increases,
- number of knots, and so on.

All of these factors have an effect on the coating process and the coating used for the staining, sealing, filling, or top coating of wood [141]. In 1992, members of RadTech met to discuss whether or not the radiation-curing, coating industry could meet the demands of the wood industry [142]. It was pointed out that the wood industry offered one of the largest potential markets for radiation-cured coatings. At that time, ultraviolet radiation curing of coatings for flat wood stock, such as is used for kitchen cabinets and wood flooring [143,144], as well as "wet look" wood finishes [145,146], had found a good niche. Again, environmental pressures were pushing the industry to effectively 100% solids systems [144], and with this impetus, the members of RadTech felt the industry could meet the challenges of many-step processes [145–150], maximum cure temperatures of 100° C, variable plant humidity, and the variability of wood itself. The spraying of 100% solids, or near that figure, and the curing of three-dimensional objects were other areas of interest and concern [151,152]. Since that time, they have been proven correct, and wood coating is an important facet of radiation-curing technology. The situation in Europe during the mid to late 1980s has been documented providing a good history of the technology [153].

During the early 1990s, many investigations were carried out in the wood industry. Work in this area included:

- Studies of waterborne radiation curable urethane acrylates [154–157],
- Testing of 21 oligomeric acrylates and numerous monomeric acrylates in finalsealer and topcoat formulations for crosshatch and nickel scrape adhesion, sandability, solvent resistance, and holdout [158,159],
- A study of actinic radiation lamps with enhanced peak irradiance values in the 350–450 nm wavelength range for heavily pigmented coatings [160],
- Formulation development of cycloaliphatic epoxide-based cationic cure coatings and the testing of these coatings for a broad range of properties including hardness, adhesion, hot-cold checking, abrasion, stain resistance, and others [161],
- And cationic formulations for high solids and water-based epoxy systems [162].

By 1996, the interest in and development of radiation curable products for the wood industry had grown to the point where the RadTech organization held a seminar dedicated to technology for the wood industry [163]. At this seminar, topics dealt with application techniques such as roll, spray, and vacuum coatings of 100% solids ultraviolet radiation cured coatings, as well as waterborne and pigmented UV coatings, UV curing equipment for flat and three-dimensional products, health and safety considerations, and end-user comments about the conversion from a conventional to a radiation-cure process. Vacuum coating is a well-known technique that has recently been applied to ultraviolet radiation curable coatings. Such coatings have low viscosity and excellent flow properties, coupled with fast air release and high cure speeds. A starting point formulation for a white vacuum applied coating and a recommended process is available [164].

During the latter half of the 1990s, and into the 21st Century, radiation-curing studies were carried out in a number of areas including:

- photoinitiator fragments released from furniture coatings based on free radical chemistry [165],
- inks and wood fillers that contained aluminum trihydroxide filler, which is transparent to ultraviolet radiation, in free radical-cured formulations [166],
- sprayable water-based wood systems [167],
- comparisons of the conversion from conventional waterborne to ultraviolet radiation cured coatings, pointing out value increases, decreases in waste, and improved physical properties [168], and the economics of roll coating and ultraviolet radiation curing [169],
- use on small wood parts such as broom handles and brush blocks where a 20% solids nitrocellulose industry standard system was replaced [170,171],
- economic and VOC-reduction advantages for flat stock, such as kitchen cabinets and wood flooring [172,173], for guitar manufacturing where productivity gains reduced finishing time per guitar [174], and how to apply low emission furniture coatings [175],
- formulation studies of and application processes for free radical systems as primers, sealers, and top coats for furniture [176] and abrasion-resistant parquet flooring clear coats, as well as methods for testing the cured coatings [177],
- technology dealing with sanding wood coatings with emphasis on ultraviolet radiation cured coatings, including coated abrasives charts for converting between various meshes and average micron size [178].

In 1999, it was reported that in Europe radiation curing of acrylates and unsaturated polyesters had taken over 10.3% of the wood coating market [179]. For the United States, regulatory items that pertain to coating in the wood industry have been summarized [180].

General

In addition to the above specific areas, there are hundreds of particular end uses in which the radiation curing industry participates, and the number and market share keep increasing [181–190]. In some areas it dominates; for example, it has been said that radiation-curable materials are used to decorate essentially all yogurt cups. The large number of end uses includes barrier coatings, book covers, bottles and bottle caps, cans and can ends, catalogues, cigarette package cartons, coated foils and films, conductive coatings, conformal coatings, cosmetic cartons, credit cards, cups and containers, decals, decorative mirrors, denture fillings and restorative dentures, dielectric coatings, electrical insulation, electronics packaging, flocked fabric, labels, liquor cartons, magazines, medical devices, name plates for cars and other transportation vehicles, orthopedic casts and splints, overprint coatings, paper coatings, perfume cartons, prostheses, sealants for syringes, shrink films, tapes—pressure sensitive and magnetic, wallpaper, and wire and wood coatings, as well as many others. Some of these are detailed by reference in the following tabulation.

END USE	REFERENCE
Bioactive substance immobilization on a membrane	191
Cabling for optical fibers	192
Can coatings—2-piece and 3-piece cans	193–197
Ceramic (gypsum) tile manufacture	198
Corrosion resistant coatings—inside ballast tanks of submarines	199
Dental filling materials	200
Flame retardant coating—halogen free	201
Flooring—vinyl sheet, vinyl tile, wood	202–204
Glass fiber reinforced laminates	205–207
Insulation—building and clothing	208, 209
Leather coatings	210, 211
Lottery tickets (scratch-off type)	212
Medical—sustained-release wound dressings	213
Metallization of paper	208, 214
Metallized products—UV-curable base- and/or top-coats	215–217
Microporous membranes and coatings	218
Molds for injection molding	219
Paper, film, foil—packaging	208, 220–222
Photography	223, 224
Pipe coatings—temporary external rust proofing	225
Plastic decoration—cup, container, and other	226–228
Polyboard	229
Polymeric film coating and decoration	230–232
Record jackets	208
Space and satellite	233, 234
Tinplate decoration	235
Transfer decoration	236
Wide web converting	209, 237, 238

The Future

Born amid an upheaval in the coating industry that was caused by environmental concerns, petrochemical shortages, and regulation in the early 1970s, radiation curing has grown into an important segment of the coating industry. The early days were difficult, and through the first decade and more since 1970, growth was slow, but market share and penetration was increasing all the time. Radiation-curing technology found market segments in the industry that could be championed, and there were people who had an interest in championing them. There was very rapid growth of clear overprint varnishes in the late 1980s and of flexographic inks in the late 1990s. These particular areas have turned some minor suppliers into major suppliers and have brought about wider acceptability of radiation-curing technology. Coatings and adhesives for electronics, wood, vinyl flooring and other plastics, automotive, pressure sensitive adhesives, and use in the other areas mentioned above are growing in importance.

Today there is a realization by many that essentially 100% solids, liquid systems that cure in an effectively instantaneous manner into coatings with functional and decorative properties, have merit. The technology has changed the way end-users think about manufacturing. They have found handling times and production or throughput markedly improved by radiation-curing technology. They have found the technology provides fast application of coatings that cure into high quality surfaces without surface flaws that are caused by handling. Pressure sensitive adhesive technology seemed a natural end use for radiation curing, and with particular people leading the area, it soon was adapted and used. In addition, radiation-cured coatings are ecologically sound and very energy efficient. Importantly, the technology allowed manufacturers to meet existing regulations.

Although radiation curing still has only a small percentage of the total coatings market, it is in a rapid growth phase. Those who use the technology are well pleased with the results—a factor that is important to growth. New photoinitiators, monomers and oligomers, equipment for application to substrates and for electron beam and ultraviolet curing is being developed and is entering the marketplace. Cross technologies such as radiation-curable powder coatings that allow powder coatings to be cured on temperature-sensitive substrates should bode well for the future. A handle is being obtained on consistent gloss control, and this will further open up the matte and semigloss coating areas. Thick-section curing that takes place in a rapid manner is becoming a reality, and this will further open the sealants area to radiation curing. Nanocomposite technology is a new arena that is beginning to receive attention. Growth can be expected in all of the market segments mentioned in this chapter with wood, automotive, electronic and optical equipment coatings, and graphic arts leading the way in the near-term future. Longer term, many other segments of the coating, adhesives, and sealant markets will turn to radiation curing as they realize the benefits that can be obtained from radiation curing—with areas such as exterior automotive and other transportation coatings, coatings for exterior plastic signs, business machines, toys, some unrealized areas, and so on, leading the way to a bright future for radiation curing.

References

[1] Lewarchik, R. J. and Hurwitz, D. A., "New Developments in UV Curables for Automotive and Product Finishes," *Proceedings of RadCure '84, AFP/SME*, Atlanta, GA, 10–13 Sept. 1984, p. 3-1.

[2] Near, William R., "Opportunities for UV-Curable Coatings on Automobile Plastics," *Proceedings of RadCure '86*, Baltimore, MD, 8–11 Sept. 1986, p. 2-1.

[3] Bradford, C., "The Evolution of UV Coatings Toward the Automotive Market," *RADTECH Report*, Vol. 13, No. 6, November/December 1999, p. 19.

[4] Cameron, C. and Zimmer, M., "UV Cure—A Cure for the Automotive Processing Dilemma," *RADTECH Report*, Vol. 13, No. 6, November/December 1999, p. 25.

[5] Brandl, C., "UV-Curing Applications for the Automotive Market," *Proceedings of RadTech—Europe '99*, Berlin, Germany, 8–10 Nov. 1999, p. 173.

[6] Meisenburg, U., Joost, K. H., Schwalm, R., and Beck, E, "Automotive Coatings and UV Technology: a Paradigm?" *Proceedings of RadTech—Europe '99*, Berlin, Germany, 8–10 Nov. 1999, p. 201; *RadTech 2000*, Baltimore, MD, 9–12 Apr. 2000, p. 1.

[7] Muzeau, E., von Stebut, J., and Magny, B., "The Scratch Resistance of Radiation Curable Coatings for Automotive Applications," *Proceedings of RadTech—Europe '99*, Berlin, Germany, 8–10 Nov. 1999, p. 193.

[8] Rekowski, V. and Frigge, E., "Progress in UV Clear Coats for Exterior Use," *Proceedings of RadTech 2000*, Baltimore, MD, 9–12 Apr. 2000, p. 29.

[9] Vu, C., Faurent, C., Eranian, A., Vincent, P., Wilhelm, D., et al., "Abrasion and Scratch Resistant UV-Cured Clearcoats Using Colloidal Silica Acrylates," *Proceedings of RadTech 2000*, Baltimore, MD, 9–12 Apr. 2000, p. 822.

[10] Willard, K., "UV/EB Curing for Automotive Coatings," *Proceedings of RadTech 2000*, Baltimore, MD, 9–12 Apr. 2000, p. 18.

[11] Fischer, W., "New Alternatives for UV-Curing Clearcoats Based on Urethane Acrylate Chemistry," *Proceedings of RadTech—Europe '99*, Berlin, Germany, 8–10 Nov. 1999, p. 179.

[12] Decker, C., "Weathering Resistance of Thermoset and UV-Cured Acrylate Clearcoats," *Proceedings of RadTech—Europe '99*, Berlin, Germany, 8–10 Nov. 1999, p. 209.

[13] Svejda, P., "Possibilities of Three-Dimensional Curing UV-Curable Clearcoats in Car Body Finish From the Point-of-View of a Coating Equipment Supplier," *Proceedings of RadTech—Europe '99*, Berlin, Germany, 8–10 Nov. 1999, p. 215; *Proceedings of RadTech 2000*, Baltimore, MD, 9–12 Apr. 2000, p. 53.

[14] Fischer, W., Meier-Westhues, U., and Hovestadt, W., "Dual-Cure, New Possibilities of Radiation Curing Coatings, *Proceedings of RadTech 2000*, Baltimore, MD, 9–12 Apr. 2000, p. 38.

[15] Löffler, H., "UV Curing System for Refinishing Application," *Proceedings of RadTech—Europe '99*, Berlin, Germany, 8–10 Nov. 1999, p. 205.

[16] Zweirs, R. J. M., "Photopolymerization in the Electronics Industry: Molecular Design to Meet High-Technology Specifications," *Proceedings of RadCure—Europe '87*, Munich, Germany, 4–7 May 1987, p. 4-1.

[17] Fisher, L. J., "UV Curable Primary Image Resists," *Proceedings of Radiation Curing VI*, Chicago, IL, 20–23 Sept. 1982, pp. 6-1.

[18] Delany, R., "Electroplated Copper Hybrids Using Photolithographic Techniques," *Proceedings of RadCure '84*, Atlanta GA, 10–13 Sept. 1984, pp. 13–23.

[19] Mathias, E., Ketley, A. D., Morgan, C. R., and Gush, D. P., "Solventless Radiation Curable Conductive Inks," *Proceedings of RadCure '84*, Atlanta GA, 10–13 Sept. 1984, p. 13-1.

[20] Jones, G., W., Russell, D., Simpson, L. L., and Rosenthal, G., "Additional Factors Affecting Step Wedge Readings in Negative Dry Film Photoresists," *Proceedings of RadCure '86*, Baltimore, MD, 8–11 Sept. 1986, p. 13-1.

[21] Sulzberg, T., Schubert, F., and Brewster, W. D., "The Interaction of UV-Curable Solder Masks with Copper Surfaces," *Proceedings of RadCure '86*, Baltimore, MD, 8–11 Sept. 1986, pp. 13–15.

[22] Lescarbeau, D.and Csuros, Z., "The Economic and Technical Advantages of UV Curable Temporary Masking," *Proceedings of RadTech '90–North America*, Chicago, IL, Vol. 2, 25–29 Mar. 1990, p. 62.

[23] Pozzolano, J. and DePoto, R. E., "Permanent Additive Resist Process and Operating Specifications," *Proceedings of RadTech '90–North America*, Chicago, IL, Vol. 2, 25–29 Mar. 1990, p. 417.

[24] Van Dover, L. K., Berg, J. C., and Foshay, ÚV Curable Epoxy Resins for Printed Circuit Board Coatings, *Presentation at the Electrical/Electronics Insulation Conference*, Boston, MA, 14 Nov. 1975; NEPCON West, Anaheim, CA, 25 Feb. 1976.

[25] Patent: Crivello, J. V., General Electric Co., Heat Curable Cationically Polymerizable Compositions and Method of Curing Same with Onium Salts and Reducing Agents, US 4,216,288. (1980); Patent: General Electric Co., Method of Making Printing Plates and Printed Circuit, US 4,193,799. (1980).

[26] Kurisu, V., "Photoimageable Solder Masks," Printed Circuit Fabrication, Vol. 13, No. 11, Nov. 1990, p. 66.

[27] Sinka, J. V. and LieBerman, R. A., "New Developments in Second Generation Radiation Curable Products for Printed Circuit Boards and Fiber Optics," *Proceedings of Radiation Curing VI*, AFP/SME, Chicago, IL, 20–23 Sept. 1982, p. 9-1.

[28] Berg, C. J., "UV Curable Epoxy Resins for Printed Circuit Board Coatings," AFP Technical Paper FC 76-511, Dearborn, MI, 1976.

[29] Grant, S. M., "UV Light Curing Compositions for Electronics Assembly Processes," *Proceedings of RadCure '87—Europe*, Munich, Germany, 4–7 May 1987, p. 7-1.

[30] Bennington, L. D., "Novel Silicone Conformal Coatings," Vol. 1, *Proceedings of RadTech '90—North America*, Chicago, IL, 25–29 Mar. 1990, p. 233.

[31] Patent: Su, W. F. A., Westinghouse Electric Corp., Ultraviolet Curable Conformal Coatings, US 5,013,631. (1991).

[32] Patent: Koleske, J. V., Union Carbide Corp., Conformal Coatings Cured with Actinic Radiation, US 5,155,143. (1992); US 5,043,221. (1991).

[33] Krawiec, S. L., "LED Protection: The UV Answer," *Proceedings of RadTech '88—North America*, New Orleans, LA, 24–28 Apr. 1988, p. 425.

[34] Sun, B. Y. and Lee, B. L., "UV Curable Adhesives Increase the Productivity for Pin Connection on Liquid Crystal Displays," *Radiation Curing*, Vol. 14, No. 3, August 1987, p. 4.

[35] Kelly, K., "Motorola Wants to Light Up Another Market," *Business Week*, 14 Oct. 1991, p. 50.

[36] Auletti, C., R., Kessler, J. L., and Spalding Jr., J. A., "Compliant Silicone Helps Control Thermal Expansion Mismatch," *Adhesives Age*, Vol. 32, No. 2, February 1989, p. 33.

[37] Pandiri, S. M., "The Behavior of Silver Flakes in Conductive Epoxy Adhesives," *Adhesives Age*, Vol. 30, No. 10, October 1987, p. 31.

[38] Hart, A. C., "New Developments in Metallic Pigments for Conductive Applications," *Polymers Paint Colour Journal*, Vol. 176, No. 4168, 1986, p. 416.

[39] Patent: Ernsberger, C. N., CTS Corp., Solderable Conductive Employing an Organic Binder, U S 4,396,666. (1983).

[40] Kunkle, R. E., "Multiwire Today," *Printed Circuit Fabrication*, Vol. 13, No. 9, September 1990, p. 94.

[41] Decker, C., "Laser Curing of Photoresist Systems for Imaging of Micro-Circuits," Technical Paper FC83-265, *RadCure '83*, Lausanne, Switzerland, 9–11 May 1983.

[42] Caffrey, K. F., "UV Curing of Dry Offset Inks for Electronic Components," *Proceedings of Radiation Curing V: A Look to the 80s*, AFP/SME, Boston, MA, 23–25 Sept. 1980, p. 521.

[43] Green, W. J., "Processing Polymer Thick Film Circuitry with Radiation Curing," pp. 11–18, *Proceedings, Radiation Curing VI*, AFP/SME, Chicago, IL, 20–23 Sept. 1982.

[44] Patent: Poulsen, V., "Method of Recording and Reproducing Sounds or Signals," US Letters 661,619. (1900).

[45] Santosusso, T. M., "Radiation Curable Polymer Blends for Magnetic Media," *Radiation Curing*, Vol. 11, No. 3, August 1984, p. 4; *Proceedings of RadCure '84*, Atlanta GA, 10–13 Sept. 1984, pp. 16–42.
[46] Zillioux, R.M., "Polyester Film—Its Use and Benefits as a Substrate in Radiation Curing Applications," *Conference Proceedings of RadCure '86*, Baltimore, MD, 8–11 Sept. 1986, p. 8-1.
[47] Huisman, H. F., "Volume Concentrations of (Magnetic) Pigments," *Journal of Coating Technology*, Vol. 57, No. 721, 1985, p. 49
[48] Rand, Jr., W. M., "Magnetic Media and the Electron," *Radiation Curing*, Vol. 11 No. 3, August 1984, p. 18.
[49] Patent: Nakajima, K., Somezawa, M., Takamizawa, M., Inoue, Y., and Yoshioka, H., Sony Corp., Magnetic Recording Medium, US 4,368,242. (1983).
[50] Winquist, D. P., "Particles for Magnetic Media–Market Analysis and Future Trends," Paper FC84-1018, *RadCure '84*, Atlanta, GA, 10–13 Sept. 1984.
[51] Patent: Kolycheck, E., G., The B. F. Goodrich Co., Curable Polyurethanes, US 4,408,020 (1983).
[52] Jurek, I. E. and Keller, D. J., "Electron Beam Curable Polyurethane Blends as Magnetic Media Binders," *Journal of Radiation Curing*, Vol. 12, No. 1, January 1985, p. 20; *Proceedings of RadCure '84*, Atlanta, GA, 10–13 Sept. 1984, pp. 16–25.
[53] Patent: Osborn, C. L. and Koleske, J. V., Union Carbide Corp., Magnetic Recording Medium, US 4,585,702. (1986).
[54] Koleske, J. V., Peacock, G., S., Kwiatkowski, G., T., Union Carbide Corp., Magnetic Recording Medium, US Patent 4,555,449. (1985).
[55] Pacansky, J., and Waltman, R. J., "Studies on the Effects of Energy Deposition and Backscattered Electrons on Magnetic Media," *Proceedings of RadCure '84*, Atlanta, GA, 10–13 Sept. 1984, p. 16–1.
[56] Tu, R. S., "Effect of Molecular Weight of Binder on Slurry Milling Efficiency and Performance of a Typical EB Curable Magnetic Tape System," *Proceedings of RadTech '90—North America*, Chicago, IL, Vol. 2, 25–29 Mar. 1990, p. 62, 405
[57] Laskin, L., Ansel, R., Murray, K. P., and Schmid, S. R., "EB Curable Wetting Resin for Magnetic Media Coatings," Paper FC84-1023, *RadCure '84*, Atlanta, GA, 10–13 Sept. 1984.
[58] Pacansky, H. and Waltman, R. J., "Electron Beam Curing of Poly(perfluorinated ethers) Used as Lubricants in the Magnetic Media Industry," *Proceedings of RadCure '86*, Baltimore, MD, 8–11 Sept. 1986, p. 6-1.
[59] Tu, R. S. and Anglin, D. L., "Performance Comparison of Electron Beam and Thermal Curable Magnetic Tapes," *Proceedings of RadCure '86*, Baltimore, MD, 8–11 Sept. 1986, pp. 6–35.
[60] Patent: Ohno, T., Yamaguchi, T., and Itoh, S., Kabushiki Kaisha Toshiba Portable Storage Medium and Apparatus for Processing the Same, US 5,570,124. (1996).
[61] Patent: Carlson, J. G., Anderson, J. T., and Rotto, N. T., Minnesota Mining and Manufacturing Co., Strong Carboxylic Acid Functional Polyurethane Polymers and Blends Thereof Used in Magnetic Recording Media, US 5,512,651. (1996).
[62] Patent: Haidos, J. C., Arudi, R. L., and Rotto, N. T., Minnesota Mining and Manufacturing Co., Magnetic Recording Medium Having Radiation Curable Binder with α-Methyl styrene Unsaturation, US 5,523,115. (1996).
[63] Blyer, Jr., L. L. and Aloisio, C. J., "Polymer Coatings for Optical Fibers," *Applied Polymer Science*, 2nd ed., ACS Symposium Series 285, American Chemical Society, 1985.
[64] Seo, N., "The Assembly of Optical Components Using UV Curable Adhesives," *Proceedings of RadCure '84*, AFP/SME, Atlanta, GA, 10–13 Sept. 1984, pp. 2–27.
[65] Rie, J., "UV-Cured Adhesives for Fiber Optics and Electronics Assembly," *Proceedings of RadCure '86*, Baltimore, MD, 8–11 Sept. 1986, p. 3-1.
[66] Norton, R. V., "UV Cure Coatings for Optical Fibers: A Formulary Review," Conference Papers, *RadTech '88—North America*, New Orleans, LA, 24–28 Apr. 1988, p. 1.

[67] Seo, N., "High Adhesional Quality Using Ultraviolet Curing Adhesives," *Proceedings of RadCure '86*, Baltimore, MD, 8–11 Sept. 1986, p. 3–37.

[68] Patent: Ansel, R. E., DeSoto, Inc., Soft and Tough Radiation-Curable Coatings for Fiber Optic Application, US 4,682,851. (1987).

[69] Patent: Zimmerman, J. M. and Bishop, T. E., DeSoto, Inc., Ultraviolet Curable Optical Glass Fiber Coatings from Acrylate Terminated, End-Branched Polyurethane Polyurea Oligomers, US 4,690,502. (1987).

[70] Patent: Pasternack, G., Bishop, T. E., and Cutler, Jr., Orvid L., DeSoto, Inc., Optical Glass Fiber Coated with Cationically Curable Polyepoxide Mixtures, US 4,585,534. (1986).

[71] Stowe, R. W., "Improvements in Efficiency of UV Curing Systems for Optical Fiber Buffer Coatings," *Proceedings of RadCure '86*, Baltimore, MD, 8–11 Sept. 1986, pp. 3–23.

[72] Bishop, T. E., DeSoto Inc., Ultraviolet Curable Buffer Coatings for Optical Fibers, US Patent 4,629,287. (1986).

[73] Reese, J. E., "UV-Curable Coatings for Cabling Optical Fibers," *Proceedings of RadCure '86*, Baltimore, MD, 8–11 Sept. 1986, pp. 3–11.

[74] Patent: Krajewski, J, J., Bishop, T. E., Coady, C, J., Zimmerman, J. M., Noren, Gerry K., and Fisher, C. E., DeSoto, Inc., Ultraviolet Curable Coatings for Optical Glass Fiber Based on a Polyfunctional Core, US 4,806,574. (1989).

[75] Leppert, H. D., Berndt, J., and Zamzow, P. E., "Interaction Between UV Coating and Optical Fiber Properties," *Proceeding of RadCure Europe '87*, Munich, Germany, 4–7 May, 1987, pp. 4–27.

[76] Cutler, Jr., O. R., "Thermal and Aging Characteristics of UV Curable Optical Fiber Coatings," Technical Paper FC84-1022, *RadCure '84*, Atlanta, GA, 10–13 Sept. 1984.

[77] Murray, R. T. and Jones, M. E., "Command Cure Precision Cements in Optics," Technical Information Brochure, ICI Resins US, 1993.

[78] Patent: Nakayama, J., Katayama, H., Takahashi, A., Ohta, K., and Van, K., Optical Recording Medium and Manufacturing Method Thereof, Sharp Kabushiki Kaisha, US 5,989,671. (1999).

[79] Patent: Amo, M. and Inouchi, M., Kitano Engineering, Co., Ltd., Method of Curing an Optical Disc, US 6,013,145. (2000).

[80] Patents: Nishizawa, Akira, Victor Company of Japan, Ltd., Optical Discs, Producing Methods and Production Apparatus of the Optical Discs, US 6,279,959. (2001); Savant, G. and Latchinian, H., Physical Optics Corp., Composition for Use in Making Optical Components, US 6,168,207 (2001).

[81] Patent: Savant, G. and Latchinian, H., Physical Optics Corp., Composition For Use in Making Optical Components, US 6,262,140. (2001).

[82] Maguire, E., "EB Units," *End User Conference of RadTech 2000*, Baltimore, MD, 9–12 Apr. 2000, p. 63.

[83] Hanrahan, B. D., Manus, P., and Eaton, R. F., "Cationic UV Curable Coatings and Inks for Metal Substrates," *Proceedings of RadTech '88—North America*, New Orleans, LA, 24–28 Apr. 1988, p. 14.

[84] Heberle, M., "The Success Story of an End User," *Proceedings of RadTech—Europe '99*, Berlin, Germany, 8–10 Nov. 1999, p. 293.

[85] Hutchinson, I., Smith, M., Grierson, W., and Devine, E., "Pigments and Oligomers for Inks—Moving Towards the Best Combination," *Proceedings of RadTech—Europe '99*, Berlin, Germany, 8–10 Nov. 1999, p. 285.

[86] Bolle, T. and Schulz, R., "High Speed UV-Printing—A Photoinitiator Selection Study," *Proceedings of RadTech Europe '99*, Berlin, Germany, 8–10 Nov. 1999, p. 255.

[87] Siegel, S., "UV Safety and the Environment," End User Conference, *RadTech 2000*, Baltimore, MD, 9–12 Apr. 2000, p. 73.

[88] Kinter, M. Y., "Environmental Issues and UV Technology," End User Conference, *RadTech 2000*, Baltimore, MD, 9–12 Apr. 2000, p. 80.

[89] Wright, R. E., "Radiometry, Why Should I Care?" End User Conference, *RadTech 2000*, Baltimore, MD, 9–12 Apr. 2000, p. 83.
[90] Jägers, D., "UV-Use in the Printing Industry: Facts and Figures," *Proceedings of RadTech '99—Europe*, Berlin, Germany, 8–10 Nov. 1999, p. 227.
[91] Faure, G., "Overview of the UV/EB Inks and Coatings European Market," *Proceedings of RadTech—Europe '99*, Berlin, Germany, 8–10 Nov. 1999, p. 239.
[92] Justice, W. H., "Getting Started—UV Curing of Conventional Inks," *Proceedings of RadCure '86*, Baltimore, MD, 8–11 Sept. 1986, p. 7-1.
[93] O'Brien, T. F., "Electron Beam Cylinder Wipe Intaglio Inks," *Proceedings of RadCure—Europe '87*, Munich, Germany, 4–7 May 1987, pp. 8–9.
[94] Sugita, K., Ishizawa, T., Tateno, H., Kushida, M., Harada, K., Saito, K., et al., "Photodegradable Toners for Electrophotography II. Accelerated Photodegradation of Matrix Resin by DUV—Exposure at an Elevated Temperature," *Conference Proceedings of RadTech Europe '99*, Berlin, Germany, 8–10 Nov. 1999, p. 325.
[95] Gupta, P., "UV Flexo, Technical Advancements," *Proceedings of RadTech 2000*, Baltimore, MD, 9–12 Apr. 2000, p. 462.
[96] Mulligan, B., "The Anilox Roll: The Key to Consistency," End User Conference, *RadTech 2000*, Baltimore, MD, 9–12 Apr. 2000, p. 97.
[97] De Micheli, P., "Flexographic Inks: The Compromise," *Conference Proceedings of RadTech Europe '99*, Berlin, Germany, 8–10 Nov. 1999, p. 261.
[98] Gringeri, F., "UV Flexo: A Cost Comparison Evaluation," *Conference Proceedings of RadTech Europe '99*, Berlin, Germany, 8–10 Nov. 1999, p. 267.
[99] Wolf, F., and Kies, J., "Print Quality with UV Flexo," *Conference Proceedings of RadTech Europe '99*, Berlin, Germany, 8–10 Nov. 1999, p. 299.
[100] Hausman, G. and Bean, A., "Radiation Curing for Flexible Packaging," *Proceedings of RadTech '90—North America*, Chicago, IL, Vol. 2, 25–29 Mar. 1990, p. 159.
[101] Midlik, E., "What You Need to Know to be a Successful Wide Web Flexo UV Coating Operation," End User Conference, *RadTech 2000*, Baltimore, MD, 9–12 Apr. 2000, p. 97, 109.
[102] Pascale, J. V., Greenslade, R. T., De, D.A. and Helsby, D. A., "An Evaluation of Ink Performance Properties for UV Versus Conventional Sheetfed Ink," *Proceedings of RadCure '86*, Baltimore, MD, 8–11 Sept. 1986, pp. 7–19.
[103] Maki, D., "UV Coating and HWO: Process Considerations," End User Conference, *RadTech 2000*, Baltimore, MD, 9–12 Apr. 2000, p. 114.
[104] Honaker, J., "The Fundamentals of In-Line, Tandem Printing," End User Conference, *RadTech 2000*, Baltimore, MD, 9–12 Apr. 2000, p. 116.
[105] Waldo, R. M., "Lithographic Inks: A Correlation of Emulsification Test Methods," *Proceedings of RadTech 2000*, Baltimore, MD, 9–12 Apr. 2000, p. 473.
[106] Van Esch, C., "Is the Waterbalance a Lithographic UV Problem," *Conference Proceedings of RadCure Europe '87*, Munich, Germany, 4–7 May 1987, pp. 8–21.
[107] Branscomb, D. N., "A 'Shotgun' Approach to Roller Covering Observations/Questions," End User Conference, *RadTech 2000*, Baltimore, MD, 9–12 Apr. 2000, p. 89.
[108] Lewis, V., "Fountain Solutions for Radiation Curable Inks," End User Conference, *RadTech 2000*, Baltimore, MD, 9–12 Apr. 2000, p. 92.
[109] Takayama, M., Ishii, H., and Hatta, S., "Studies on Materials of UV Curable Lithographic Inks," *Proceedings of RadTech '90—North America*, Chicago, IL, Vol. 2, 25–29 Mar. 1990, p. 167.
[110] Bargenquest, B., "Retrofitting Sheetfed," End User Conference, *RadTech 2000*, Baltimore, MD, 9–12 Apr. 2000, p. 120.
[111] Kaufman, Martin, "UV Rotary Letterpress and UV Screen Printing," *Conference Papers Addendum of RadTech '88–North America*, New Orleans, LA, 24–28 Apr. 1988, p. 66.
[112] Lendle, E., "New Aspects of UV-Drying of Screen Printing Inks," *Conference Proceedings of RadTech—Europe '99*, Berlin, Germany, 8–10 Nov. 1999, p. 317.

[113] Gibbons, T. J., "UV Coated Products and Web Offset," *Conference Papers Addendum of RadTech '88–North America*, New Orleans, LA, 24–28 Apr. 1988, p. 68.
[114] Anderson, U., "Banknote Printing with EB- and UV-Cured Inks," *Conference Proceedings of RadTech—Europe '99*, Berlin, Germany, 8–10 Nov. 1999, p. 249.
[115] Walker, J., "The Future of Money—The Way Forward is Clear," *Conference Proceedings of RadTech—Europe '99*, Berlin, Germany, 8–10 Nov. 1999, p. 311.
[116] Helsby, D., "UV Curing Printing Inks for Food Packaging," *Conference Proceedings of RadTech—Europe '99*, Berlin, Germany, 8–10 Nov. 1999, p. 279.
[117] Carter, J. W. and Jupina, M. J., "Cationic UV Ink Migration and Safety Assessment," *Proceedings of RadTech—Europe '97*, Lyon, France, 16–18 June 1997, p. 250; *Proceedings of RadTech—Asia '97* Conference, Yokohama, Japan, 4–7 Nov. 1997, p. 514.
[118] Bishop, J. A., "Parameters Affecting U.V. Ink Performance in High Speed Aluminum Can Decoration," *Proceedings of RadTech '88—North America*, New Orleans, LA, 24–28 Apr. 1988, p. 257.
[119] Anon, "Polyrad® UV-8 Coating Delivers for U. S. Postal Service," Brochure, Morton Thiokol, Inc., Morton Chemical Div., 1987.
[120] Wolinski, L. E. and Sadowski, G. E., "UV Light Cured Coatings Which Prevent Fading of Inks Without Loss of Adhesion," *Conference Proceedings of RadCure '84*, AFP/SME, Atlanta, GA, 10–13 Sept. 1984, pp. 6–10.
[121] Bean, A. J., "The Technology of Using UV & EB Curable Inks and Coatings for Decorative and Protective Packaging," *Conference Proceedings of RadCure '84*, AFP/SME, Atlanta, GA, 10–13 Sept. 1984, p. 9-1.
[122] Braddock, J. K., Carter, J. W., and Lamb, K. T., "Cationic UV Curable Inks for Rigid Packaging," *Proceedings of RadTech '98—North America*, Chicago, IL, 19–22 Apr. 1998, p. 545.
[123] Simpson, D. L., "An Ideal Marriage—Radiation Curing and Polycarbonate Plastic," *Conference Proceedings of RadCure '86*, Baltimore, MD, 8–11 Sept. 1986, pp. 7–13.
[124] Manus, P. J. M., "Cationic UV-Curable Coatings for Metal and Plastic Substrates," *Polymers, Paints, and Colour Journal*, Vol. 179, No. 4242, 1989, p. 524.
[125] Cattanel, M., Viconti, M., and Filpa, R., "LFC 1001 in Printing Inks," *Proceedings of RadTech 2000*, Baltimore, MD, 9–12 Apr. 2000, p. 1048.
[126] Rudolph, A., "Radiation Curable Coatings for the Graphic Arts Enhanced by Aqueous Anchor Coat," *Conference Papers of RadTech '88—North America*, New Orleans, LA, 24–28 Apr. 1988, p. 345.
[127] Mehnert, R., "Excimer UV Curing in Printing," *Proceedings of RadTech—Europe '99*, Berlin, Germany, 8–10 Nov. 1999, p. 303.
[128] Patent: Hall, C. W., UVP, Inc., Apparatus for Production of Three-Dimensional Objects by Stereolithography, US 4,575,330. (1986).
[129] Johnson, D. L., "A Review of Stereolithography," *RADTECH Report*, Vol. 14, No. 5, September/October 2000 p. 11.
[130] "Stereolithography (SLA)," Tech, Inc., Internet, *www.techok.com/sla*, 2000.
[131] "The Stereolithography Process: How Does the Process Work?" Internet, *www.cs.cmu.edu/People/rapidproto/students.98/master*, 2000.
[132] Anon, "Rapid Prototyping—Solid Model Building," *RADTECH Report*, Vol. 5, No. 3, May/June 1991, p. 12.
[133] Patent: Murphy, E. J., Ansel, R. E., and Krajewski, J. J., DeSoto, Inc., Investment Casting Utilizing Patterns Produced by Stereolithography, US 4,844,144. (1989).
[134] Patent: Hull, C. W., 3D Systems, Inc., Method for Production of Three-Dimensional Objects by Stereolithography," US 4,929,402. (1990).
[135] Krajewski, J. J. and Murphy, E. J., "Investment Castable Photofabricated Parts," *Proceedings of RadTech '90–North America*, Chicago, IL, Vol. 1, 25–29 Mar. 1990, p. 211.
[136] Murphy, E. J. and Krajewski, J. J., "Some Characteristics of Steric Polymerization," *Proceedings of RadTech '90—North America*, Chicago, IL, 25–29 Mar. 1990, p. 217.

[137] Patent: Steinmann, B., Wolf, J. P., Schulthess, A., and Hunziker, M., Ciba-Geigy Corp., (Cyclo)aliphatic Epoxy Compounds, US 5,599,651. (1997).

[138] Herskowits, V., "Stereolithographic Materials–Real and Imaginary Requirements," *Proceedings of RadTech '90—North America*, Chicago, IL, Vol.1, 25–29 Mar. 1990, p. 227.

[139] Flach, L. and Chartoff, R. P., "An Analysis of Laser Photopolymerization (Applied to Stereolithography)," *Proceedings of RadTech '90—North America*, Chicago, IL, Vol. 2, 25–29 Mar. 1990, p. 52.

[140] Patent: Rex, G. C., Union Carbide Chemicals & Plastics Tech. Corp., Distortion Control Additives for Ultraviolet-Curable Compositions, US 5,498,782. (1996); US 5,801,392. (1998).

[141] Feist, W. C. and Williams, R. S., "Exterior Wood Coatings: Current Needs and Changes for the Future," *Paint & Coatings Industry*, Vol. IX, No. 3, April 1993, p. 28.

[142] Rayball, J., "RadTech Members Discuss UV/EB Curing Technology," *RADTECH Report*, Vol. 6, No. 4, July/August, 1992, p. 13.

[143] Stranges, A. J., "Why European Style U.V. Flat Lines?" *Proceedings of RadTech '90—North America*, Chicago, IL, Vol. 1, 25–29 Mar. 1990, p. 117.

[144] Garratt, P. G., "The Use of Unsaturated Polyester Resins in UV Curable Paint Formulations for Use in the Furniture Industry," *Proceedings of RadCure '84*, AFP/SME, Atlanta, GA, 10–13 Sept. 1984, pp. 3–13.

[145] Riedell, A. W., "Some Contribution of Radiation Curable Coatings to the Production of High Gloss "Wet" Look Wood Finishes, *Proceedings of RadTech '90—North America*, Chicago, IL, Vol. 1, 25–29 Mar. 1990, p. 119.

[146] Anon, "EB/UV Cure of 100% - Solids 'Wet-Look' Pigmented Coatings," *Industrial Finishing*, Vol. 18, No. 12, December 1987, p. 34; Danneman, J., *Modern Paint and Coatings*, Vol. 78, No. 2, February 1988, p. 28.

[147] Cox, R., "UV Curable Coatings Technology in the Wood Industry: Compliance for the '90s," *RADTECH Report*, Vol. 6, No. 4, July/August 1992, p. 24.

[148] Gruber, G. W., "UV Curing of Coatings," *Radiation Curing*, Vol. 13, No. 1, 1986, p. 4.

[149] Schrantz, J., "Wood Finishing Needs 'Flexibility'," *Industrial Finishing*, Vol. 14, No. 4, April 1983, p. 14.

[150] Thelander, L. and Hoel, O., "Experience of Ultraviolet Curing on Hardwood Floors," *Proceedings of RadCure '84*, AFP/SME, Atlanta, GA, 10–13 Sept. 1984, p. 3–29.

[151] Riedell, A., "3-Dimensional Finishing of Wood Furniture," *Proceedings of RadTech '88—North America*, New Orleans, LA, 24–28 Apr. 1988, p. 556.

[152] Arnold, Sr., H. S., "3D UV Curing in Furniture and Wood Products Now and in the Future," *Conference Papers Addendum of RadTech '88–North America*, New Orleans, LA, 24–28 Apr. 1988, p. 80.

[153] Chiocchetti, P., "Radiation Curing Coatings in the Wood Industry," *Proceedings of RadCure—Europe '87*, Munich, Germany, 4–7 May 1987, p. 8-1.

[154] Stenson, P. H., "Radiation Curable Water-Borne Urethanes for the Wood Industry," *Proceedings of RadTech '90—North America*, Chicago, IL, Vol. 1, 25–29 Mar. 1990, p. 114; *Modern Paint and Coatings*, Vol. 80, No. 6, June 1990, p. 44.

[155] Stenson, P. H., "New Developments in Water-Borne Radiation Curable Resins Enhance Opportunities in Wood Finishing," *Paint and Coatings Industry*, Vol. VIII, No. 9, September 1992, p. 22.

[156] Dvorchak, M. J. and Riberi, B. H., "Water Reducible, Unsaturated Polyesters as Binders and Clear Coatings for UV-Curable Furniture Coatings," *Journal of Coatings Technology*, Vol. 64, No. 818, May 1992, p. 41.

[157] Dvorchak, M. J., "UV Curing of Pigmented High-Build Wood Coatings Based on Non-Air-Inhibited Unsaturated Polyesters," *Journal of Coatings Technology*, Vol. 67, No. 842, March 1995, p. 49.

[158] Mahon, W. F. and Nason, D. L., "Testing UV-Cure Coatings Systems for Wood," *Modern Paint and Coatings*, Vol. 82, No. 6, June 1992, p. 44.

[159] Mahon, W. F. and Nason, D. L., "UV Cure Finishing Systems for Wood," *Proceedings of RadTech '92—North America*, Boston, MA, 26–30 Apr. 1992, p. 190.
[160] Schaeffer, W.R., "Enhanced UV Curing of Heavily Pigmented Coatings for the Wood Industry Through Increased Peak Irradiance and Careful Wavelength Selection," *Proceeding of RadTech '92—North America*, Boston, MA, 26–30 Apr. 1992, p. 201.
[161] Technical Bulletin, "Starting Formulations for Wood Substrates," Bulletin No. 1, Union Carbide Chemicals and Plastics Co. Inc., 16 Mar. 1990.
[162] Eaton, R. F., Hanrahan, B. D., and Braddock, J. K., "Formulating Concepts for Both High Solids and Water Based UV Cured Cycloaliphatic Epoxy Wood Coatings," *Proceedings of RadTech '90—North America*, Chicago, IL, Vol. 1, 25–29 Mar. 1990, p. 384.
[163] Anon, "UV/EB Curing: Practical Solutions for the Wood Industry," *RadTech International North America*, Atlanta, GA, 21 Aug. 1996.
[164] Valdes-Aguilera, O., "UV Curing: Vacuum Coatings," *Modern Paint & Coatings*, Vol. 90, No. 12, Dec. 2000, p.17.
[165] Salthammer, T., "Release of Photoinitiator Fragments from UV-Cured Furniture Coatings," *Journal of Coatings Technology*, Vol. 68, No. 856, May 1996, p. 41.
[166] Dando, N. R., Kolek, P. L., Martin, E. S., and Clever, T. R., "Performance Optimization of 100% Solids, UV-Cure Inks and Wood Fillers Using Aluminum Trihydroxide (ATH) Filler," *Journal of Coatings Technology*, Vol. 68, No. 859, August 1996, p. 67.
[167] Wang, J. Z., Arceneaux, J. A., and Hall, J., "UV Curable Aqueous Dispersions for Wood Coatings," *Modern Paint and Coatings*, Vol. 86, No. 8, 1996, p. 24.
[168] Modjewski, R. J., "UV Curing for Wood Applications," *RADTECH Report*, Vol. 13, No. 3, May/June 1999, p. 45.
[169] Schlatter, H., "Economic Surface Treatment by Means of Roller Coating Technique and UV Curing," *Proceedings of RadTech—Europe '99*, Berlin, Germany, 8–10 Nov. 1999, p. 153.
[170] Berejka, A. and Larsen, S., "A Clean Sweep—UV Curing of Small Wood Products," *RADTECH Report*, Vol. 13, No. 3, May/June 1999, p. 20.
[171] Berejka, A. J., "UV Curing of Small Wood Products—An Industrial Demonstration Product, *RADTECH Report*, Vol. 11, No. 4, July/August 1999, p. 30.
[172] Derbas, J., "UV Coatings as an Alternative Option for the Changing Environment in Wood Finishes," *RADTECH Report*, Vol. 13, No. 3, May/June 1999, p. 31.
[173] Bankowsky, H. H., Enenkel, P., Beck, E., Lokai, M., and Sass, K., " The Formulation and Testing of Radiation Curing Binders for Abrasion-Resistant Parquet Varnishes," *Proceedings of End User Conference at RadTech 2000*, Baltimore, MD, 9–12 Apr. 2000, p. 210.
[174] Osborne, A., "Guitar Makers Create Beautiful Music with UV Coatings," *RADTECH Report*, Vol. 13, No. 3, May/June 1999, p. 16.
[175] Salthammer, T., Hofmockel, U., Lokai, M., Prieto, J., Hansemann, W., "Application of UV-Curing for Low Emission Furniture Coatings," *Proceedings of RadTech—Europe '99*, Berlin, Germany, 8–10 Nov. 1999, p. 159.
[176] Van den Branden, S., "Wood Coatings for the Furniture Industry," *Proceedings of RadTech—Europe '99*, Berlin, Germany, 8–10 Nov. 1999, p. 147.
[177] Bankowsky, H. H., Enenkel, P., Beck, E., Lokai, M, and Sass, K., "The Formulation and Testing of Radiation Curing Binders for Abrasion-Resistant Parquet Varnishes," *Proceedings of End User Conference at RadTech 2000*, Baltimore, MD, 9–12 Apr. 2000, p. 210.
[178] Beaty, B., "Sanding Wood Substrates and UV Coatings," *Proceedings of End User Conference at RadTech 2000*, Baltimore, MD, 9–12 Apr. 2000, p. 199.
[179] Bankowsky, H. H., Enenkel, P., Lokai, M., and Menzel, K., "Radiation Curing of Wood Coatings," *Proceedings of RadTech—Europe '99*, Berlin, Germany, 8–10 Nov. 1999, p. 131.
[180] Loof, R. M., "Regulatory Issues in the Wood Coating Industry," *Proceedings of End User Conference at RadTech 2000*, Baltimore, MD, 9–12 Apr. 2000 p. 223.
[181] Pelgrims, J., "Cationic UV Polymerization of CYRACURE® Cycloaliphatic Epoxies: A Review of Technology and Markets," Technical Paper presented at the VILF Association, Krefeld, Germany, November 1986.

[182] Laughlin, C., "The Solution is in the Cure," *RADTECH Report*, Vol. 4, No. 2, March/April 1990, p. 12.

[183] Elias, P., "UV/EB Curing Technology: Making Headway in Technology and Applications," *Paint and Coatings Industry*, Vol. X, No. 5, May 1994, p. 60.

[184] Anon, "Radiation Curing 1996," *Modern Paint and Coatings*, Vol. 86, No. 4, April 1996, p. 29.

[185] Busato, F., "Waterborne, High Solids, and Rad-Cure Technologies: Progress in the U.S., Europe and Asia," *Modern Paint and Coatings*, Vol. 87, No. 3, March 1997, p. 30.

[186] Koleske, J. V., "Radiation-Curable Coatings on the Upsweep of a Strong Growth Curve," *Paint and Coating Industry*, Vol. XVI, No. 6, June 2000, p. 34.

[187] Walker, J. H., "Radiation Curing: A Future Force to be Reckoned With?" *Modern Paint and Coatings*, Vol. 90, No. 10, October 2000, p. 15.

[188] Hess, J., "Still Misunderstood but Making Progress: Rad-Cure Coatings," *Coatings World*, Vol. 6, No. 4, April 2001, p. 38.

[189] Willard, K., "Driving Innovation: UV/EB Technologies Find New Uses in Automotive Industry," *Modern Paint and Coatings*, Vol. 91, No. 4, April 2001, p. 29.

[190] Joesel, K. H., "Automotive Applications for UV-Cure," *Industrial Paint and Powder*, Vol. 77, No. 4, April 2001, p. 26.

[191] Jin, C., Ye, Y., and Zongxian, S., "Immobilization of Bioactive Substance on Membranes by Radiation Technique," *Proceedings of RadTech '90—North America*, Chicago, IL, Vol. 2, 25–29 May 1990, p. 154.

[192] Pasternack, "Ultraviolet Light (UV) Curable Resins for Cabling Optical Fibers," *Proceedings of Conference on Radiation Curing—Asia*, Tokyo, Japan, 20–22 Oct. 1986, p. 138.

[193] Crabtree, T. A., "UV Curing of Two-Piece Cans: An Update," *Conference Papers of RadTech '88—North America*, New Orleans, LA, 24–28 Apr. 1988, p. 231.

[194] Iimure, T., "UV-Cured White Coat for 3-Piece Beverage Cans," *Proceedings of RadTech '88—North America*, New Orleans, LA, 24–28 Apr. 1988, p. 240.

[195] Matyska, K. J., "UV Curing Matures at U. S. Can," *RADTECH Report*, Vol. 11, No. 6, November/December 1997, p. 20.

[196] Thompson, R. J., "Advances in UV Curing of Two-Piece Cans," Technical Paper FC86-856, *RadCure '86*, Baltimore, MD, 8–11 Sept. 1986.

[197] Patent: Huemmer, T. F. and Plooy, R. J., The O'Brien Corp., Radiation Curable Can Coating Composition, US 3,912,670 (1975)

[198] Maruyama, T., Ogawa, M., and Sugimoto, K., "Electron Beam Curing Coating for Gypsum Tile," *Proceedings of Conference on Radiation Curing—Asia*, Tokyo, Japan, 20–22 Oct. 1986, p. 275.

[199] Mejiritski, A., Marino, T., Martin, D., "Corrosion Resistant Visible Light Curable Coatings, Part II," *Proceedings of RadTech 2000*, Baltimore, MD, 9–12 Apr. 2000, p. 439.

[200] Patent: Bowen, R. L., United States of America, Dental Filling Material Comprising Vinyl Silane Treated Fused Silica and a Binder Consisting of the Reaction Product of Bisphenol and Glycidyl Acrylate, US 3,066,112. (1962).

[201] Smit, C. N., Hennink, W. E., de Ruiter, B., Luiken, A. H., and Marsman, M. P. W., "Radiation Cured Halogen Free Flame Retardant Coatings, *Proceedings of RadTech '90—North America*, Chicago, IL, Vol. 2, 25–29 May 1990, p. 148.

[202] Ross, J. S., Leininger, L. W., Sigel, G. A., and Tian, D., "A Brief Review of Radiation Cure Systems Used in Flooring," *Proceedings of RadTech 2000*, Baltimore, MD, 9–12 Apr. 2000, p. 241.

[203] Tian, D., Ross, J. S., and Sigel, G. A., "A Comparison of Caprolactone Based Polyols in Flooring Formulations," *Proceedings of RadTech 2000*, Baltimore, MD, 9–12 Apr. 2000, p. 251.

[204] Barclay, R., "Uses of Radiation Curing in Flooring," *Radiation Curing*, Vol. 13, No. 5, May 1986, p. 4.

[205] Jung, T., Koehler, M., and Trovato, D., "Preparation of Glass Fiber Reinforced Materials—An Unconventional Application for UV-Curing," *Proceeding of RadTech '99*—Europe, Berlin, Germany, 8–10 Nov. 1999, p. 649.

[206] Scholz, D. and Koser, W., "Photocuring 'Prepregs': New Opportunities for Efficient RP-Processing," *Radiation Curing*, Vol. 11, No. 11, November 1984, p. 10.

[207] Cordts, H. P. and Karloske, J. E., "UV Cured Gelcoat for Translucent Panels," Session 22A of 38th Annual Conference, Reinforced Plastics/Composites Institute, The Society of the Plastics Industry, Inc., 7–11 Feb. 1983, p. 1.

[208] Keough, A. H., "Functional Coated Products from Electron Beam Curing," *Proceedings of RadCure '84*, Atlanta, GA, 10–13 Sept. 1984, pp. 10–38.

[209] Keough, A. H. and Beaupre, P. M., "Radiation Curing on Wide Web Substrates for Insulation," *Proceedings of Radiation Curing VI*, Chicago, IL, 20–23 Sept. 1982, pp. 5–15.

[210] Scholnick, F. and Buechler, P. R., "The Use of UV and EB Curing for Leather Coatings," *Proceedings of RadCure '86*, Baltimore, MD, 8–11 Sept. 1986, pp. 18–23.

[211] Kronick, P. and Scholnick, F., "Conference on Radiation Curing Leather Finishing," US Department of Agriculture, Eastern Regional Research Center, Philadelphia, PA, 13–14 Dec. 1988.

[212] Anon, "High-Volume Security Printer Masters UV Curing Setting World-Class Standards for Lottery Tickets," High Volume Printing, June 2000, pp. 48–52.

[213] Szycher, M., Battistone, G. C., Vincent, J., and Borowski, R. S., "Medical Applications of UV-Curable Elastomers," Technical Paper, FC84-1026, *RadCure '84*, Atlanta, GA, 10–13 Sept. 1984.

[214] O'Neill, J. R., "Electron Curing & Vacuum Metallization of Paper," *Proceedings of RadTech '88—North America*, New Orleans, LA, 24–28 Apr. 1988, p. 245.

[215] Sato, M., "UV Curable Coatings for Metallized Products," *Proceedings of Conference on Radiation Curing—Asia*, Tokyo, Japan, 20–22 Oct. 1986, p. 249.

[216] Costanza, J. R. and Kuzma, E. J., "UV Curable Coatings for Metal Coated Plastics," Technical Paper FC84-1021, *RadCure '84*, Atlanta, GA, 10–13 Sept. 1984.

[217] Dorfner, K and Ohngemach, J., "Novel Radiation Curable Pigmented Systems Exhibiting Metallic Effects," Technical Paper FC84-1029, *RadCure '84*, Atlanta, GA, 10–13 Sept. 1984.

[218] Tanny, G. B., Shchori, E., and Kenigsberg, Y., "UV/EB Polymerized Microporous Membranes and Coatings," *RadCure '86*, Baltimore, MD, 8–11 Sept. 1986.

[219] Yamamura, T., Tanabe, T., Ukachi, T., and Morohoshi, K., " Injection Molding by Utilizing Photofabricated Resin Mold," *Proceedings of RadTech 2000*, Baltimore, MD, 9–12 Apr. 2000, p. 427.

[220] Aurin, W., "Some Samples of Radiation Curing in the Paper, Film and Foil Converting Industry," *Proceedings of RadCure '84*, Atlanta, GA, 10–13 Sept. 1984, p. 10-1.

[221] Visser, J.D., "Functional or Decorative? EB-Curable Coatings for Film, and Paper," *Proceedings of RadCure '86*, Baltimore, MD, 8–11 Sept. 1986, p. 5-1.

[222] Rudolph, A. C., "Ferrotype Electron-Beam Cured Coatings," *Proceedings of RadCure '86*, Baltimore, MD, 8–11 Sept. 1986, pp. 5–7.

[223] Patent: Harasta, L. P., Leszyk, G. M., and Morrison, E. D., Eastman Kodak Co., Radiation-Curable Compositons for Restorative and/or Protective Treatment of Photographic Elements," US 4,426,431. (1984).

[224] Patent: Leszyk, G. M., Eastman Kodak Co., Radiation-Curable Composition for Restorative and/or Protective Treatment of Photographic Elements, US 4,333,998. (1982).

[225] Fuga, T., Murao, A., Enomoto, Y., and Kimazuka, T., "Applications of UV Resins to the Temporary Rust-Proofing for the External Surface of Steel Pipes," *Proceedings of Conference on Radiation Curing—Asia*, Tokyo, Japan, 20–22 Oct. 1986, p. 269.

[226] Dominico, J., "A Discussion on the Impact of UV Curing in the Printed Cup Manufacturing Operation," *Proceedings of RadTech '90—North America*, Chicago, IL, Vol. 2, 25–29 May 1990, p. 69.

[227] Frazier, J. F., "Offset Container Decorating with UV Coating," *Proceedings of Radiation Curing VI*, Chicago, IL, 20–23 Sept. 1982, p. 13-1.

[228] Fleischer, J. E., "Coatings for Plastics," *Modern Paint & Coatings*, Vol. 91, No. 21, April 2001.

[229] Pascale, J. V., "An Evaluation of UV & EB Clear Coating Cured on Polyboard," *Proceedings of RadCure '84*, Atlanta, GA, 10–13 Sept. 1984, p. 10-9.

[230] Kerr, III, S. R., "Graphic Arts Quality for Polycarbonate Film with Combined Chemical Resistance and topcoatability," *Proceedings of RadTech '88—North America*, New Orleans, LA, 24–28 Apr. 1988, p. 326.

[231] Guenther, D., "A Production System for Continuous U.V. Coating of Clear Extensible Films," *Proceedings of RadTech '88—North America*, New Orleans, LA, 24–28 Apr. 1988, p. 337.

[232] Patent: Miller, J. E., The Sherwin-Williams Co., Coating Polycarbonates with UV Curable Coatings, US 4,908,230. (1990).

[233] Anderson, E. A. and Rawls, R. M., "UV Cured Coatings and Adhesives for Space a nd Satellite Applications," *Proceedings of RadCure '84*, Atlanta, GA, 10–13 Sept. 1984, pp. 11–30.

[234] Anon, "New Developments Include UV-Cure Aerospace Coatings," *Modern Paint and Coatings*, Vol. 79, No. 9, September 1989, p. 40.

[235] van Neerbos, A., Hoefs, C. A. M., and Giezen, H., "UV Curing for Tinplate Decoration," XVth FATIPEC Congress, Amsterdam, The Netherlands, 8–13 June 1980, p. I-319.

[236] Yamada, H. and Sumita, M., "Novel Transfer Decoration System by EB Curing," *Proceedings of Conference on Radiation Curing—Asia*, Tokyo, Japan, 20–22 Oct. 1986 p. 281.

[237] McIntyre, F.S. "One Pass "Inline" EB/UV Silicone and PSA Coating Process for Tape/Label and Wider Web Products," *Proceedings of RadCure '84*, Atlanta, GA, 10–13 Sept. 1984, p. 10–27.

[238] Warren, J. B., "A Comparison of the Curing of Thin Films by Ultraviolet and Electron Beam Radiation," *Proceedings of Radiation Curing VI*, Chicago, IL, 20–23 Sept. 1982, p. 5-1.

Glossary of Terms*

Terminology Used for Ultraviolet (UV) Curing Process Design and Measurement

---

This glossary of terms has been assembled in order to provide users, formulators, suppliers, and researchers with terms that are used in the design and measurement of UV curing systems. It was prompted by the scattered and sometimes incorrect terms used in industrial UV curing technologies. It is intended to provide common and technical meanings as used in and appropriate for *UV process design, measurement, and specification*. General scientific terms are included only where they relate to UV measurements. The object is to be "user-friendly," with descriptions and comments on meaning and usage, and minimum use of mathematical and strict definitions, but technically correct. Occasionally, where two or more terms are used similarly, notes will indicate the preferred term.

For historical and other reasons, terms applicable to UV curing may vary slightly in their usage from other sciences. This glossary is intended to "close the gap" in technical language, and is recommended for authors, suppliers, and designers in UV curing technologies.

---

absorbance. An index of the light absorbed by a medium compared to the light transmitted through it. Numerically, it is the logarithm of the ratio of incident spectral irradiance to the transmitted spectral irradiance. It is unitless number. Absorbance implies monochromatic radiation, although it is sometimes used as an *average* applied over a specified wavelength range.

absorptivity (absorption coefficient). Absorbance per unit thickness of a medium.

additive lamps. Medium pressure mercury vapor UV lamps (arc or microwave) that have had small amounts of metal halide(s) added to the mer-

* Published with permission of RadTech International North America.

cury within the bulb. These materials will emit their characteristic wavelengths in addition to the mercury emissions. (This term is preferred over **doped lamps**).

bandwidth. The range of wavelengths between two identified limits, expressed in the same units as wavelength (nm).

cosine response. Description of the spatial response to incident energy where response is proportional to the cosine of the incident angle. A radiometer with a diffuser or a photo-responsive coating will exhibit nearly cosine response.

dichroic. Exhibiting significantly different reflection or transmission in two different wavelength ranges. Dichroic reflectors which have reduced reflectance to long wavelengths (IR) are also called "cold mirrors."

diffuse. A characteristic of a surface that reflects or scatters light equally in all directions (often confused with *spread reflectance*).

doped lamps. Term applied to UV lamps having metal halide additives to the mercury to alter the emission spectrum of the lamp. (Historically, this term has been used by UV arc lamp manufacturers. It is a slightly imprecise usage, as the added chemical does not alter the properties of the other). (The preferred term is **additive lamps**).

dose. A common, but loosely used, term for **energy density**, or radiant flux density, at a surface. (It is a precisely defined term in EB curing: 1 Gray (Gy) = 1 J/kg, a measure of absorbed energy per unit mass). In other technologies, the term usually applies to energy absorbed *within* the medium of interest, but in UV curing, is equated only to **irradiant energy density** *arriving at the surface* of the medium of interest. (The preferred shortened term is **energy density**, expressed in J/cm^2 or mJ/cm^2).

dynamic exposure. Exposure to a varying irradiance, such as when a lamp passes over a surface, or a surface passes under a lamp, or lamps. In that case, energy density is the time-integral of the irradiance profile.

dynamic range. The span between the *minimum* irradiance and the *maximum* irradiance to which a radiometer will accurately respond. Expressed as a ratio, or in measured units (e.g., $watts/cm^2$).

effective energy density. Radiant energy, *within a specified wavelength range*, arriving at a surface per unit area, usually expressed in joules per square centimeter or millijoules per square centimeter (J/cm^2 or mJ/cm^2). Is expressed in a specified wavelength range (without wavelength specification, it is essentially meaningless). Commonly accepted abbreviations are W_λ or E_λ.

effective irradiance. Radiant power, *within a specified wavelength range*, arriving at a surface per unit area. It is expressed in watts or milliwatts per square centimeter (W/cm^2 or mW/cm^2) in a specified wavelength range (without wavelength specification, it is essentially meaningless). For brevity, when the wavelength range is *clearly* understood, the term is shortened to **irradiance**. Commonly accepted abbreviations are E_λ or I_λ. Compare **spectral irradiance**.

emission spectra. Radiation from an atom or atoms in an excited state, usually displayed as radiant power vs wavelength. Emission spectra are unique to each atom or molecule. The spectra may be observed as narrow line emission (as in atomic emission spectra), or as quasi-continuous emission (as in molecular emission spectra). A mercury plasma emits both line spectra and continuum simultaneously.

energy density. Radiant energy arriving at a surface per unit area, usually expressed in joules or millijoules per square centimeter (J/cm^2 or mJ/cm^2). It is the time-integral of irradiance. (Terms applied in other technologies include "radiant exposure," "light dose," and "total effective dosage"). Compare **fluence, dose.**

fluence. The time-integral of fluence rate (J/m^2 or J/cm^2). For a parallel and perpendicularly incident beam, not scattered or reflected, **energy density** and **fluence** become identical.

fluence rate. The radiant power of all wavelengths passing from all directions through an infinitesimally small sphere of cross-sectional area dA, divided by dA. For a parallel and perpendicularly incident beam, not scattered or reflected, **irradiance** and **fluence rate** become identical. (W/cm^2 or mW/cm^2).

flux (radiant flux). The flow of photons, in einstein/second; one einstein = one mole of photons.

intensity. A generic term, with a variety of meanings; undefined, but commonly used to mean **irradiance**. Generally misapplied in UV curing. Its *precise* optical meaning is flux/steradian (W/sr), applied to *emission* of light; not useful in UV curing. (The preferred terms are **irradiance** or **effective irradiance**).

irradiance. Radiant power arriving at a surface from all forward angles, per unit area. It is expressed in watts per square centimeter or milliwatts per square centimeter (W/cm^2 or mW/cm^2). Compare **effective irradiance, spectral irradiance,** and **fluence rate**.

irradiance profile. The irradiance pattern of a lamp; or, in the case of dynamic exposure, the varying irradiance at a point on a surface that passes through the field of illumination of a lamp or lamps; irradiance versus time.

joule (millijoule). A unit of work or energy (a newton-meter). The time-integral of power. Abbreviated ***J*** or ***mJ***. (Although derived from a proper name, the term is *not* capitalized, while its abbreviation *is* capitalized).

line emission. Narrow lines of emission from an atom in an excited state. These are the "spikes" observed in spectrometry. Low-pressure sources exhibit finely distinguished line emission; higher pressure sources exhibit more continuous spectra.

monochromatic. Light radiated from a source that is concentrated in only a very narrow wavelength range (**bandwidth**). This may be accomplished either by filters or by narrow-band emission.

monochromator. An instrument that separates incoming radiant energy into its component wavelengths for measurement. Two methods are used for dispersing the radiation: diffraction grating or prism. The typical resolution may be 1 nanometer or less.

nanometer. Unit of length. Abbreviated nm. Equals 10^{-9} meter, = 10^{-3} micron, = 10 Å (ångstrom). Commonly used unit to define wavelength of light, particularly in the UV and visible ranges of the electromagnetic spectrum. An older equivalent term, millimicron, is rarely used today.

optical density. The logarithm of the reciprocal of reflectance or transmittance. A dimensionless number. In printing and color, it is the log of the ratio of visible light absorbed by an "absolute white" to the light absorbed by the measured ink.

peak irradiance. The intense peak of focused power directly under a lamp. The maximum point of the **irradiance profile.** Measured in irradiance units (W/cm^2).

photometer. An instrument for measuring visible light, usually filtered or corrected to match the human eye response.

power (radiant) (see **radiant power).** The rate of radiant energy or total radiant power (W) emitted in *all* directions by a source.

power (UV lamp). Tubular UV lamps are commonly described by their operating power in "watts per inch" or "watts per centimeter." This is derived simply from the electrical power input divided by the effective length of the bulb. (It does not have a direct meaning to the output efficiency of a lamp system, to the spectral conversion efficiency, to the curing performance, or to the UV irradiance delivered to a work surface).

polychromatic, or **polychromic.** Consisting of many wavelengths.

quantum yield. A measure of the photon efficiency of a photochemical reaction. The ratio of the number of chemical events per unit time to the number of photons absorbed per unit time. It is a unitless measure.

radiachromic. Exhibiting a change of color or optical density with exposure to light. A character of films whose color or density change can be correlated to exposure to UV energy.

radiance. Generally refers to the radiant *output* of a source. It is radiant flux per unit area per steradian ($W/cm^2/sr$). In UV curing, it is used generically rather than as a precise optical term.

radiant power. Rate of energy transfer, expressed in watts or joules/second ($W = J/sec$).

radiant intensity. Power per unit of solid angle from a source, expressed in watts/steradian (W/sr).

radiant energy. Energy transfer, expressed in joules or watt-seconds ($J = W \times sec$).

radiometer. A device that senses irradiance incident on its sensor element. Its construction may incorporate either a thermal detector or a photonic detector. The instantaneous signal output will usually have a linear proportionality to radiant flux, and will depend on incident wavelength(s). The resulting characteristic response to irradiance versus wavelength is called responsivity.

responsivity (spectral sensitivity). The response or sensitivity of any system in terms of incident wavelength. In radiometry, it is the output of a device versus wavelength.

spectral output. The radiant output of a lamp versus wavelength. It is displayed in a variety of ways, but commonly a graph or chart of output watts plotted against wavelength. The appearance of the plot will vary dramatically, depending on the wavelength resolution used. A technique of normalizing is to integrate energy over 10-nanometer bands, to reduce the difficulty of quantifying the effects of line emission spectra.

spectral absorbance (absorbance spectrum). Absorbance described as a function of wavelength.

spectral irradiance. Irradiance at a given wavelength per unit area per unit wavelength interval. Expressed in $W/cm^2/nm$. Usually measured with a spectroradiometer. Compare **effective irradiance.**

spectroradiometer. An instrument that combines the functions of a radiometer and a monochromator to measure irradiance in finely divided wavelength bands.

static exposure. Exposure to a constant irradiance for a controlled period of time. Contrast with **dynamic exposure**.

UV. Ultraviolet. Radiant energy in the 100–450 nm range. 100–200 nm is generally called *vacuum UV (VUV)*, because it does not transmit in air. There is no precisely defined boundary between UV and visible light, and may be considered about 400–450 nm.

UVA, UVB, UVC. Designations of UV wavelength ranges, originally for distinction of physiological effects of UV, and establishment of safe exposure limits. The generally accepted ranges are:

VUV: 100–200 nm
UVC: 200–280 nm
UVB: 280–315 nm
UVA: 315–400 nm

UVA is commonly referred to as *long UV wavelengths*; while **UVC** is considered *short UV wavelengths*. **VUV** stands for "vacuum UV." Measurement of specific ranges may be defined by the responsivity of a radiometer. It should be made clear, when referring to these ranges, *exactly* what wavelengths they represent. Specific manufacturers of radiometers will use uniquely specified ranges.

watt (milliwatt). The absolute meter-kilogram-second unit of power equal to the work done at the rate of one joule per second or to the power produced by a current of one ampere across a potential difference of one volt : 1/746 horsepower. Abbreviated W or mW. In optics, a measure of radiant or irradiant power. (Even though the term is derived from a proper name, it is *not* capitalized, while the abbreviation *is* capitalized).

wavelength. A fundamental descriptor of electromagnetic energy, including light. It is the distance between corresponding points of a propagated wave. It is the velocity of light divided by equivalent frequency of oscillation associated with a photon. UV wavelengths are currently measured in nanometers (10^{-9} meter). An older term, Ångstroms (Å= 10^{-10} meter) is rarely used today. The typical symbol for wavelength is λ(lambda).

INDEX

P

R

S

W

X

Y